W9-DEV-594

Experience Forms

World Anthropology

MOUTON PUBLISHERS · THE HAGUE · PARIS · NEW YORK

Experience Forms

Their Cultural and Individual Place and Function

Editor

GEORGE G. HAYDU

MOUTON PUBLISHERS · THE HAGUE · PARIS · NEW YORK

ISBN 90-279-7840-9 (Mouton)
ISBN 0-202-90088-6 (AVC Inc.)
Indexes by Society of Indexers, Great Britain
Jacket photo by Béla Bácsi
Cover and jacket design by Jurriaan Schrofer
Phototypeset in V.I.P. Times by
Western Printing Services Ltd, Bristol
Printed in Great Britain

General Editor's Preface

However else humans see our species in contrast to others, attention usually focuses on mental aspects of our individual experiences, and their collective representations. This book discusses, in a new context, the interrelations of forms of human experience and the symbols, meanings, and values as these are developed in psycho-social life. The concept of "experience forms" unifies papers from a variety of specialities and backgrounds which were assembled for a Congress which encouraged such combinations, and the consequent flow of ideas.

Like most contemporary sciences, anthropology is a product of the European tradition. Some argue that it is a product of colonialism, with one small and self-interested part of the species dominating the study of the whole. If we are to understand the species, our science needs substantial input from scholars who represent a variety of the world's cultures. It was a deliberate purpose of the IXth International Congress of Anthropological and Ethnological Sciences to provide impetus in this direction. The *World Anthropology* volumes, therefore, offer a first glimpse of a human science in which members from all societies have played an active role. Each of the books is designed to be self-contained; each is an attempt to update its particular sector of scientific knowledge and is written by specialists from all parts of the world. Each volume should be read and reviewed individually as a separate volume on its own given subject. The set as a whole will indicate what changes are in store for anthropology as scholars from the developing countries join in studying the species of which we are all a part.

The IXth Congress was planned from the beginning not only to include as many of the scholars from every part of the world as possible, but also with a view toward the eventual publication of the papers in high-quality volumes. At previous Congresses scholars were invited to bring papers which were then read out loud. They were necessarily limited in length; many were only summarized; there was little time for discussion; and the sparse discussion could only be in one language. The IXth Congress was an experiment aimed at changing this. Papers were written with the intention of exchanging them before the Congress, particularly in extensive pre-Congress sessions; they were not intended to be read aloud at the Congress, that time being devoted to discussions — discussions which were simultaneously and professionally translated into five languages. The method for eliciting the papers was structured to make as representative a sample as was allowable when scholarly creativity — hence self-selection — was critically important. Scholars were asked both to propose papers of their own and to suggest topics for sessions of the Congress which they might edit into volumes. All were then informed of the suggestions and encouraged to re-think their own papers and the topics. The process, therefore, was a continuous one of feedback and exchange and it has continued to be so even after the Congress. The some two thousand papers comprising *World Anthropology* certainly then offer a substantial sample of world anthropology. It has been said that anthropology is at a turning point; if this is so, these volumes will be the historical direction-markers.

As might have been foreseen in the first post-colonial generation, the large majority of the Congress papers (82 percent) are the work of scholars identified with the industrialized world which fathered our traditional discipline and the institution of the Congress itself: Eastern Europe (15 percent); Western Europe (16 percent); North America (47 percent); Japan, South Africa, Australia, and New Zealand (4 percent). Only 18 percent of the papers are from developing areas: Africa (4 percent); Asia-Oceania (9 percent); Latin America. (5 percent). Aside from the substantial representation from the U.S.S.R. and the nations of Eastern Europe, a significant difference between this corpus of written material and that of other Congresses is the addition of the large proportion of contributions from Africa, Asia, and Latin America. "Only 18 percent" is two to four times as great a proportion as that of other Congresses; moreover, 18 percent of 2,000 papers is 360 papers, 10 times the number of "Third World" papers presented at previous Congresses. In fact, these 360 papers are more than the total of *all* papers published after the last International Congress of Anthropological and Ethnological Sciences which was held in the United States (Philadelphia, 1956).

The significance of the increase is not simply quantitative. The input of scholars from areas which have until recently been no more than subject matter for anthropology represents both feedback and also long-awaited theoretical contributions from the perspectives of very different cultural, social, and historical traditions. Many who attended the IXth Congress were convinced that anthropology would not be the same in the future.

The fact that the Xth Congress (India, 1978) was our first in the "Third World" may be symbolic of the change. Meanwhile, sober consideration of the present set of books will show how much, and just where and how, our discipline is being revolutionized.

Readers of the present volume will find in this series of books on world anthropology many related ideas from linguistic, psychological, and cultural anthropology, and data from the gamut of historic cultures against which to test them.

Chicago, Illinois
May 22, 1979

SOL TAX

Table of Contents

Introduction

GEORGE G. HAYDU

Cultural and personal event patterns are organically related. Culture is a system of skills, instrumentalities, and certainties through which members of a society form and pursue their needs. Up to the nineteeth century the individual was considered mainly a recipient of cultural forms. Absorption in the tribal ways, enculturation as a molding of the individual to patterns altogether inflexible, these were the basic concepts. As culture change and the succession of culture patterns became recognizable, the power of persons on cultural patterning and pattern transformations could be considered. The hypothesis that the experiences and transformations of the individual have profound effects on cultural transformations appeared to be increasingly tenable. It has become evident that we are considering a two-way process. The study of general traits and vectors are imprecise in this respect. Without regarding the configuration of individuality of these interrelated structures, the two-way dynamics of culture and individual transformations cannot be understood.

Experience is not a continuous stream. It is a sequence of events, and each event is an experience entity. The experience entity is an integration of many components, chiefly an aroused need that seeks and finds its instrumentalities, concepts, images and action. Action is then fed back to indicate results and consequences. This multicomponential integration is the smallest unity that is still a full-fledged psychological happening. It has its particular shape or form quality. Experience entities cluster and combine chiefly according to their most significant features, mainly around significant persons or powerful events. Such structures are the experience forms. They can be expressed and they can be communicated. They are basic to interpersonal relations. They depend on and can be encoded in the semiotic ecology of which the person is member. Relations, objects, and intentions are viewed and weighed through the outcome of these events: The outcome-oriented aspect of experience forms constitute axiological value. Values can be divided in two rather artificial categories according to outcome: drive object outcome and drive integration outcome.

Experience forms, both experimentally and conceptually, can elucidate the interrelationships of individual growth, on the one hand, and cultural transformations, on the other — their continual interplay, their limiting, normative, as well as transforming, innovative dynamics. The reality of experience forms also accommodates the fullness of human happenings and their expressions as realized through the creative activities of man without the unsatisfactory reductionism of the recent past. In addition, experience forms can be operationally described and used in empirical studies and in organizing suitable data, thereby bridging theory and experimental work.

Experience forms, as integrative patterns, serve important functions in establishing the meaning of any event of which people are aware. Their stability as well as modifiability indicate how cultural and individual continuities are structured, how old creative forms may be superseded and new ones relevant and useful. To discover these regularities in the many related fields is a difficult undertaking. The following papers indicate the range and hazards of such an enterprise.

The Neural Representation of Experience as Time

ROBERT W. THATCHER

The original title of this paper was simply "Time." However, after some thought it became clear that time is merely an abstraction and, in a concrete sense, does not exist. To avoid the confusion which stems from the concept time, when it is erroneously used as a concrete entity, the word "change" should instead be used since this word is fundamental to time. Change reflects a particular category of relationships between events. In this sense time is an abstraction based upon the succession of events. Event A is succeeded by event B. The regular occurrence of such relationships provides the basis by which objective time can be understood. It is believed that the subjective perception of time (e.g., duration, time passing, temporal perspective, simultaneity, etc.) can best be understood by first examining the means by which objective or external time is measured. The last statement may at first seem naive. However, an understanding of subjective time must ultimately stem from the fact that a living organism operates on a world that exhibits succession and change. How is succession and change represented in the brain? What are the mechanisms by which the brain manipulates and makes use of time? What are some of the basic processes that underly time estimation?

EXTERNAL OR OBJECTIVE TIME

This paper will not attempt to answer all these questions in detail. Rather, basic experimental data will be reviewed which deal with the elementary aspects of neural time representation. The general question that will be asked is: What is the nature of the internal references used for the

The author gratefully acknowledges the contributions arising from many enjoyable discussions with Alfredo Toro, Rebecca Thatcher, and Kim Hopper.

subjective estimation of time? It is felt that the answer to this question will provide a basis for understanding many of the other aspects of time. However, in order to determine the internal references for subjective time, it is necessary to first consider the role of a reference in the measurement of objective or external time.

For example, it is clear that humans exist in a universe that exhibits ordered change. "Time's arrow" is an inexorable movement of events which follows the laws of cause and effect and exhibits a specific direction. But can time go backwards? Modern physicists seriously explore this possibility, though with little success. It is remarkable to consider the evolution of human intelligence which can reflect on the order of events, anticipate their future occurrence, and manipulate relationships with predictable consequence. Modern technological society puts a high premium on the precise accounting of the relationship between events. But how is this accomplished in today's sophisticated societies? Very simply, by establishing a reference (i.e., a clock) so that the events that constitute "time's arrow" can be anticipated and quantified. Today, the cesium crystal which oscillates between two energy levels, precisely 9,192,631,770 cycles per second (or Hertz, "hz") is the clock reference. Thus, the elementary unit of time, the second, is defined by a highly regular or ordered event change. Of course, not all events are regular. Some events occur only once in an infinity of time, while others are frequent but random with respect to the clock reference. The important point is that objective time depends on a reference. That is, measurements are made by comparing one set of events with another set.

The idea of a reference system is of fundamental importance in physics. The early laws of motion, as posited by Newton, assumed that all bodies were either at rest or moving with respect to three-dimensional space, which was an unchanging and absolute reference. Gottfried Leibnitz, Newton's rival, rejected the theory of absolute space and argued that space was merely a system of relationships between the bodies in it and had no existence independent of these bodies. The statement "this body is in motion" does not imply some kind of intrinsic motion but only a change of position of the body relative to another body. The latter constitutes a reference system that is conveniently thought of as a set of three mutually perpendicular coordinate axes.

Albert Einstein generalized the above notion of relativity to pertain to all natural phenomena. Einstein focused his theoretical discussions on the concept of simultaneity of two events taking place at different locations. He was able to demonstrate that the simultaneity of two events occurring at different places would be interpreted differently by observers in two coordinate systems moving relative to each other. Measurements of simultaneity depended entirely on the relative motions of the reference systems. Einstein stressed that there was no absolute space, but rather a coordinate reference system of space-time. Also, he abandoned the idea of instaneous transmission of effects and replaced it with the constant of the speed of light.

In summary, relativity theory leaves us with an environment constituted by events whose relations are changing. Concepts of space and time are understood in terms of specific frames of reference. Space is understood in terms of the relations between objects, and time is understood in terms of the relative change of events.

THE CONSTRUCTION OF SUBJECTIVE TIME

Studies of the developing child show that subjective time, parallel to the development of the awareness of space, is constructed little by little and involves the elaboration of a system of relations (Piaget 1971a). More specifically, time sense develops ontogenetically and involves learning and the construction of a set of operations by which one makes comparisons between events. The title of this section was chosen to emphasize that time sense is an adaptive, creative process that enables an individual to manipulate and extend his control over the environment.

Obviously, cultural factors play an important role in determining the way in which an individual conceives and uses time. But again this fact helps emphasize the adaptive aspect of subjective time.

Now, what is meant by the "construction" of subjective time? It is suggested that subjective time is constructed through the action of an internal reference or signal. This internal signal develops ontogenetically and parallels the development of memory and the self (Piaget 1971b). A clearer perspective of the nature of this process can be obtained by considering the various aspects of subjective time operations. For convenience, subjective time can be divided into at least four (nonmutually exclusive) categories (Ornstein 1969):

1. Succession (change) and simultaneity
2. Awareness of time passing (the specious present)
3. Estimates of duration
4. Time ordering (temporal perspective)

The first category is actually fundamental to the others. At the root of succession is the concept of change and, as discussed earlier, this involves the representation of a relationship between events. Event A can be discriminated from succeeding event A′ when a minimum interval of time elapses (in the case of the human observer this is approximately twenty to sixty milliseconds). The relationship between simultaneity and succession is clear since two events will be perceived as one if they occur successively at a short interval of time. It is important to emphasize that both succession and simultaneity involve a comparison, namely, comparisons between the neural representations of events. In 1962 Halliday and Mingay (see Efron 1963) measured the latency of cortical evoked potentials following stimulation of the toe and index finger. They found that the evoked potentials elicited by the toe stimulus occurred approximately twenty milliseconds later than the potentials evoked by the finger stimulus. Psychophysical experiments showed that the stimuli delivered to the toe and finger were not perceived as simultaneous unless the toe stimulus preceded the finger stimulus by nine to seventeen milliseconds. Halliday and Mingay concluded that the central nervous system does not correct for the time error produced by conduction differences and that the perception of simultaneity occurs when two sensory messages reach some point in the nervous system at the same time. Robert Efron (1963) extended these findings by demonstrating a consistent time error for the awareness of simultaneity of two to six milliseconds when stimuli are delivered to left and right symmetrical parts of the body. For left-handed individuals the delay was in stimuli delivered to the right half of the body, while for right-handed individuals the delay was in stimuli delivered to the left half of the body. This led Efron (1963) to conclude that the sensory messages received by the nondominant hemisphere are transferred by a pathway (probably the corpus callosum) to the dominant hemisphere for language where the conscious comparison of the time of occurrence of two sensory stimuli occurs.

Awareness of time passing also appears to involve an active comparison. The comparison in this case is between short or long term memory and information processing of immediate events. It is closely linked to the awareness of succession but as William James (1890) suggests, time passing involves an interplay between the immediate past and the present. Awareness of time passing seems to involve consciousness of the occurrence of events. Using Morris Halle and Kenneth Steven's (1962) model of speech recognition, it is possible to argue that consciousness is a creative process involving the synthesis of an internally generated signal that is continually being matched and mismatched with a constellation of sensory input, memories, feelings, and other constitutive factors occurring in the present. Consciousness thus would be a process whereby a reference signal, which is part of the constructed self (developed gradually over many years), is continually compared, on a multidimensional basis, with sensory, affective, and endogenous factors. If such occurs then there should be a time delay for awareness which is the time required for the process to complete itself. This view is one where consciousness and the awareness of time passing are a continual, creative process that ceases during slow wave sleep and starts again when we are awake or dreaming.

Estimates of duration can be considered in terms of both short durations and long durations. Estimates of durations seem to involve memory and other processes similar to those operating during awareness of time passing. Memories of durations from the past may be compared with durations occuring in the present. An important and distinctive feature of duration is that this process involves the use of time "markers" or tags to label the beginning and end of an interval. It is important to ask: How is the beginning and end of an interval represented in the brain? Or more precisely: What is the nature of the time markers? Examples of the neural representation of duration will be discussed in detail in a later section. However, suffice it to say that change in a stimulus event invariably results in the excitation of a set of neurons and in the inhibition of others. One consequence of this action is synchronization of neural ensembles. The synchronization process appears to be of fundamental importance in the creation of neural time markers. Furthermore, synchronization, which has neuroanatomical correlates, may be involved in initiating construction of representational systems. Neurophysiological data show that synaptic inhibition plays a basic role in the synchronization process (Purpura 1969, 1970). For example, intracellular analyses of visual cortical neurons show that synaptic inhibition disappears, while excitation is left largely unaltered, near the interstimulus interval for flicker fusion (Kuhnt and Creutzfeldt 1971). This finding further supports the notion that inhibition is important in the segregation of neural representational systems in time.

Recurrent and lateral inhibition appear to be integral parts of the dynamics of nearly all brain activity. Recurrent inhibition has been shown to be of vital importance in thalamic synchronization processes (Thatcher and Purpura 1972, 1973) and is believed to underlie the production of alpha rhythms (Andersen and Andersson 1968) and cortical EEG (Creutzfeldt, Watanabe, and Lux 1966). Both lateral and recurrent inhibition are most likely of fundamental importance in the isolation and protection of neural representational systems in space and time.

If memories are retrieved through the use of tags or markers created by the onset of the event, then estimates of duration may involve an active comparison of memory segments from the past and stimulus input segments occurring in the present (consciousness referent). A change of brain state, such as by drugs or natural arousal, may affect the consciousness referent (created in the present) in a manner that could cause underestimation or overestimation of a standard interval.

Temporal perspective is determined to a large extent by social factors. By temporal perspective is meant the ordering of time based on a particular view of events that a given individual may choose. Such perspective, however, also involves references. The references may be past experiences or a specific plan or an idea that occurs at the moment. By perspective is also meant the manipulation of past events or the reorganization of ideas and relations to enable an individual to internally order either his own actions or events in time. The adaptive and constructive manipulation of events in terms of time, as well as the general use of knowledge to predict the future are all consumed under the term "time perspective."

The ability to have time perspective is not given at birth. Rather, a gradual emergence of intellectual capacities parallels the development of time perspective (Piaget 1971b). The development of concepts of space, of the relation of the self to others and objects develops hand in hand with time sense (Piaget 1971b).

It is beyond the scope of this paper to discuss in detail the nature of the various and many referents used in time sense. However, suffice it to say that it can be argued that match-mismatch processes involving internal references underlie many aspects of time sense. These match-mismatch processes all appear to involve the interaction and manipulation of neural representational systems. In the remainder of this paper experimental data will be discussed in light of the previous discussion of time duration. Data will be presented showing the existence of a neural representational system inextricably bound to the reconstruction of an interval of time. The purpose of this presentation is an attempt to derive some understanding of the basic neural elements involved in time representation. What is the basic element of neural time representation? Is it a movement through neural space, or the pulse character of a synchronizing process, or a set of such processes? Based on the previous discussions, a good starting point on the journey to answering these questions would be to first inquire into the nature of neural representational systems.

NEURAL REPRESENTATIONAL SYSTEMS

What is an internal representation of experience and how is a neural representation of experience translated into time? Recent neurophysiological research indicates that information from the external world is represented in terms of specific spatiotemporal organizations of neural activity (see Thatcher and John 1977). These internal organizations which reflect information from the external world are called "representational systems." A representational system is defined as "any structure of which the features symbolize or correspond in some sense to some other structure" and "for any given structure there may be several equivalent representations" (Mackay 1969). This brief discussion allows us to define information as that which constitutes a change in a representational system. According to this definition, in its broadest sense is that which creates, adds to, or changes a representation (Mackay 1969).

The precise mechanism by which neural representational systems are created are currently unknown. However, the fact that neural representational systems exist and can be described in (at least) four dimensions has been clearly established (John 1967; Thatcher 1970; Freeman 1977; Disterhoft and Olds 1972; Morrel and Jasper 1956; John and Killam 1960; Libersen and Ellen 1960).

Recent behavioral experiments have demonstrated functional relationships between representational systems and time estimation. These experiments have led Robert Ornstein (1969) to postulate what he calls the "storage size" hypothesis of time estimation. This hypothesis argues that the size of a representational system (which is related to the amount of information it contains) is directly related to the estimation of time. Neurophysiological experiments conducted in the late 1960's (Thatcher and Cadell 1969; Thatcher 1970) tend to support Ornstein's hypothesis by showing that a neural representational system can expand or contract (spatially) depending on the amount of information the system represents. The latter studies are important since they provide the only empirical evidence to date showing a relationship between the amount of neural space occupied by an experience (which is directly related to the amount of information) and the reconstruction of an interval of time. For this reason, the remainder of this chapter will be devoted largely to a discussion of the Thatcher (1970) and Thatcher and Cadell (1969) experiments. The aim of this discussion is to provide experimental evidence that hopefully helps illuminate the basic mechanisms by which time is represented in the brain.

LABELLED RHYTHMS AND TIME RECONSTRUCTION

A very powerful method to study neural representational systems involves the use of what are called "labelled rhythms." This method was first discovered by Russian workers (Livanov and Poliakov 1945) and subsequently developed and formalized as an experimental technique by many scientists (Chow 1964; Stern, Viett, and Sines 1960; Libersen and Ellen 1960; Morrel and Jasper 1956; Yoshii, Shimokochi, and Yamaguchi 1960; John and Killam 1960). The technique involves conditioning animals to a very specific frequency of a stimulus. The electrical activity of the brain is then analyzed in order to detect the presence of frequency specific neural activity (labelled activity) which occurs as a function of the animals past experience.[1] For instance, animals can be

[1] The technique of conditioning an animal to a frequency specific stimulus (tracer stimulus) and then, either simultaneously or at some later date, detecting the presence of neuroelectrical activity at the frequency of that tracer stimulus is called the "tracer technique." The "labelled frequency response" is defined as a neural response that is produced by the tracer stimulus and occurs at a fundamental or a harmonic of the frequency of that stimulus. The retrieval of labelled frequency information is demonstrated when the presence of tracer frequency specific neural activity is shown to be *produced* by a *past* tracer stimulus experience.

trained to make differential responses to different frequencies of a flickering light (e.g., pressing a bar on the left to frequency one and pressing a bar on the right to frequency two). When this discrimination is well established an intermediate frequency can be presented and the animal will be observed to generalize. Experiments by John and Killam (1960) show that when the animal interprets the intermediate stimulus as frequency one then rhythms specific to frequency one appear in the brain. On the other hand, when the animal interprets the intermediate stimulus as frequency two, rhythms specific to frequency two appear in the brain. Thus, the labelled rhythm technique is a method by which brain processes (representational systems) are labelled or tagged by a past experience. Much like radioactive tracers, labelled rhythms allow researchers to detect the presence of neural representational systems (from the past) and to trace their appearance (development) throughout regions of the brain.

The experiments by Thatcher and Cadell (1969) and Thatcher (1970) involved conditioning two groups of rats to two different frequencies of a flickering light conditioned stimulus (a painful footshock paired with either a 3.3hz or an 8.2hz CS). Each group was given several classical conditioning experiences with only one frequency of flicker. The rats were subsequently tested for their memory of the experience by presenting an intermediate frequency of flicker (5.8hz) while the animals' bar pressed for food. When the rats saw the intermediate flicker they immediately froze (and exhibited fear responses), demonstrating recognition of the past association between the flicker and a painful footshock (UCS). The electrical activity recorded while the animals were "freezing" was subjected to a power spectral analysis. This analysis allowed quantification of the amount of energy in each of several areas of the brain at specific frequencies. The study involved four separate experiments (and over 190 animals) using four different CS-UCS intervals (40, 60, 80, and 120 sec.). The amount of power (in microvolts/cycle/sec) in the EEG at 3.3 ± 0.25hz and 8.2 ± 0.25hz (and at harmonic frequencies, 4.lhz, 6.6hz, 16.4hz) was calculated for each animal for each of several brain structures. The relative amounts of power (percent of total) at these labelled frequencies were then compared for the two groups of animals. It was found that animals conditioned at 3.3hz exhibited significantly greater percent power at 3.3 ± 0.25hz than did animals conditioned at 8.2hz. On the other hand, animals conditioned at 8.2hz exhibited significantly greater power at 8.2 ± 0.25hz and at 4.1 ± 0.25hz than animals conditioned at 3.3hz. Furthermore, it was shown that these representational systems (labelled rhythms) occurred in particular structures at particular times and reflected the CS-UCS interval. The results of this study are shown in Table 1. This table shows the intervals (following the 5.8hz test flicker onset) at which significant differences between groups occurred at the

Table 1. Intervals during which statistical significance occurred ($P < 0.025$)

CS duration	Seconds 0–20	20–40	40–60	60–80	80–100	100–120	120–140
120 seconds	VC, 8.2 MLTh, 3.3	MTh, 3.3 PLTh, 4.1	----	----	MRF, 3.3	VC, 3.3 MTh, 3.3 PLTh, 3.3	MRF, 3.3 MRF, 4.1 PLTh, 3.3 PLTh, 6.6
80 seconds	VC, 8.2	PLTh, 3.3	MLTh, 3.3	----	VC, 3.3 VC, 4.1 MRF, 3.3	----	----
60 seconds	----	MRF, 8.2	----	VC, 3.3 MTh, 3.3 PLTh, 3.3 MLTh, 3.3	----	----	----
40 seconds	PLTh, 3.3 PLTh, 4.1	----	PLTh, 3.3 PLTh, 6.6	----	----	----	----

Key: MRF = Mesencephalic reticular formation; VC = Visual cortex; MLTh = Medial lateral thalamus; MTh = Midline thalamus; PLTh = Posterior lateral thalamus.

various labelled frequencies. The largest number of significant differences occurred in the 120-sec. CS experiment and the smallest number occurred in the 40-sec. CS experiment. Note that no significant differences occurred beyond 40 to 60-sec. with a 40-sec. CS, or beyond 60 to 80-sec. with a 60-sec. CS or beyond 80 to 100-sec. with an 80-sec. CS. This shows that an interval of time specific to the CS-UCS interval was consistently reconstructed. A clear picture of the anatomical and temporal evolution of the 3.3hz phenomenon (which was the strongest of the labelled rhythms) is shown in Figure 1. This figure shows the difference in

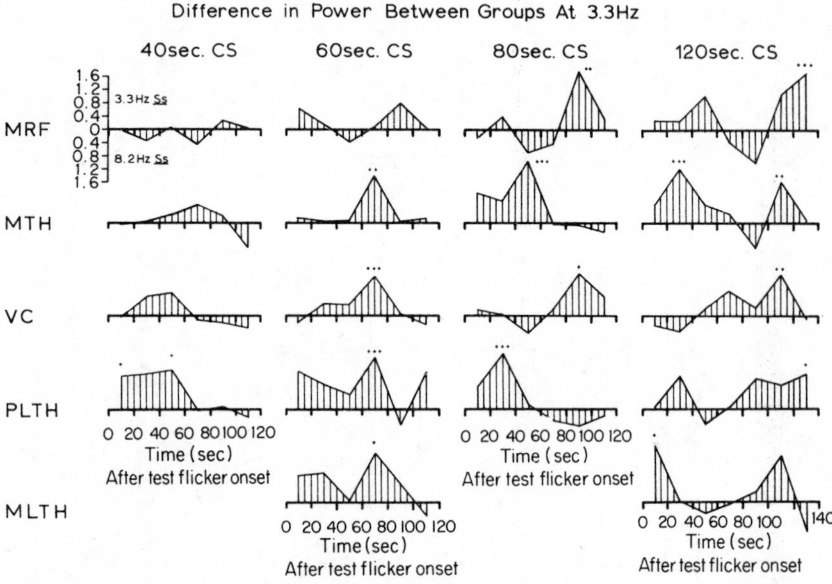

Figure 1. The ordinate is the absolute difference in mean percent power at 3.3 ± 0.25hz for the two groups of subjects. Points above the zero line (see MRF, 40 sec. CS) represents greater power at 3.3hz in animals conditioned with a 3.3hz CS than in animals conditioned with a 8.2hz CS. Points below the zero line represent greater 3.3hz power in subjects that were conditioned with an 8.2hz CS. Generally, there was more power at 3.3hz in subjects conditioned at 3.3hz as evidenced by the fact most of the points are above the zero line. Statistically significant differences appeared frequently during intervals associated with the time footshock would normally occur. For instance, follow the changes in the time of significant differences in the VC (visual cortex) and MRF (reticular formation) with different CS durations. Also, note that progressively more structures exhibited significant differences as CS duration was lengthened

Key: $=+++P<0.005$, $++= P=<0.01$, $+=P=<0.025$

power between groups of 3.3 ± 0.25hz during successive 20-sec. epochs beginning with the onset of the intermediate (5.8hz) flicker. All points above the zero line (see MRF 40-sec. CS) represent greater power at 3.3hz in animals conditioned at 3.3hz than in animals conditioned at 8.2hz. All points below the zero line represent greater power at 3.3hz in animals conditioned with an 8.2hz flicker. Note that most of the points are above the zero line.

Several important features can be seen. First, progressively more structures exhibited significant differences as CS duration was lengthened. Second, the rhythms consistently represented the reconstruction of the CS-UCS interval. These two features provide the first evidence that the size of a representational system can expand as a function of time. Third, there is a specific space-time organization of the rhythms. That is, particular structures exhibit labelled rhythms at specific times following the onset of the test flicker.[2]

The only variable manipulated in this study was CS duration. As CS duration was lengthened, the flicker representational system became anatomically more widespread and temporally more complex. What is the significance of this phenomenon? Why is the CS-UCS interval reconstructed? And what do these evolving anatomical-temporal organizations mean? One answer is that these phenomena represent the action of a system involved in predicting or anticipating the time of expected footshock. It has been mathematically proven that the best predictor of the future is based on an optimally succinct description of the past (Van Heerden 1968). It is reasonable to argue, therefore, that the best way to predict footshock is to reconstruct those events that, in the past, led to footshock. That is, the results reflect the fact that the flicker CS was the most consistent event preceding footshock. This interpretation is also consistent with the fact footshock is the most significant past event determining the animals' behaviors.

The results of this series of experiments are very complex. Differences are noted between 8.2 and 3.3 and the subharmonic 4.1. Each of these rhythms may reflect a local representational system which, in combination, constitutes a global or very large representational system.

Is it possible to reduce this complexity to the operation of a set of very elementary operations? To do so would be desirable since simple models generally are the most comprehensive and provide the most readily testable predictions. In the paragraphs to follow a very simple loop model of time representation will be presented which is capable of explaining most of the major findings of the Thatcher (1970) study.

Neural Loop Model of Time Representation

How is an interval of time represented in the brain? A clue to the answer

[2] It is important to note that these differences only occurred in animals that suppressed responding and exhibited recognition of the past conditioning experiences. Animals that failed to suppress responding (55 percent with a 2 min. CS) did not show statistically significant differences.

is provided by the well-documented fact that the brain is composed, anatomically, of a large number of neural loops (Lorente de No 1938; Scheibel and Scheibel 1958, 1965, 1966). These loops seem to be of various sizes nested within and between each other (Lorente de No 1938; Verzeano 1972). Evidence is available indicating that individual neurons may be members of many loops, i.e., one neuron may be shared by more than one loop (Lorente de No 1938; MacGregor et al, 1973; John and Morgades 1969) and in this way serve as nodal points within oscillatory systems. Thus, it would be consistent with anatomy and physiology to argue that time is represented by the circulation of activity within neural loops. That is, the basic functional element of time estimation is a loop. The larger the loop or the greater the delay between elements in the loop, the longer the interval of time. The following is a formalized loop model of time representation.

Consider that time is represented by the circulation of activity in a loop of cells (see Figure 2). A given interval of time can be represented by one circulation in a large loop or by a number of circulations in a smaller loop. In Figure 2 assume that a signal initiated at the entering axon passes through the loop C_1, C_2, C_3, ... C_n. Transit time (t_L) of the signal through the loop is the number of neurons (n) in the loop times the time for activity to pass through an individual neuron (a), i.e., $t_L = nxa$. For simplicity consider that transmission time *a* is represented by: (1) axonal conduction time; (2) synaptic delay (≈ 0.5 msec.); and (3) soma-dendritic integration time (≈ 0.2 msec.). Assume that the neurons in the loop are similar; that is, they have the same transmission properties such that a is the same for all neurons and equal to one msec. The time (t_L) required for transmission through a loop that consists of n neurons is given by $t_L = na = n$ (msec.). If the CS interval b is 100 msec., then n is related to b, such that $b = t_L K$ for some integer K. In other words, b equals the time t_L for circulation within one loop times the number of circulations in that loop K. Since $t_L = na = n$, then $n = b/K = 100/K$. Thus if $K = 1$ then $n = 100$, i.e., there are 100 neurons in the loop. If $K = 10$, then $n = 10$, i.e., there are 10 neurons in the loop. Since $b = t_L K = nk$ for some integer K, then K is the loop frequency, that is, the number of times the loop has to be traversed in order to match the interval b. The relationship of K to the interval b is given by $K = b/na$. That is, the number of circulations K that represent the interval b is inversely related to na.

The formula $K = b/na$ can apply to multiple loop systems of considerably greater complexity than that represented in Figure 2. For instance, circulation of activity can occur in a system of loops arranged in such a way that completion of circulation in one loop activates a second, anatomically different loop which then activates a third anatomically different loop, etc. In other words K can represent the circulation of activity within a single loop or a series of loops or both.

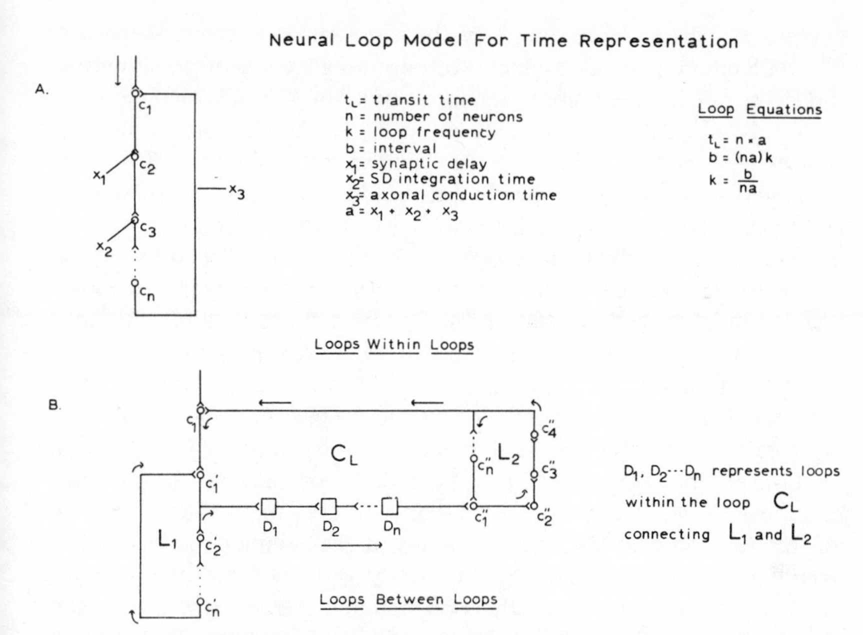

Figure 2. In this model time is represented by the circulation of activity within reiterant neural loops. The relationship between loop size, delay factors, and transit time is given in A. The loop equation k = b/na expresses a relationship between reiteration, loop size, and an interval of time (see text for details). The simple loop system in A can be expanded by coupling loops across distributed regions, see B. In B a diagrammatic representation of loops nested within loops is shown. The purpose of this is to illustrate complexity possible with such a system. No formalization is provided for the control of coupling or switching of loops.

Application of the Model to the Labelled Frequency Findings

Since the labelled frequency phenomenon was strongest and most reliable at 3.3hz the loop equation k = b/na will be applied to these data first (refer to Figure 1). Consider the 40-sec. and 60-sec. CS experiments. In the 40 and 60-sec. CS experiment, 3.3hz labelled activity was confined (in the structures investigated) to the PLTh but extended through the 0 to 20-sec. to the 40 to 60-sec. interval. In the 60-sec. CS experiment labelled activity was anatomically more extensive but the temporal extent within a structure was reduced. That is, the peak at 0 to 20-sec. In the PLTh just reached the 0.05 level and there was a clear cut dip in power to 40 to 60 sec. The finding that the within-structure temporal extent of the labelled activity was reduced when anatomical distribution was increased is explained by the loop model. That is, if time is represented by the circulation of activity in a loop system, then as the size of the system expands the number of circulations necessary to represent an interval decreases. This is represented by the equation $k = b/na$ where b = CS–UCS interval and na is a factor representing the size of the system. In the 40-sec. CS experiment, reconstruction of the CS–UCS interval involved the circulation of activity within a large number of anatomically confined loops. When the system expanded with a 60-sec. CS, activity circulated across an anatomically distributed system (i.e., increased na) and the number of circulations (k) within a structure decreased. According to this model, labelled frequency loops were established by experience with the CS, and the reconstruction of the CS–UCS interval involved reactivating the coupling relationships between the loops and/or spirals.[3]

The model can be extended to the 80-sec. CS experiment (Figure 1). With this duration of CS a further anatomical expansion of the system occurred. Correspondingly labelled frequency peaks appeared only during single 20-sec. epochs. However, with this duration CS temporal relationships between structures appeared. According to the model, such relationships represent circulation between structures. Thus, with an 80-sec. CS, increased K is an increase in the number of circulations extending across structures. An important consequence of the model is that in order for k to increase the loop system representing the interval b must be anatomically confined. That is, na is inversely related to k; thus, if k increases, na must be restricted.

The results of the 120-sec. CS experiment are entirely consistent with this aspect of the loop model. As can be seen in Figure 1, a 120-sec. CS results in an increased temporal extent of labelled activity within a structure as well as increased temporal relations between structures. At this duration dual peaks (iteration) occurred within structures. These findings suggest an increased amount of circulation both within a structure and between structures (i.e., increased k). The labelled frequency phenomenon begins to resemble that seen in the anatomically confined 40-sec. CS experiment (Figure 1). This indicates that the loop system is restricted with CS durations, that exceed 60 sec. The results of the 120-sec CS experiment are consistent with the notion that a loop system involved in reconstructing the CS–UCS interval begins to involute or reiterate when it is spatially restricted and that this restriction occurs with CS durations greater than approximately 60 sec. A specific experimental prediction arises from this model: Further increases in the CS–UCS

[3] The equation $K = b/na$ is a linearization of a much more general noneuclidean model. Using the notation of differential geometry, neural time representation can be described as a loop C' of n elements where $n \in C'$. The elements are neurons or groups of neurons. The relationship between CS frequency and the CS–UCS interval can be represented as d x $(n \in C') = b \in K)$, where d is the interelement delay, n is an element of the loop C' b = time for activity related to the CS (the CS frequency) to circle once in a loop and K = the CS–UCS interval. According to this more general formulation the CS interflicker interval is represented as a unit disk (or circle) and the CS–UCS interval is a larger Riemannian manifold upon which successive unit disks are mapped.

interval should result in further increases in the temporal extent of activity within a structure until some maximum level is reached.

A parsimonous feature of the loop model is that it can explain differences between 3.3hz and 8.2hz. If b = 303 msec. (for 3.3hz) and k = 1 then n = 303 neurons. If b = 123 msec. (for 8.2hz) the n = 123 neurons. This means that the loop system representing 8.2hz is smaller than that for 3.3hz, and thus for a given structure, there are a smaller number of neurons involved in the representation of 8.2hz. Given the signal-to-noise nature of this spectral technique one therefore would expect 8.2hz to be more difficult to detect than 3.3hz. This is what was observed in this study.

The loop model can also be applied to some features of the 8.2hz–4.1hz phenomenon. A prominent 8.2hz–4.1hz phenomenon occurred in the 40 and 120-sec. CS experiments. This is expected, based on the loop model, because k is greatest with a 40 and 120-sec. CS. That is, since there is a large number of circulations within anatomically confined systems in the 40-sec. and 120-sec. CS experiments 8.2hz and 4.1hz is easy to detect.

This model of the neural representation of time explains several aspects of phenomena observed in the Thatcher (1970) study and emphasizes, in particular, that long intervals of time may be represented by the circulation of neural activity within loops of cells distributed across widespread regions of the brain. It emphasizes that much of the phenomenology of time can be reduced to coupled loop systems in the brain. The application of the loop model emphasizes that time is directly related to spatial distribution or in other words, a space-time transformation. Anatomical order in the brain is translated into time. The data indicate that anatomy and time are indissolubly interrelated and that time reconstruction involves the reproduction of a geometrical representation.

The loop model also has limitations. For instance, no account of inhibitory synaptic control was provided. This is a serious omission since inhibition is believed to be of fundamental importance in neural control processes (Purpura 1970) and may, in fact, sculpture the flow of excitation in space-time (Thatcher and Purpura 1972, 1973). The model is also linear. A great many oscillations in the brain are nonlinear. This emphasizes another drawback of the model since it has been demonstrated that a dynamic time structure common to living systems can be described using nonlinear oscillators and linearization of the same equations abolishes the time structure (Goodwin 1963). The model, however, does give rise to several important and testable predictions. For example, the model suggests that a three-minute CS would result in very powerful (but not in anatomically more distributed) labelled rhythms. Also, it suggests that rats with smaller brains would exhibit (e.g., in strains selected for small cranial size) between structure reiteration with shorter CS durations than in rats with larger brains. The latter prediction arises from the postulate that time intervals are represented by a finite system of neurons and that an eventual limit in the spatial extent of the representational system can be reached. When the limit is attained other processes, such as biochemical rate reactions or oscillator subharmonics, may operate to represent very long intervals of time. Finally, if time indeed, is, represented by the circulation of neural activity in loops, then cooling of the total brain or selected fiber systems should affect the space-time aspects of the labelled rhythms in a predictable manner.

The rather slow circulation times of very large loops postulated by the model are consistent with measured circulation velocities in hypothetical neural loops. For instance, Verzeano and coworkers (Verzeano and Negishi 1960, 1961; Verzeano 1955, 1963, 1970, 1972; Verzeano et al. 1965) have extensively examined circulation velocities in groups of cells. The velocities they observed ranged from 0.5 to 8mm./sec. Experiments recently reviewed by Verzeano (1972) indicate that neural activity moves within both spirals and closed loops. The loop model, both conceptually and in terms of velocity valves, is consistent with the results of Verzeano.

The results of the Thatcher (1970) and Thatcher and Cadell (1969) studies are also consistent with recent findings by Olds and coworkers (Disterhoft and Olds 1972; Olds et al. 1972; Segal and Olds 1972) demonstrating differentiated anatomical and temporal neural organizations in the rat. The latter studies also indicated that the posterior thalamus (see Figure 1) is an important structure in the possible initiation and organization of developing representational systems.

CONCLUSIONS

There is considerable anatomical physiological evidence demonstrating the existence of loops in the central nervous system. Such systems are both very large and very small and capable of forming a hierarchy of organizations. The evidence presented in this chapter argues that the representation of experience occupies space in the brain, and that, in fact, such representations possess a specific shape or geometry in space-time. The latter conclusion is also suggested by the results of others (Disterhoft and Olds 1972; Segal and Olds 1972; and Olds et al. 1972). The results of the experiments presented in this chapter appear to demonstrate a representational system which exhibits a development in space as a function of time. The successful application of a loop model (k=b/na) to these data suggests that space is translated into time through the sequential operation of neuroanatomically distributed loops. These loops appear to be part of a highly stable space-time structure.

The larger question of the subjective experience of time passing requires further analysis. However, data by Ornstein (1969) and others (Frankenhaeuser 1959; Creelman 1962; Fraisse 1963; Michon 1966) show that subjective estimates of time are related to information processing. Ornstein (1969) presents evidence indicating that the size of a stored representation of a past experience can directly influence time estimation. This suggests that the storage size hypothesis as well as the relationship between information and time presented here may be understood in terms of expanding, contracting, and differentiating representational systems.

But the question still remains: What gives rise to the awareness of time sense and time passing? Humans most likely recall experiences of time passing from their past and use this as a reference to compare and estimate time passing in the present. In this sense a matching of information processing with an "internal reference signal" can serve as the basis for time estimation. In Thatcher's experiment (1970), described earlier, it is possible that the labelled rhythms represent the actual synthesis of the reference signal. The loop model may reflect the basic physiological mechanism by which the internal signal is generated. In humans, in contrast to the rat, very different parameters may operate so that smaller loops or a simple cascading of time "chunks" are used to reconstruct an interval of time. Such a process would require more elaborate rules, but also, less space in the brain. In any case, the main assertion of these arguments is that time and space are interrelated in the events of the brain and that time sense is intimately related to information processing.

REFERENCES

ANDERSEN, P., S. A. ANDERSSON
1968 *Physiological basis of the alpha rhythm.* New York: Appleton-Century-Crofts.

CHOW, K. L.
1964 Bioelectrical activity of isolated cortex, III. Conditioned electrographic responses in chronically isolated cortex. *Neuropsychologia* 2: 175–187.

CREELMAN, C. D.
1962 Human discrimination of auditory duration. *Journal of the Acoustical Society of America* 34: 582–593.

CREUTZFELDT, O. D., S. WATANABE, H. D. LUX
1966 Relations between EEG phenomena and potentials of single cortical cells, II. Spontaneous and convulsoid activity. *Electroencephalography and Clinical Neurophysiology* 20: 19–37.

DISTERHOFT, J. F., J. OLDS
1972 Differential development of conditioned unit changes in thalamus and cortex of rat. *Journal of Neurophysiology* 35: 665–679.

EFRON, R.
1963 The effect of handedness on the perception of simultaneity and temporal order. *Brain* 86: 261–284.

FRAISSE, P.
1963 *The psychology of time*. New York: Harper and Row.
FRANKENHAEUSER, M.
1959 *Estimation of time, an experimental study*. Amsterdam: Almqvist and Wiksell.
FREEMAN, W. J.
1977 Spatial patterns of EEG activity with odors at surface of olfactory bulb. *Society for Neuroscience Abstract* 726.
GOODWIN, B. C.
1963 *Temporal organization in cells*. New York: Academic Press.
HALLE, M., K. STEVENS
1962 Speech recognition: a model and program for research. *IRE Transactions on Information Theory*, IT–8, 155–159.
JAMES, W.
1890 *The principles of psychology*. New York: Henry Holt and Co.
JOHN, E. R.
1967 *Mechanisms of memory*. New York: Academic Press.
JOHN, E. R., K. F. KILLAM
1960 Electrophysiological correlates of differential approach-avoidance conditioning in the cat. *Journal of Nervous Mental Disease* 131: 183–201.
JOHN, E. R., P. P. MORGADES
1969 The pattern and anatomical distribution of evoked potentials and multiple unit activity elicited by conditioned stimuli in trained cats. *Communications in Behavioral Biology* 3: 187–207.
KUHNT, U., O. D. CREUTZFELDT
1971 Decreased post-synaptic inhibition in the visual cortex during flicker stimulation. *Electroencephalography and Clinical Neurophysiology* 30: 79–82.
LIBERSON, W. T., P. ELLEN
1960 "Conditioning of the driven brain wave rhythm in the cortex and the hippocampus of the rat," in *Recent advances in biological psychiatry*. Edited by J. Wortis. New York: Grune and Stratton.
LIVANOV, M. N., K. L. POLIAKOV
1945 The electrical reactions of the cerebral cortex of a rabbit during the formation of a conditioned defense reflex by means of rhythmic stimulation. *Izvestiya Akademiya Navk. USSR Series Biology* 3: 287–306.
LORENTE DE NO, R.
1938 Analysis of the activity of the chains of internuncial neurons. *Journal of Neurophysiology* 1: 207–244.
MACKAY, D. M.
1969 *Information, mechanism and meaning*. London: The MIT Press.
MACGREGOR, R. J., R. PRIETO-DIAZ, S. W. MILLER, P. M. GROVES
1973 Statistical properties of neurons in the rat reticular formation: evidence for reverberating loops, widespread rhythmicities, and functional reorganization. *Brain Research* 64: 167–187.
MICHON, J.
1966 Tapping regularity as a measure of perceptual motor load. *Ergonomics* 9: 401–412.
MORRELL, F., H. JASPER
1956 Electrographic studies of the formation of temporary connections in the brain. *Electroencephology and Clinical Neurophysiology* 8: 201–215.

OLDS, J., M. SEGAL, R. HIRSH, J. F. DISTERHOFT, C. L. KORNBLITH
1972 Learning centers of rat brain mapped by measuring latencies of conditioned unit responses. *Journal of Neurophysiology* 35: 202–219.

ORNSTEIN, R.
1969 *On the experience of time.* Baltimore: Penguin.

PIAGET, J.
1971a *The construction of reality in the child*. New York: Ballantine Books.
1971b *Biology and knowledge*. Chicago: University of Chicago Press.

PURPURA, D. P.
1969 "Interneuronal mechanisms in synchronization and desynchronization of thalamic activity," in *The interneuron*. Edited by M. A. B. Brazier. Los Angeles: UCLA Forum in Medical Sciences.
1970 "Operations and processes in thalamic and synaptically related neural subsystems," in *The neurosciences II* pp. 458–470. Rockefeller University Press.

RUCHKIN, D., E. R. JOHN
1966 Evoked potential correlates of generalization. *Science* 153: 209–211.

SCHEIBEL, M. E., A. B. SCHEIBEL
1958 "Structural substrates for integrative patterns in the brain stem reticular core," in *Reticular formation of the brain.* Edited by H. H. Jasper. 31–55. Boston: Little, Brown.
1965 Periodic sensory nonresponsiveness in reticular neurons. *Archives of Italian Biology* 103: 279–299.
1966 "Patterns of organization in specific and nonspecific thalamic fields," in *The thalamus*. Edited by D. P. Purpural and M. D. Yahr, 13–46. New York: Columbia University Press.

SEGAL, M., J. OLDS
1972 Behavior of units in hippocampal circuit of the rat during learning. *Journal of Neurophysiology* 35: 680–717.

STERN, J. A., G. A. VIETT, J. O. SINES
1960 "Electrocortical changes during conditioning," in *Recent advances in biological psychiatry*. Edited by J. Wortis. New York: Grune and Stratton.

THATCHER, R. W.
1970 "A demonstration of anatomical and temporal changes in the organization of the brain which occur as a function of CS duration." Unpublished Ph.D. Dissertation, University of Waterloo, Ontario, Canada.

THATCHER, R. W., T. E. CADELL
1969 "A demonstration of time dependent processes associated with recall." *Proceedings of the 77th annual meeting of the American Psychological Association*.

THATCHER, R. W., E. R. JOHN
1977 *Foundations of cognitive processes.* Hillsdale, N. J.: Erlbaum.

THATCHER, R. W., D. P. PURPURA
1972 Maturational status of inhibitory and excitatory synaptic activities of thalamic neurons in neonatal kitten. *Brain Research* 44: 661–665.
1973 Postnatal development of thalamic synaptic events underlying evoked recruiting responses and electrocortical activation. *Brain Research* 60: 21–34.

VAN HEERDEN, P. J.
1968 *The foundation of empirical knowledge*. Wassenaar, Netherlands: N. V. Uitgeverij Wistik.

VERZEANO, M.
1955 Sequential activity of cerebral neurons. *Archives of International Physiology and Biochemistry* 63: 458–476.
1963 The synchronization of brain waves. *Acta Neurologica Latinoamerica* 9: 297–307.
1970 "Evoked responses and network dynamics," in *The neural control of behavior*. Edited by R. E. Whalen, R. F. Thompson, M. Verzeano, and N. M. Weinberger. New York: Academic Press.
1972 "Pacemakers, synchronization and epilepsy," in *Synchronization of EEG activity in epilepsies*. Edited by H. Petsche and M. A. B. Brazier. New York: Springer-Verlag.

VERZEANO, M., K. NEGISHI
1960 Neuronal activity in cortical and thalamic networks. A study with multiple microelectrodes. *Journal of General Physiology* 43 (suppl.): 177–195.
1961 "Neuronal activity in wakefulness and in sleep," in *The nature of sleep* Edited by G. Wolstenholme and M. O'Connor. London: Ciba Symposium.

VERZEANO, M., M. LAUFER, P. SPEAR, S. McDONALD
1965 "L'activité des reseaux neuroniques dans le thalamus du singe," in *Actualites neurophysiologigues* volume 6. Edited by A Monnier 223–253. Paris: Masson.

YOSHII, N., M. SHIMOKOCHI, Y. YAMAGUCHI
1960 Conditioning frequency characteristic repetitive response with electrical stimulation at some thalamic structures. *Medical Journal Osaka University* 10: 375–381.

A Comparative Analysis of U.S. and Slovenian Sociopolitical Frames of Reference

LORAND B. SZALAY and VID PECJAK

> The sources of tragedy in international life lie in the differences of outlook that divide the human race; and it seems to me that our purposes prosper only when something happens in the mind of another person, and perhaps in our mind as well, which makes it easier for all of us to see each other's problems and prejudices with detachment and to live peaceably side by side
>
> (George Kennan 1964)

The following psychocultural analysis represents an indepth study with apparently conflicting objectives. It is directed toward intangible cultural dispositions which at times are not able to be verbalized (i.e. images, meanings, beliefs) which rely on inferences drawn from verbal associations. It aims at an emic yet comparative analysis. It strives to be apolitical while dealing with timely political topics. It introduces findings of methodological and theoretical consequences which have been gained through extensive data on specifics.

As the method has been discussed at length in several recent publications (Szalay and Brent 1967; Szalay and D'Andrade 1972; Szalay and Bryson 1973; Szalay and Maday 1973), the present article will simply describe a few details indispensable to an understanding of the results.

In line with the original objective of providing an apolitical, emic analysis, no effort was made to formulate hypotheses or to anticipate findings; rather, a heuristic approach has been taken, assuming that the findings are sufficiently coherent to speak for themselves.

The investigation reported here represents cooperative research which explores the effects of the processes of political socialization on comparable student groups in the United States and Yugoslavia, representing countries with pragmatic or ideological orientations respectively. More specifically, it focuses on this question: Do the pragmatically and ideologically oriented processes of political socialization each produce distinct and clearly identifiable effects on the political frame of reference? This question has a direct bearing on our lack of knowledge concerning man's beliefs and values as products of different sociopolitical systems of East and West. This gap has been extensively commented upon recently by several leading anthropologists, such as Margaret Mead and Edward Hall, at the hearings of the Senate Foreign Relations Subcommittee (United States Senate 1969).

Compared to the United States, a leading "Western democracy," Slovenia represents a relatively small central European country, a republic of the Yugoslav Federation with a Marxist-socialistic constitution. Of the six republics of Yugoslavia, Slovenia is the most Western; it has a distinct sense of national identity, its own language, and the highest average level of income and education. Slovenia was a part of Austria until 1918 and its history is characterized by repeated independence movements and, during World War II, by strong partisan activity.

The data reported in this article represent part of a broader comparative study of American and Slovenian students that involved a variety of domains concerning their social, moral, and political frames of reference. With special reference to the meanings, images, and attitudes generated by the varied processes of political socialization, the present article focuses on the group frame of reference in three major political domains: (1) *isms*, *ideology*, (2) *politics and political institutions*, and (3) *values, ideals.*

Comparative analytic studies of pragmatic and ideological political orientations offer insights into the development and dynamics of belief systems which investigations limited to single social systems would hardly provide. Very few studies have explored the question of how ideological systems affect people in countries where political socialization has a strong ideological foundation. The most outstanding in many respects is the Harvard study conducted on Soviet citizens (Inkeles and Bauer 1961; Bauer 1959). Some major trends that emerged from this study were the emphasis on order, acceptance of the authority principle, rejection of certain manifestations of freedom, collectivistic orientation, heavy emphasis on society and social justice, and faith in Communist ideals despite disillusionment with many practical aspects and timely realities. In considering the case of Yugoslavia, the relevance of observations based on Soviet citizens is naturally limited, but it is interesting to keep these findings in mind.

The present study deals with the following questions: Are there consistent trends in perceptions, attitudes, and beliefs which are characteristic of Slovenians and which differentiate them from Americans? How do the actual beliefs of these two groups relate to the prevalent sociopolitical ideologies associated with the United States and Yugoslavia? Does ideology influence beliefs only in the domain of *isms*, or does it also influence other domains as well? What is the level of ideological polarization; in other words, to what extent do they reject ideologies associated with the sociopolitical system of the other group? For example, to what extent do Slovenians approve of socialism and at the same time reject capitalism? Finally, it is intriguing to ask whether it is possible to identify elements of the Slovenian belief system that are characteristic of the broadly emphasized, individual Yugoslav approach to socialism.

METHOD

Data Collection

In the framework of a broader investigation, data were collected from American and Slovenian student groups during the summer of 1969. The U.S. group was composed of fifty University of Maryland students, twenty-five male and twenty-five female. The Slovenian student group had a similar composition and was tested at the University of Ljubljana in Yugoslavia.

The original testing, with the objective of a comparative cultural analysis, consisted of three consecutive steps in which continued verbal association tasks were used with small modifications. First, the high-priority domains of both cultural groups were identified. Second, the four most dominant representative themes within these domains were assessed for both cultural groups. Third, a combined list of representative themes was compiled, translated into both languages, and then administered to members of both cultural groups.

This article presents the results of an analysis obtained on a sample of twelve themes used in the representation of three political domains: *ideologies*, *institutions*, and *ideals*.

The research method, the Associative Group Analysis technique (AGA), elicits native-language, free verbal associations by using selected themes as stimulus words. The subjects are asked to give as many responses as they can think of in one minute in the context of each stimulus word (theme) presented on separate cards.

Each response is given a score which indicates the relative importance of this element to the meaning of the stimulus for the particular group. Scores consist of frequencies within fifty-member groups weighted by the order of occurrence. The weights assigned to responses, beginning with the first in the sequence, are: 6, 5, 4, 3, 3, 3, 3, 2, 2, 1, 1. . . . These weights have been empirically derived from the differential stability of rank place assessed by the test-retest method in previous investigations (see Szalay and Brent 1967).

All shared responses to a particular theme are compiled into a group-response list that describes the meaning that a particular theme has for a particular group. Meaning is used here not in a linguistic but in a psychological sense, as a subjective coding reaction to a word learned by a particular person (see Osgood, Suci, and Tannenbaum 1957).

Methods of Analysis

To extract the information provided by these response lists, various analytic procedures have been developed. These analytic methods enable the investigator to determine how the understandings of a theme by one group relates to that of another group, how dominant a theme is for each group, to what extent the groups agree on the interpretation of a particular theme, what are the major components of their agreement and disagreement, and so on.

Based on the responses in the group-response lists, two analytic procedures were used to obtain the desired categories of information.

IDENTIFICATION OF MAIN MEANING COMPONENTS. An initial major objective is to organize the many different responses into a more concise and manageable form, focusing on a smaller number of the major components of group meaning.

The method developed for identifying the primary meaning components relies on a content analysis of the responses. Two or more independent judges establish eight to sixteen categories that they feel contain all the responses in meaningful groupings relevant to the stimulus word. They then assign the responses to these categories based on their similarities in content. Responses that do not seem to fit into any of the categories are put into a miscellaneous category. In previous investigations, the interjudge reliability calculated by correlations among four judges across categories averaged 0.7. In the present study two judges were used: one Slovenian and one American.

ASSESSMENT OF GROUP PRIORITIES. The psychological priorities characteristic of a particular group or culture can be inferred from dominance scores. The dominance score is a modified version of Noble's measure of "meaningfulness" (1952). It is based on the number of responses produced in common by the members of the group and weighted by the sequence in which they were produced. Responses in common are those associations that were given by at least two members of the group. The dominance scores indicate how meaningful and how important a theme (stimulus word) is for a particular group. Group-dominance scores show group-specific priorities not only for single themes, but also for clusters of themes representing broader domains.

Other analytic methods developed for assessing the similarity of group meanings of selected themes, for mapping the relationship of themes as well as domains, and for assessing group attitudes are described elsewhere (see Szalay and Brent 1967; Szalay and Lysne 1970).

RESULTS AND DISCUSSION

1. *Isms, Ideology*

A comparison of the political frames of reference of the American and Slovenian students has, by nature, an ideological dimension. Whether this dimension is important or not, whether the differences are small or large are presently debatable, academic questions. In an effort to make empirical comparisons, four of the most frequently used *isms* have been selected for analysis: *communism, Marxism, socialism,* and *capitalism* (see Figures 1, 2, 3, and 4).

COMMUNISM (KOMUNIZEM). The main components of this theme (Figure 1) are described below:

1. *Positive characteristics* (U.S. 23, S. 160). This predominantly Slovenian component expresses positive evaluation. The most weighty Slovenian positive attributes were *equality* (68), *peace* (21), *justice* (15), and *freedom* (14); the only sizable U.S. response was *good* (17).
2. *Political isms* (U.S. 152, S. 153). The largest common meaning elements were *socialism* (U.S. 63, S. 76), *capitalism* (U.S. 30, S. 37), and *Marxism* (U.S. 30, S. 9). Some smaller U.S. references — *Nazi* (5), *Fascist* (9), and *democracy* (15) — were not mentioned by the Slovenian group. Nonetheless, the two groups apparently perceive the same degree of relationship of communism with other isms, especially socialism.
3. *Theory, ideology* (U.S. 19, S. 111). This was an overwhelmingly Slovenian category. The largest Slovenian responses were *utopia* (35), which seems to express some skepticism, and *future* (40), which may reflect the belief that communism is the system of the future.
4. *Party, system* (U.S. 34, S. 76). The largest American responses were *party* (18) and *government* (12); the largest Slovenian responses were *society* (32), *party* (24), and *system* (13). Certain responses such as *society* and *class* have fairly specific interpretations in Marxist theory.
5. *Symbolic references* (U.S. 66, S. 35). The largest American response was *red* (41).
6. *Miscellaneous* (U.S. ??) ?

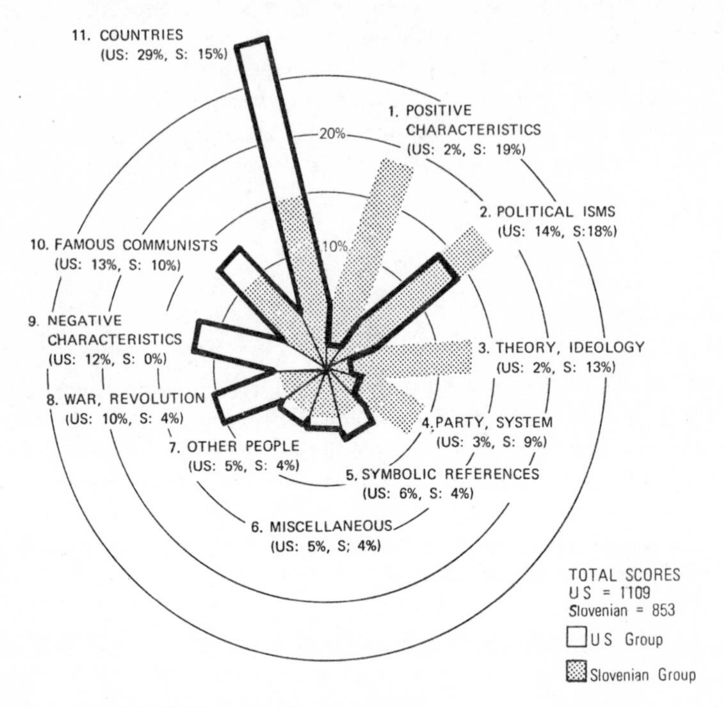

Figure 1. *Communism*: main meaning components

U.S. GROUP: The American image of communism includes as the most salient element foreign *countries* and political systems, represented primarily by the Soviet Union and personified by Marx, Krushchev, and other *famous Communists*. It contains distinct references to *isms*, especially to socialism; it is aggressive, expansionistic, promoting *wars* and *revolution*. The general evaluation of communism is strongly *negative*.

SLOVENIAN GROUP: The Slovenian image of communism is composed of somewhat different elements. There is less emphasis on *countries*, that is, on the systematic aspect of communism which in this case includes their own sociopolitical system. As a major component, *ideology* is in the foreground of interest. The list of *famous Communists* includes their own leader Tito, and the role of the *party* is more central. The general evaluation of communism is strongly *positive;* it is depicted as an ideal and just system.

7. *Other people* (U.S. 60, S. 37). The American group concentrated on *dictator* (19) and *people* (22); the largest Slovenian responses were *worker* (21) and *proletariat* (9), both of which carry distinct ideological connotations.

8. *War, revolution* (U.S. 106, S. 37). This meaning component shows that the American group perceives communism as an aggressive, expansive, and oppressive force: *war* (25), *take over* (13), *spread* (12), and *oppression* (11).

9. *Negative characteristics* (U.S. 128, S. 4). The negative evaluations were sizable and came mainly from the American group. The main attributes, *bad* (51), *poor* (19), *evil* (10), *threat* (15), and *hate* (14), express strong negative evaluation.

10. *Famous Communists* (U.S. 149, S. 82). The Slovenian group focused primarily on two well-known Communists, *Tito* (29) and *Lenin* (46). The American responses were more diverse, with emphasis on *Marx* (74) and *Kruschev* (31). This component reflects a strong American emphasis on Marxist-Soviet leadership and a more Leninist and Yugoslav trend by the Slovenian group.

11. *Countries* (U.S. 319, S. 130). For the American group, communism is primarily a system characteristic of the *Soviet Union* (55) or *Russia* (141), *China* (43), and *Vietnam* (39). This "systemic" aspect of communism had lower salience for the Slovenians. The Slovenians' references to *Yugoslavia* (23) closely follow their references to *Soviet Union* (36) and *China* (26), which suggest that this group places its own country at practically the same level of communism as the two Communist giants.

MARXISM (MARKSIZEM) The main components of this theme (Figure 2) are described below:

1. *Famous representatives* (U.S. 236, S. 292). The two foremost representatives of Marxism were *Marx* (U.S. 104, S. 171) and *Lenin* (U.S. 93, S. 63). The U.S. group made more references to *Stalin* (17); the Slovenian more to *Engels* (43) and a few to *Tito* (7). This component was somewhat stronger for the Slovenian group.

2. *Philosophy, ideology* (U.S. 83, S. 158). This primarily Slovenian category reflects that *Marxism* implies a strong philosophical and ideological orientation. The main elements were *philosophy* (U.S. 16, S. 36), *history* (U.S. 12, S. 21), *science* (S. 21), *book* (U.S. 8, S. 20), and *ideology* (S. 18).

3. *Future, remote* (U.S. 0, S. 42). This purely Slovenian component suggests a view that *Marxism* is the trend of the *future* (18), but that views on such a development are not free of skepticism (*Utopian* 11).

4. *Positive characteristics* (U.S. 13, S. 49). Such values as *equality* (24) and *humanity* (20) were emphasized by the Slovenians. The American responses were small and nonspecific.

6. *Society, people* (U.S. 54, S. 51). This social aspect received an about equal attention from the American and Slovenian groups. The two main elements were *workers* (U.S. 17, S. 27) and *man* (U.S. 3, S. 11).

7. *Negative characteristics* (U.S. 28, S. 0). This small, purely American component reflects generally negative evaluation (*bad* 12, *poor* 6).

8. *Economic references* (U.S. 40, S. 24). This small but articulate component was somewhat stronger for the American group. It was com-

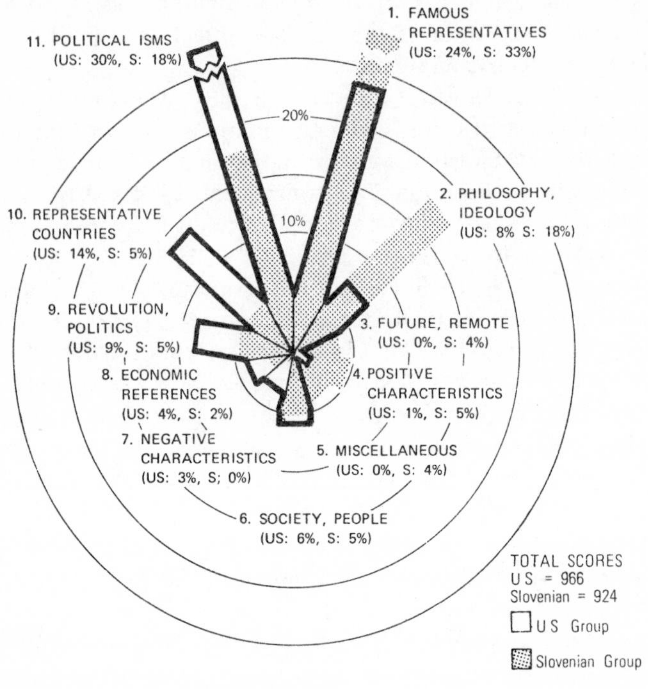

Figure 2. *Marxism*: main meaning components

U.S. GROUP: The U.S. image of Marxism contains an especially salient component involving *political isms*, especially communism. This systemic focus, which probably derives from a close association with communism, is also expressed by the emphasis placed on *representative countries*. There is a strong focus on *famous representatives* and a certain weight on *revolution* as a practical objective. The evaluation of Marxism is primarily *negative*.

SLOVENIAN GROUP: The Slovenian image of Marxism contains a larger component on *famous representatives* than does the U.S. image; this component emphasizes the role of Karl Marx and other ideologues. Similarly, the larger score on *philosophy, ideology* reflects this primarily theoretical focus, suggesting that for Slovenians Marxism implies more an ideological system than etatism or an action program. References to *future* may suggest that Marxism as an ideal has not materialized. The *societal* implications of Marxism receive more attention than the *economic* ones. The evaluation of Marxism by the Slovenian group is *positive*.

posed of diverse elements such as *economy* (14), *industry* (8), and *economics* (8), while the only Slovenian response was *capital* (24).

9. *Revolution, politics* (U.S. 84, S. 47). Both groups referred to *revolution* (U.S. 23, S. 20) and to *war* (U.S. 10, S. 9). The American group placed more emphasis on *government* (16) and *dictatorship* (11) and expressed concern with centralized *power* (4).

10. *Representative countries* (U.S. 134, S. 50). *Russia* was the largest

American referent by far (76); for the Slovenian it scored much lower (18). For the Slovenians, *Yugoslavia* was the main representative (19); *China* received more attention from the Americans (19) than from the Slovenians (7). This component was much stronger for the American group than for the Slovenian, while the representative ideologues and ideologies were more strongly emphasized by the Slovenians. This suggests that for the Slovenian group *Marxism* is associated more with people than with countries and more with political ideology than political systems.

11. *Political isms* (U.S. 190, S. 173). The two largest elements were *communism* (U.S. 188, S. 67) and *socialism* (U.S. 84, S. 46). The American emphasis on this component was distinctly greater than the Slovenian.

SOCIALISM (SOCIALIZEM). The main components of this theme (Figure 3) are described below:

1. *Ideals* (U.S. 89, S. 163). Ideals and principles appeared to play a greater role for the Slovenians than for the Americans: *freedom* (U.S. 28, S. 19), *equality* (U.S. 19, S. 57), and *peace* (U.S. 10, S. 25). This category reflects specific values as well as general positive evaluation.

2. *Representative people* (U.S. 88, S. 148). The Slovenians assign a greater role to leading representatives and ideologues. Most of the American references were to *Marx* (U.S. 82, S. 86). For the Slovenian group *Tito* (46) was the most prominent representative.

3. *Society, state* (U.S. 83, S. 143). This primarily Slovenian component reflects an emphasis on large social organizations: *society* (U.S. 20, S. 45), *social order* and *order* (S. 26).

4. *Negative characteristics* (U.S. 19, S. 54). The Slovenian responses suggest more negative evaluation than the American responses: *bad* (U.S. 12, S. 10), *exploitation* (S. 10), *deception* (S. 8). These negative elements were outweighed by the positive ones mentioned under ideals, but in combination suggest a certain ambivalence.

5. *Economy* (U.S. 45, S. 24). In this small category, both groups agree on the relevance of *work* (U.S. 14, S. 15), and other American responses (*economy* 11, *money* 7) imply a focus on economic variables.

7. *Government, power* (U.S. 76, S. 41). The many American references to government (U.S. 52, S. 8) and the larger category score suggest a focus on the role of government and on socialism as a political system. The largest Slovenian response was *self-government* (15), a characteristically Yugoslavian solution central to the "Yugoslav way to socialism."

8. *Programs, principles* (U.S. 80, S. 7). This American component involves such action programs as *welfare* (U.S. 27) and *medicare* (U.S. 22).

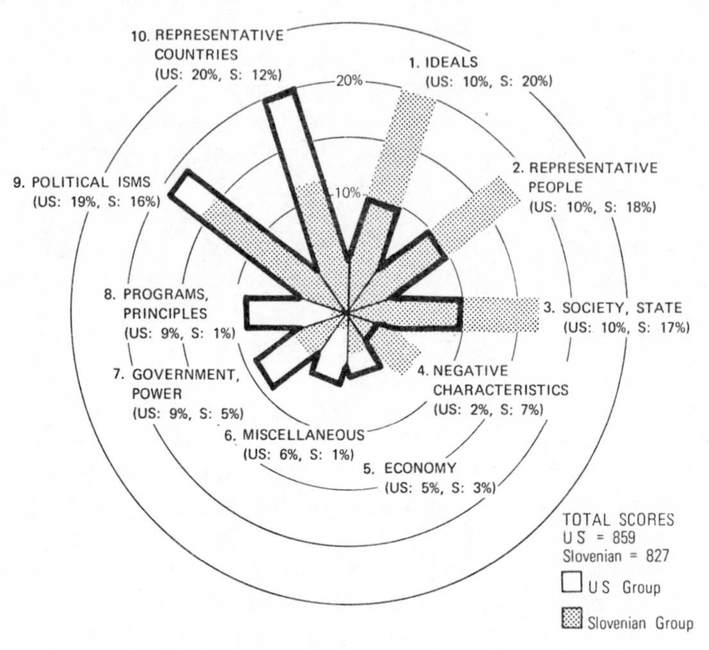

Figure 3. *Socialism*: main meaning components

U.S. GROUP: The American image of socialism is apparently a complex one. It includes two different political systems as shown by *representative countries*: one which is closely synonymous with the Soviet system and another which is represented by Western democracies such as England and Sweden. The American interpretation of socialism is less theoretical but has a strong focus on the organization of *power* and on *government*. As a related trend, distinct attention is given to practical *programs*: welfare, medicare. There are numerous *ideals* attached to socialism, suggesting positive evaluation.

SLOVENIAN GROUP: The Slovenian image of socialism is less complex but perhaps emotionally more ambivalent. Socialism stands for the theory and reality of their own *country* and for the ideology of their own leader (see *representative people*). Slovenians appear to be strongly preoccupied with characteristics of the *society*, with the high *ideals* it should represent, and with apparent disappointment over certain *negative characteristics* which, however, are outweighed by the positive ones.

9. *Political isms* (U.S. 160, S. 129). The isms may refer to both political systems and ideologies. *Communism* (113) was the largest American response, but there were also references to *democracy* (17), indicating Western democracies. The Slovenians referred mainly to *communism* (71) and *capitalism* (63) as opposites. The response *capitalism* suggests an ideological focus, emphasizing the polarity of socialism and capitalism.

10. *Representative countries* (U.S. 169, S. 101). This category shows

what socialism denotes for both groups: that is, particular countries or systems. The political system aspect appears to be more dominant for the American than for the Slovenian group. For the American group, there were two main categories of political systems representative of socialism: (a) *Soviet Russia* (58), which may be labeled Soviet socialism, and (b) a group of Western countries —*England* (37), *Sweden* (23), and the *United States* (20). The Slovenians do not mention Western countries, only their own country *Yugoslavia* (84) and *Soviet Russia* (17).

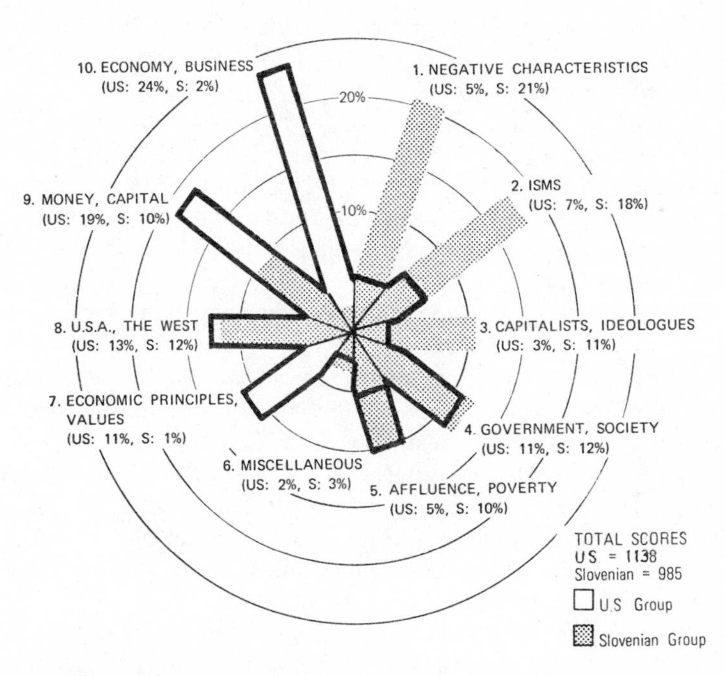

Figure 4. *Capitalism*: main meaning components

U.S. GROUP: The American image of capitalism reflects primarily an economic-financial system founded on *economy, business* and on *money, capital*. In their image of this system, social considerations find little representation; political considerations are also subordinate, mainly limited to *government. Economic principles*, however, occupy a central position. The system is viewed as representative of the *United States*. Its evaluation contains references to positive *values* as well as critical attitudes, *negative characteristics*.

SLOVENIAN GROUP: The Slovenian image of capitalism, in contrast to the American, contains remarkably few economic, financial elements. Capitalism is perceived as *negative*, exploitative, unjust. It is conceived as a social and political system represented by the *United States* and *the West* and characterized by a polarized contrast between *affluence* and *poverty*. This image shows strong similarities with the Marxist analysis; it reflects the polarity of *isms* and the position of Marxist *ideologues*.

CAPITALISM (KAPITALIZEM). The main components of this theme (Figure 4) are described below:

1. *Negative characteristics* (U.S. 56, S. 211). The responses *exploitation* (U.S. 9, S. 85), *war* (U.S. 7, S. 26), *inequality* (S. 19), and *decay* (S. 15) have predominantly sociocritical, ideological overtones which are stronger for the Slovenians.

2. *Isms* (U.S. 78, S. 173). The total Slovenian response score is more than twice as large as the American score. The largest elements, *socialism* (U.S. 51, S. 131) and *communism* (S. 27), suggest a strong ideological focus by the Slovenian group.

3. *Capitalists, ideologues* (U.S. 36, S. 113). The main responses were *Marx* (S. 34), *worker* (S. 29), *capitalist* (S. 22), *Nixon* (U.S. 13), and *Engels* (S. 10). The American group mentioned a few representatives of capitalism, while the Slovenians referred to Marxist thinkers and to people in Marxist categories.

4. *Government, society* (U.S. 126, S. 118). The American group focused on *government* (U.S. 58) and *system* (U.S. 45, S. 27); the Slovenian group focused on *society* (U.S. 6, S. 25), *class* (S. 23), and *order* (S. 12). This implies that for the American group capitalism applies more to the government, while for the Slovenians it is a characteristic of the society or social order.

5. *Affluence, poverty* (U.S. 58, S. 94). The main responses were *wealth* (U.S. 25), *rich* (U.S. 18, S. 46), *poor, poverty* (U.S. 11, S. 37), and *famine* (S. 10). Although both groups emphasized wealth, the Slovenians placed an equal emphasis on poverty and famine, which gives a contrast between richness and poverty with obvious social implications.

7. *Economic principles, values* (U.S. 124, S. 14). The responses *free enterprise* (U.S. 21), *private enterprise* (U.S. 22), *good* (U.S. 22), and *competition* (U.S. 14) indicate further American emphasis on economic considerations and address operational characteristics, which frequently carry positive connotations.

8. *U.S.A., the West* (U.S. 146, S. 118). The responses *U.S.A.* (U.S. 146, S. 65), *West* (S. 21), and *abroad* (S. 23) indicate that for both groups the United States is the country most representative of capitalism.

9. *Money, capital* (U.S. 216, S. 95). The main elements of this sizable category were financial references: *money* (U.S. 158, S. 66), *capital* (U.S. 14, S. 29), and *profit* (U.S. 23). The heavy American references suggest a strong financial component in the American interpretation of capitalism.

10. *Economy, business* (U.S. 273, S. 15). This very sizable category shows that for the American group capitalism is a primarily economic concept, while for the Slovenians it has only negligible economic connotations. The major American elements involved *economy* (73), *business* (58), *company* (41), *industry* (32), and *trade* (22).

COMMON TRENDS OF INTERPRETATION: ISMS, IDEOLOGY. The two cultural groups studied show considerable consistency in capitalizing on the same or similar elements, priorities, and evaluations across the four themes studied. Despite the apparent dangers of oversimplification, some of the major trends can be presented in a tabular form (see Table 1). The main trends are the following (numbers refer to Table 1):

Table 1. Common trends of interpretation in the domain of *isms*, *ideology*

Trend of interpretation	Communism	Marxism	Socialism	Capitalism
1. Focus on systems, countries	U.S.	U.S.	U.S.	U.S.
2. Focus on theory, ideology	S.	S.	S.	S.
3. Greater emphasis on economic references (money)	—	U.S.	U.S.	U.S.
4. Greater emphasis on social references (society)	S.	S.	S.	S.
5. More weight on government	U.S.	U.S.	U.S.	U.S.
6. More weight on state	S.	S.	S.	S.
7. More negative evaluation	U.S.	U.S.	—	S.
8. More positive evaluation	S.	S.	S.	U.S.
9. Most representative person — U.S.	Marx	Marx	Marx	Nixon
10. Most representative person — S.	Lenin	Marx	Tito	Marx
11. Most representative country — U.S.	U.S.S.R.	U.S.S.R.	U.S.S.R.	U.S.A.
12. Most representative country — S.	U.S.S.R.	Yugo.	Yugo.	U.S.A.

1–2. It is a general trend that for the American group all four *isms* mean, to a greater extent, systems associated with particular countries in contrast to the Slovenian interpretation, which capitalizes more heavily on the theoretical, ideological interpretations. In other words, for the Americans, *isms* refer more to concrete, political systems; for the Slovenians they refer more to abstract, theoretical systems.

3–4. In the interpretation of all *isms* except communism, the American group assigns greater importance to economic factors. At the same time, the Slovenian group stresses more heavily the social and societal aspects. The systems designated by the various *isms* are viewed and evaluated primarily in the context of their perceived, or anticipated, social implications. Slovenian references to society, classes, workers, and the proletariat are consistently higher than American. Similarly, social values, especially equality, receive consistently more attention.

5–6. The American group makes consistently more references to government and uses politicalisms primarily in reference to political organizations and to the characteristics of the government. Similarly, from a political angle, the Slovenian group stresses the role of the state and the role of certain social strata, especially workers.

7–8. On three isms — *communism, Marxism*, and *capitalism* — the American and Slovenian groups show conflicting trends in their evaluations. On *socialism* both groups show positive and negative elements of evaluation whereby the positive outweigh the negative attitudes. There is also a certain contrast in the intensity of evaluation. *Communism* is the most positive for the Slovenian and most negative for the American group. On *capitalism* the American group also shows distinct negative evaluations, but the negative are outweighed by the positive. On the other hand, the Slovenian group's evaluation of *capitalism* is intensively negative.

9–10. In respect to representative people, for the Slovenian group Lenin is the major representative of *communism*; Marx, of *Marxism*; Tito, of *socialism*; and again Marx has been most frequently mentioned in the context of *capitalism*, probably as its most famous critic. In the case of American students Marx stands for *communism, Marxism*, and *socialism*, suggesting a less differentiated interpretation of these three major theories with Marxist foundation.

11–12. A similar trend is observable in respect to the most representative country mentioned by the American and Slovenian groups. The Slovenians most strongly associate the Soviet Union and Yugoslavia with *communism,* and Yugoslavia with *Marxism* and *socialism.* The Americans most strongly associate the Soviet Union with all three of these *isms.* Even if the U.S. group has two separate concepts of *socialism* — Soviet and Western — this group perceives *communism, Marxism,* and (Marxist) *socialism* with less differentiation than the Slovenian group.

2. *Politics, Political Institutions*

In the representation of this second domain, the image of a few roles and institutions — *president, government, politics,* and *political party* — were submitted to comparative analysis. Compared to the previous domain focusing on *isms, ideology*, the political institutions were expected to show not only different theories and principles but also different political realities and organizations.

PRESIDENT (PREDSEDNIK). This theme was analyzed in terms (Figure 5) of the following components:

1. *Country, state* (U.S. 75, S. 193). This component was more weighty for the Slovenian than for the American group. The Slovenian references focused on *state* (80) and *republic*[1] (54). These responses suggest a close connection between the presidency and the political structure for the Slovenian group. The Slovenians emphasized that the president is the top

[1] The Slovenian response *drzava* primarily means "state" but as a second reference also "country." The Slovenian response *republic* refers to the units of the Yugoslav Federation, which are comparable to states in the United States.

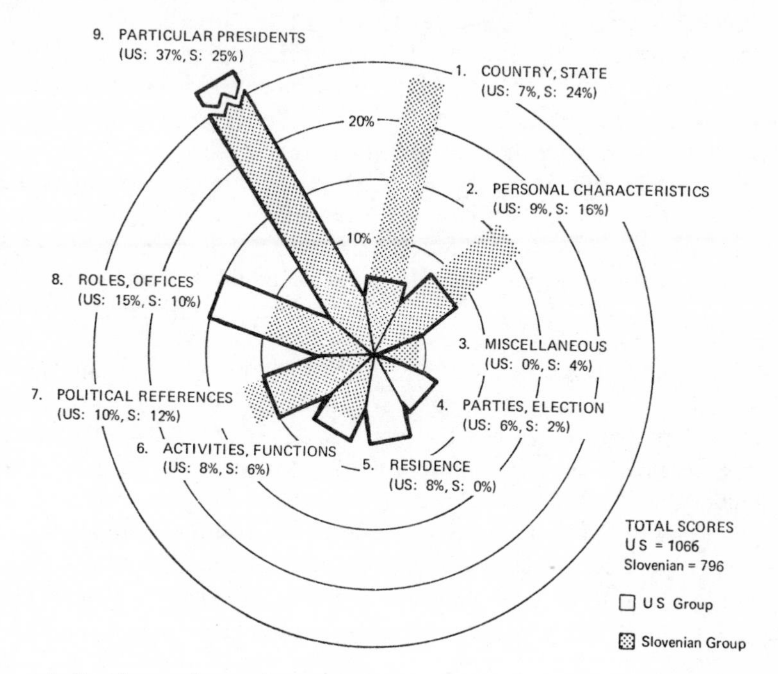

Figure 5. *President*: main meaning components

U.S. GROUP: The American image of the president is personified by the incumbent President Nixon as well as by several of his predecessors. As a *role* or *office*, it represents top leadership, civilian as well as military. The power carried by the president receives distinct attention, and from his numerous *activities* and *functions*, decisions in respect to war are emphasized. The foundation of the office on *elections* is stressed. The president is viewed in terms of *personal characteristics*, which reflect egalitarian undertones and criticism.

SLOVENIAN GROUP: The Slovenian image of the president probably reflects components determined mainly by a single person: President Tito. Accordingly, the president is viewed in his *role, office* as the head of the political structure, the state (i.e. the federation). His policy in respect to war and peace is given special attention. His *personal characteristics* are predominantly positive but contain casual notes as well so that the general impression is free of the flavor of personality cult.

man, the head of state. The main American response, *America* (U.S. 48, S. 12), is probably not free from patriotic undertones.

2. *Personal characteristics* (U.S. 100, S. 127). The Slovenian responses were more weighty and more positive: *good* (S. 40), and *responsibility* (U.S. 15, S. 15). The Slovenians also referred more to *authority* (U.S. 6, S. 25) and corpulance —*big* (S. 11) and *fat* (S. 11). The U.S. characterizations of the president involved references to *money* (11) and *richness* (8) and were more negative — *bad* (15) and *liar* (8).

4. *Parties, election* (U.S. 60, S. 19). The main American responses,

elect, election (U.S. 52, S. 6), reflect a strong emphasis on the process by which the president obtains his office. Both groups made only negligible references to *political parties* (U.S. 8, S. 8).

5. *Residence* (U.S. 79, S. 0). The U.S. references to the residence of the president — *White House* (53), *Washington* (22) — were sizable and suggest that the presidency in the United States has a stronger, more tradition-based, institutionalized foundation.

6. *Activities, functions* (U.S. 85, S. 51). In addition to a variety of small references to diverse activities — decision, visit, speak — the main responses were: *war* (U.S. 25, S. 11), *peace* (S. 15), and *rule* (U.S. 15). There was considerable similarity in the activities and functions attributed to the president by the two groups. The most weighty responses were related to war and peace.

7. *Political references* (U.S. 112, S. 92). Many of the American and Slovenian elements were similar: *government* (U.S. 33, S. 33), *politics* (U.S. 15, S. 19), *parliament* (S. 12), and *Congress* (U.S. 14). The American group paid somewhat more attention to *power* (U.S. 34, S. 9) and *democracy* (U.S. 16).

8. *Roles, offices* (U.S. 164, S. 21). This meaning component was stronger for the American group. The main roles mentioned were *leader* (U.S. 34, S. 33), *chief, commander in chief* (U.S. 36), *vice-president* (U.S. 15, S. 6), *office* (U.S. 15), and *politician* (U.S. 9, S. 15). This component has distinct military connotations (e.g. *commander in chief*).

9. *Particular presidents* (U.S. 386, S. 197). The incumbent presidents *Nixon* (U.S. 216, S. 10) and *Tito* (S. 155) received about the same attention. However, the American category total was much larger than the Slovenian as an obvious consequence of the Americans mentioning numerous preceding presidents as well: *Johnson* (U.S. 69), *Kennedy* (U.S. 67, S. 26), and *Eisenhower* (U.S. 12). This reflects that the presidential office in the United States involves change, succession, and past history. The Slovenian references to Kennedy are impressive.

GOVERNMENT (VLADA). This theme was analyzed in terms (Figure 6) of the following main components:

1. *State, political institutions* (U.S. 106, S. 180). This was the largest Slovenian category. It expressed an especially strong focus on the *state* (U.S. 10, S. 92), *étatism*. The American references to *state* have a different referent, namely the member states in the United States. The primary U.S. focus was on the legislative bodies: *Congress* (U.S. 27) and the *Senate* (U.S. 25).

2. *Statesmen* (U.S. 92, S. 131). The most salient references were to *president* (U.S. 48, S. 49), *Nixon* (U.S. 25), *Tito* (S. 29), *king* (S. 29), *ruler* (S. 13), and *emperor* (S. 11). The American group referred mainly to

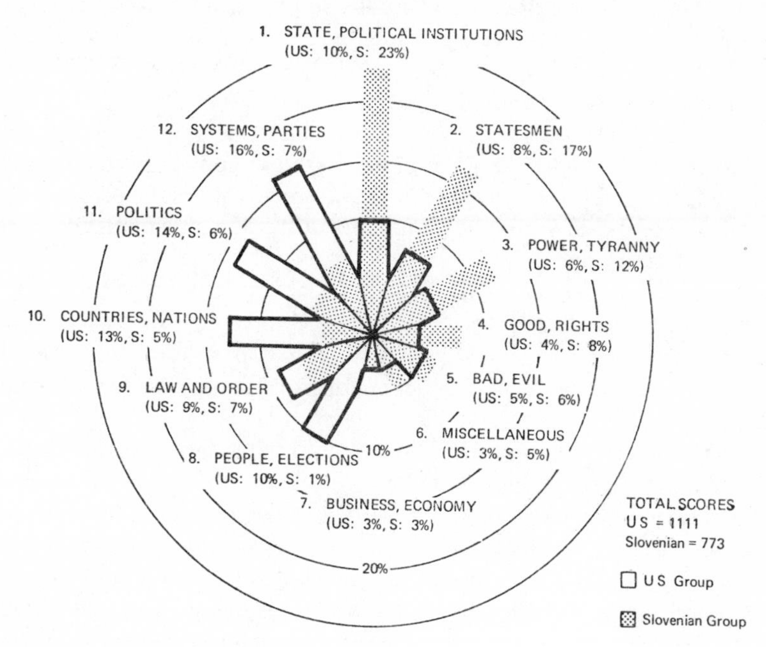

Figure 6. *Government*: main meaning components

U.S. GROUP: The American image of government is that of a political institution whose form depends a great deal on the particular *system* and specific *country*. It is apparently a function of *politics* and is regulated by *law*. Its central element is *power*, which should derive from the *people*, whom it should serve as its main function. Its legality is based on the people's will expressed by *elections*. Its evaluation is somewhat critical.

SLOVENIAN GROUP: The Slovenian image of government as a political institution is viewed in close connection with the *state* and more traditional autocrats (for example, kings), and with the leadership of *statesmen*. *Power* is a central issue. The Slovenian group pays less attention to the variations of government among *systems* and *countries*.

presidents; the Slovenian group referred to more traditional, autocratic rulers.

3. *Power, tyranny* (U.S. 63, S. 92). This component reflects concern with the use and abuse of power. The main responses were *power, powerful* (U.S. 17, S. 11), *tyranny* (S. 14), *fall* (S. 13), *war* (U.S. 13, S. 11), and *dictator* (U.S. 12, S. 10). The Slovenian responses were a little more numerous and more extreme.

4. *Good, rights* (U.S. 39, S. 59). The main elements expressing positive attitudes and evaluations were *justice* (U.S. 6, S. 10) and *good* (S. 17). This category was about as strong as the following one expressing negative attitudes.

5. *Bad, evil* (U.S. 55, S. 50). The main responses were *bad* (U.S. 10, S. 18), *lying* (S. 17), *exploitation* (S. 15), *corruption* (U.S. 13), and *bureaucracy* (U.S. 14). The negative attitudes and undertones of criticism and hostility expressed in this category are apparently shared by both groups.

7. *Business, economy* (U.S. 31, S. 26). This category represents a very small component for both groups.

8. *People, elections* (U.S. 104, S. 6). This American component expresses the philosophy that government derives its power from the people and should stand for the people. Furthermore, it reflects the principle that the power of the government is founded on the outcome of elections, the choice of the people.

9. *Law and order* (U.S. 103, S. 55). The main elements of this component were *rule* (U.S. 39), *law* (U.S. 39), and *authority* (S. 35). There was more weight on the legal and leadership functions of the government by the American group, while the Slovenian group emphasized the principle of authority.

10. *Countries, nations* (U.S. 145, S. 39). This large, predominantly American component indicates a strong focus on their own government. The main countries mentioned were: *U.S., America* (U.S. 87, S. 9) and *Yugoslavia* (S. 18), along with the capitals, *Washington* (U.S. 39) and *Belgrade* (S. 12), respectively. As the ratio of positive and negative references suggests (components 4 and 5), there is considerable ambivalence in feelings related to government. Thus, heavy references to one's own government do not automatically reflect positive identification, but merely strong concern with its characteristics.

11. *Politics* (U.S. 157, S. 43). Heavy references were made to *politics* (U.S. 145) and *policy* (U.S. 12, S. 43). Political considerations appeared to be more pervasive in the context of government for the American group than for the Slovenian.

12. *Systems, parties* (U.S. 184, S. 52). This was the largest American meaning component, with numerous references to *democracy* (U.S. 95), as well as to *communist* (U.S. 35, S. 12) and *socialist* (U.S. 16, S. 11) governments. This suggests that for the U.S. group, the system associated with the government is especially important.

POLITICS (POLITIKA). The analysis of politics (Figure 7) revealed the following main components:

1. *Negative characteristics* (U.S. 169, S. 154). The two groups gave an approximately equal number of responses, suggesting strongly critical attitudes or negative evaluations. The largest American response was *corruption* (U.S. 29); the strongest Slovenian response was *dirty* (U.S. 22, S. 37). Both mentioned *bad* (U.S. 18, S. 13).

2. *Activities, communications* (U.S. 58, S. 105). Compared with the

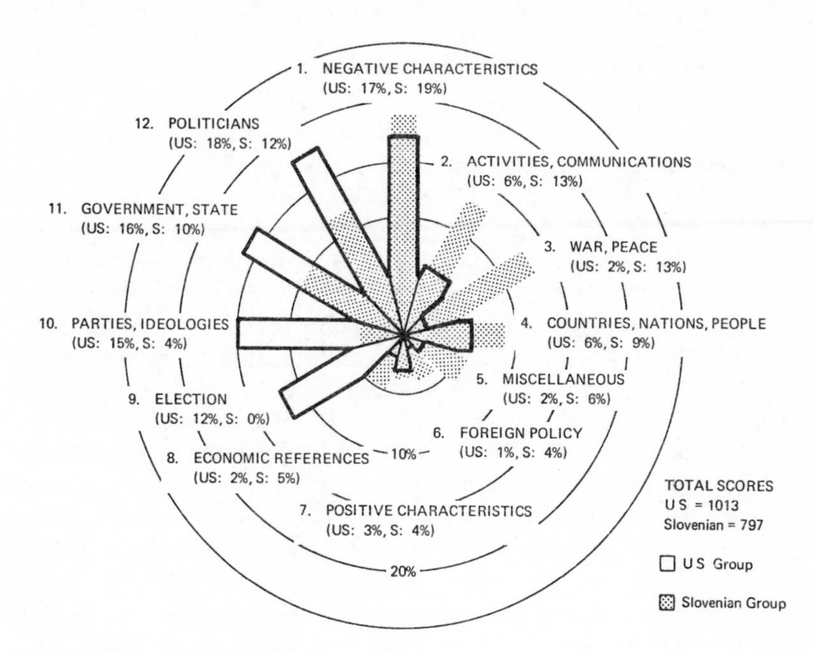

Figure 7. *Politics*: main meaning components

U.S. GROUP: The American image of politics is composed of activities conducted by three different categories of actors: (a) *politicians*, with the president as their uppermost representative; (b) *parties*, especially those that are most influential on the domestic scene; and (c) the *government*, including the legislative branch. *Elections* represent the point of departure as well as a major objective of politics. From the American group, politics receives a primarily *negative* evaluation.

SLOVENIAN GROUP: The Slovenian image of politics has *politicians* and the *state* as the two major actors. *Parties* receive negligible attention and there is no reference to *elections*, implying that they are not so important in the Slovenian interpretation of politics. Decisions concerning *war* and *peace* attract a great deal of interest. Among the various *activities, communications* play an especially important role. For this group politics has a stronger *foreign policy* component.

American interest in *rule* (11), *power* (11), and *maneuvering* (10) as political activities, the Slovenians showed a strong interest in communications. Their references to *newspaper* were especially numerous (51), suggesting that they consider newspapers to be an important instrument of politics.

3. *War, peace* (U.S. 22, S. 104). This large Slovenian category contained numerous references to *war* (U.S. 22, S. 80). The Slovenian reference to *peace* (24) was smaller but distinct. This suggests that the Slovenian group perceives politics more in a foreign, international framework than in terms of domestic activities, as may be the case with

the Americans. Their references to *diplomat, diplomacy,* and *coexistence* reinforce this impression.

4. *Countries, nations, people* (U.S. 65, S. 66). The American responses were limited exclusively to domestic references (e.g. *U.S.A.* 14, *Washington* 13), while the Slovenian group referred to *world* (11) and to various foreign countries.

6. *Foreign policy* (U.S. 7, S. 36). This was a small but distinct component. The largest Slovenian response was *foreign* (24).

7. *Positive characteristics* (U.S. 27, S. 32). This was a small category for both groups.

8. *Economic references* (U.S. 23, S. 29). A small group of Slovenian responses suggests that for this group economics has some political connotations. The American group referred only to *money* (U.S. 23, S. 8).

9. *Election* (U.S. 127, S. 0). This exclusively American response category includes references to the *election* process (45) starting with *campaign* (14) and ending with *winning* (17). There were no Slovenian responses in this category. For Slovenians elections apparently represent political events of lesser importance.

10. *Parties, ideologies* (U.S. 152, S. 33). The references to *party* (65) in general and to specific parties — *Democrat* (40) and *Republican* (23) — are purely American.

11. *Government, state* (U.S. 162, S. 84). The American focus is on *government* (U.S. 121, S. 27) as the main political organization, which includes the *Congress* (U.S. 21) and the *Senate* (U.S. 12) as well as the office of the *president* (see the theme government [Figure 6]). The Slovenians refer especially frequently to *state* (52).

12. *Politicians* (U.S. 178, S. 98). The main responses were *Nixon* (U.S. 64), *president* (U.S. 60, S. 10), and *Kennedy* (U.S. 13). This is the largest American meaning component. It shows that the office of the president is considered as characteristically political. The Slovenian responses were less numerous and only a small reference was made to President *Tito*.

POLITICAL PARTY (STRANKA). The following components of American and Slovenian interpretation of political party (Figure 8) were identified in the analysis:

1. *Political principles, organizations* (U.S. 113, S. 154). The Slovenians showed stronger political-ideological connotations. *Politics* (U.S. 27, S. 88) was the largest response. Although the Americans emphasized *government* (U.S. 25), the Slovenians referred more to *state* (U.S. 11, S. 17).

2. *Economics, finances* (U.S. 5, S. 133). The most frequent responses were *trade* (S. 41), *buy* (S. 75), *apartment* (S. 13), and *money* (U.S. 5, S. 10). The strong economic connotation attached by the Slovenian group to

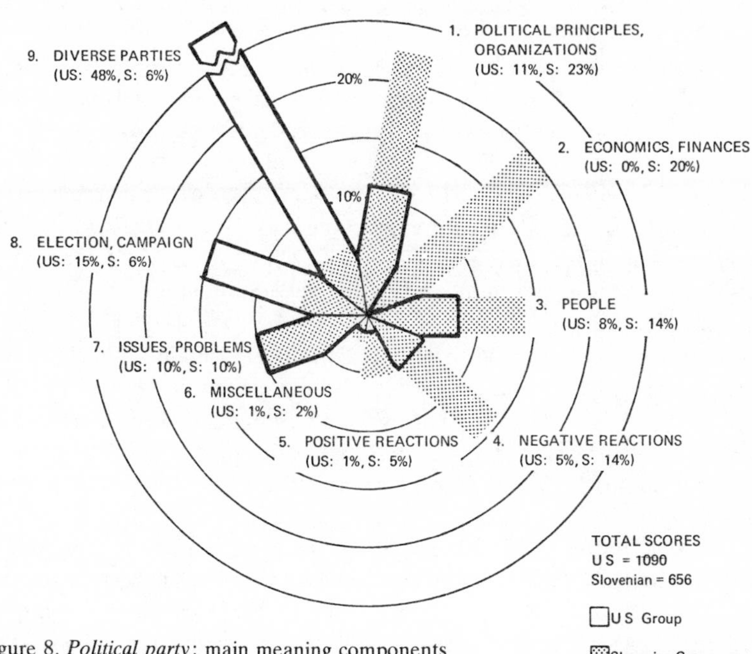

Figure 8. *Political party*: main meaning components

U.S. GROUP: The American image of political parties is first of all a pluralistic image involving a diversity of parties: Democratic, Republican, Communist. Their main activity focuses on *elections, campaigns*. Political parties are run by *people* who are leading politicians. They are the main instruments of *politics*. Their mildly *positive* evaluation is outweighed by heavier *negative* evaluations. They are associated with a variety of *issues, problems* ranging from sex to war.

SLOVENIAN GROUP: The Slovenian image of political parties is a distinctly simpler and less meaningful one. (Compare the Slovenian total score of 656 with the much larger U.S. score of 1090.) There is little diversity; the *Communist party* is the only party mentioned and a small reference is made to the opposition. References to *elections, campaigns* are negligible. The largest component deals with *political principles, organizations*, whereby the organization described has a strong connection with the state. The *economic* references produced by a second meaning of the Slovenian stimulus are weighty.

political party is explicable by the second meaning of this Slovenian word: consumer, client.

3. *People* (U.S. 86, S. 89). The three most frequent responses were *president* (U.S. 24), *Nixon* (U.S. 21), and *people* (U.S. 19). The American references to various presidents were numerous, suggesting that for this group presidents are characteristic representatives of politics.

4. *Negative reactions* (U.S. 60, S. 89). The main thrust of the American

negative attitudes appears to relate to the idea of illegal economic advantages; the Slovenians emphasized confrontation more: fight, quarrel. The Slovenian reactions may be partially due to the second meaning of consumer, client. The main responses were *quarrel* (S. 27), *fight* (S. 13), *bother* (S. 15), and *bad* (U.S. 20).

5. *Positive reactions* (U.S. 16, S. 32). In comparison with the Americans, the Slovenians were stronger in their negative as well as their positive evaluations of political party, which suggests a stronger ambivalence of feelings.

7. *Issues, problems* (U.S. 110, S. 68). Some of the American responses also belong to the election, campaign category (e.g. *convention* 52). The largest Slovenian response was *office* (18), which may be a reference to bureaucracy or a reaction elicited by the second meaning of this Slovenian word: consumer, client.

8. *Election, campaign* (U.S. 161, S. 39). This is again a nearly exclusively American response category. The main responses were *election* (U.S. 55, S. 13), *candidate* (U.S. 25), *campaign* (U.S. 26), and *vote, voter* (U.S. 25). These responses describe political activities, the parties' participation in the political process and their competition for being elected.

9. *Diverse parties* (U.S. 525, S. 39). This is an extremely strong and nearly exclusively American response category. The largest responses were *Democratic* (U.S. 215), *Republican* (U.S. 187), and *Communist* (U.S. 20, S. 15). These responses suggest that, in the American view, the main actors or forces in politics are the parties. The Slovenians mentioned only the Communist Party, and even these responses were not very frequent.

COMMON TRENDS OF INTERPRETATION: POLITICAL INSTITUTIONS. Table 2 summarizes some of the main response trends that were found with considerable consistency across the four themes used in the representation of this domain. The following comments refer to the numbers in Table 2:

1–2. In the context of all four concepts studied, the *state* emerged as the central political institution emphasized by the Slovenian group. The Slovenian word means *state* more in the sense of the French *état* than the American usage. As a second meaning, it may also refer to country. The comparable American trend to emphasize the concept of government emerged with less consistency from this domain. The exact nature of these differences requires further experimental clarification; nonetheless, the focus on state suggests that the Slovenians may be predisposed to think in terms of a relatively stable, permanent structure, which involves sovereignty in a traditional sense, while the American concept of government stresses more the idea of an elected and changing body of leadership.

3–4. As a related trend, the Slovenians made more references to politics and politicians, emphasizing politics in general and particular professional activity — domestic and international. At the same time the Americans emphasized campaigns, elections, parties: that is, the political process, which represents the foundation of these institutions, explains their changing nature, their dependence on popularity and public opinion.

Table 2. Common trends of interpretation in the domain of *political institutions*

Trend of interpretation	Politics	Political party	Government	President
1. More emphasis on state	S.	S.	S.	S.
2. More emphasis on government	U.S.	U.S.	—	U.S.
3. More weight on political process, campaign, election, parties	U.S.	U.S.	U.S.	U.S.
4. More weight on politics, politician	S.	S.	U.S.	U.S.
5. General focus on people, individuals	U.S.	U.S.	U.S.	S.
6. General focus on society	S.	S.	S.	S.
7. Greater emphasis on legislature	U.S.	U.S.	U.S.	U.S.
8. Greater emphasis on democracy	U.S.	U.S.	U.S.	U.S.
9. More weight on power	U.S.	U.S.	U.S.	U.S.
10. More weight on authority	—	S.	S.	S.
11. More emphasis on dominantly positive evaluation	—	U.S.	S.	S.
12. More emphasis on dominantly negative evaluation	U.S.	S.	S.	—
13. Primary attention to individual behavior with economic consequences, e.g. corruption	U.S.	—	U.S.	U.S.
14. Primary attention to behavior with social collective consequences, e.g. exploitation	S.	S.	S.	—

5–6. In all these contexts Slovenians made more frequent references to society, social class, which expresses a heavy societal focus and a collectivistic orientation. The ultimate criteria for the evaluation of political institutions obviously derive from these collectivistic, social considerations. On the other hand, Americans emphasized people, interpersonal relations where the individual rather than the collective is the center of interest.

7. The United States group consistently placed more weight on the legislature, the Congress, and the Senate as elected bodies. In contrast with this, the Slovenian group placed less emphasis on differentiating executive, legislative, and judiciary power. This Slovenian trend appears to be in logical agreement with the previously observed emphasis on the state.

8. From the political ideologies, *democracy* was the most frequent American reference; the common Slovenian emphasis on socialism does not become apparent in the context of this domain. The American focus on *democracy* was stronger and its main reference was the democratic process. *Socialism* for the Slovenian group was found to refer primarily to high ideals and social order.

9–10. There was a strong American interest expressed in *power*, its allocation, organization, use, and misuse as a matter of concern in respect to *political institutions.* The Slovenian focus was on *authority*, which may be conceived as institutionalized power, the mandate to decide and to rule being viewed as intrinsic to these institutions.

11–12. In the evaluation of *political institutions*, the Slovenian group displayed more emotionality: it was more negative in its reactions to *political party* and *government* and more positive toward the *president* than the American group. On the other hand, *politics* received a stronger negative general evaluation from the American group.

13–14. In the evaluation of *political institutions*, the Slovenian group emphasized the idea of exploitation, which implies a broad collective problem, in the sense elaborated by Marxism. The American concern was primarily with corruption and other negative characteristics, which suggest personal weaknesses related to economic motives.

3. *Values, Ideals*

For the third domain, the interpretation of a few sociopolitical ideals were compared. As Edward Sapir (1962) has pointed out, there are some universally approved values which are deceptive and do a dangerous disservice to mankind. They are characterized by a generally positive unifying feeling-tone, which makes people readily forget that their meanings not only do not harmonize but are in part contradictory. "Often enough," he has said, "we agree on the particular value of the label" and fail to recognize that we "disagree on the value of things and the realities of things." The following four universally approved "labels" were analyzed to assess the extent to which Americans and Slovenians actually agree about their interpretations: *peace*, *freedom*, *equality*, and *security*.

PEACE (MIR). The comparative analysis (Figure 9) relied on the following main components:

1. *War, fighting* (U.S. 210, S. 267). Both the American and the Slovenian groups were strong on this category. The two major responses were *war* and *Vietnam.* The Slovenians outscored the Americans in their references to *war* in general (U.S. 113, S. 168), while the Americans scored higher than the Slovenians on *Vietnam* (U.S. 71, S. 45).

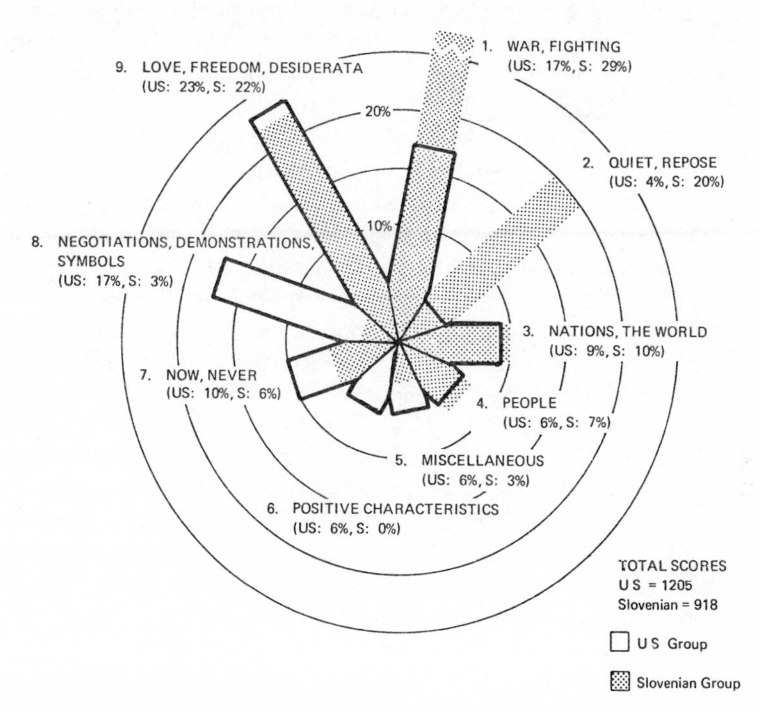

Figure 9. *Peace*: main meaning components

U.S. GROUP: The American interpretation of peace suggests a high ideal with strong emotional content involving *love, freedom*, and other desiderata. Its antithesis is *war*, with special timely connotations derived from Vietnam. In the promotion of peace, considerable weight was placed on *negotiations* and *demonstrations*. Impatience and skepticism were expressed by the component *now, never*. The reactions show strong involvement and express timely concerns.

SLOVENIAN GROUP: The Slovenian interpretation of peace is firmly anchored in the emphatic rejection of *war, fighting*. Among the ideals and values attached to *peace, freedom* is especially strong. The word has apparently a second, nonpolitical connotation as well, implying peaceful life conditions: *quiet* and *repose*. The concern with *world peace* is emphatic and articulate.

2. *Quiet, repose* (U.S. 55, S. 187). This is a mainly Slovenian category, which includes such responses as *quiet* (U.S. 28, S. 67), *noise* (73), *sleep* (23), *night* (10), and *calm* (U.S. 17). The sizable Slovenian responses stress the idea of quiet. The Slovenian word *mir* is a homonym; it means peace, but also quietness.

3. *Nations, the world* (U.S. 110, S. 90). This is a slightly stronger American than Slovenian component. The highest scoring responses were *world* (U.S. 45, S. 37), *U.S.* (U.S. 28), *nation* (U.S. 7, S. 21), *America* (U.S. 13, S. 8), and *country* (U.S. 12).

4. *People* (U.S. 80, S. 64). The most frequent responses were *friend* (U.S. 13), *people* (U.S. 13), *man* (S. 21), and *family* (S. 14).

6. *Positive characteristics* (U.S. 76, S. 0). This small but purely American component expresses positive evaluation. The highest scoring responses were *good* (20), *mind* (13), *great* (11), and *loving* (10).

7. *Now, never* (U.S. 117, S. 51). The highest scoring responses were *now* (U.S. 50), *forever* (U.S. 15), *never* (U.S. 15, S. 12), and *wish* (S. 25). The total American category score was more than twice as large as the Slovenian. Both the American and Slovenian responses express a great deal of skepticism; the American responses express impatience as well.

8. *Negotiations, demonstrations, symbols* (U.S. 203, S. 25). This is a predominantly American component which deals with activities and events which aim to achieve and maintain peace. The largest responses were *symbol* (U.S. 44), *dove* (U.S. 31), *moratorium* (U.S. 26), *march* (U.S. 17), *Christmas* (U.S. 13), and *pigeon* (S. 16).

9. *Love, freedom, desiderata* (U.S. 278, S. 203). Among the ideals associated with peace, the American group scored especially high on *love* (97, S. 11), *happiness* (43, S. 21), *hope* (27), and *tranquility* (18). The Slovenian group attributed greater importance to *freedom* (60, U.S. 10), *progress* (16), *satisfaction* (14), and *welfare* (14). These ideals suggest that peace may have a stronger personal, sentimental connotation for the American group and a more social-political connotation for the Slovenian group.

FREEDOM (SVOBODA). The following main components (Figure 10) were used in the analysis:

1. *Values, desiderata* (U.S. 234, S. 232). This was the strongest component for both the American and Slovenian group. The main American emphasis was on *liberty* (U.S. 44), *independence* (U.S. 15), and *happiness* (U.S. 28, S. 15), *love* (U.S. 23, S. 10), and *security* (U.S. 17). The Slovenians stressed such ideals as *peace* (U.S. 38, S. 73), *equality* (U.S. 11, S. 47), *joy* (U.S. 6, S. 36), *friendship* (S. 17), *brotherhood* (S. 16). The American ideas were those of individualism and liberalism, while the Slovenians were more socialistic and collectivistic, frequently reminiscent of the French revolution.

2. *Fight, war* (U.S. 48, S. 98). In this distinctly stronger Slovenian category, the main elements were *war* (S. 38), *fight* (U.S. 26, S. 28), *Vietnam* (U.S. 6, S. 13), and *revolution* (U.S. 16). As the responses suggest, freedom has a stronger connection with war and fighting for the Slovenians, a connection which may be explained by their relatively recent history. This suggests a concern with freedom at the national level.

3. *Positive attitudes* (U.S. 44, S. 33). The American and Slovenian components were about equal in weight. The highest-scoring responses were *good* (U.S. 10), *golden* (S. 13), and *wish* (S. 14).

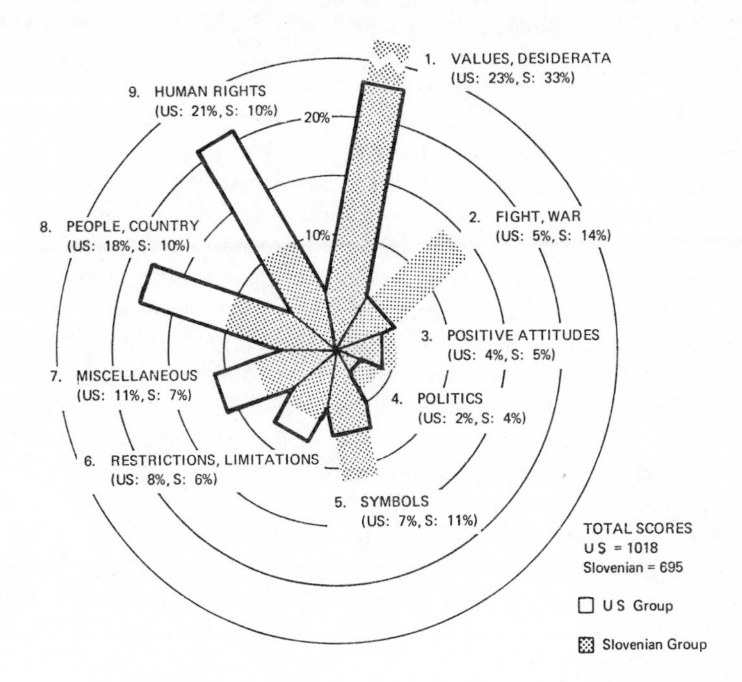

Figure 10. *Freedom*: main meaning components

U.S. GROUP: The American interpretation of freedom involves especially strong emphasis on related *values*, *ideals* such as *liberty* and *peace* and also on *human rights* such as the right of the individual to free speech and free choice. Americans especially emphasize the freedom of the *individual* and also the freedom of the *country*. America represents the ideal of freedom at an international level and democracy is the political principle most closely associated with it. There is a distinct concern with diverse *restrictions* and *limitations* of freedom.

SLOVENIAN GROUP: The Slovenian interpretation of freedom involves an especially strong emphasis on such related *values* and *ideals* as *peace* and *equality*. *Human rights*, the rights of the individual, receive relatively little attention; the emphasis is more on national freedom, a collective concern with historical foundation. Among *people*, the *partisans* and *President Tito* personify this ideal. *Fight*, *war* appears to be frequently indispensable to its realization. Among the political *isms*, *socialism* is the strongest, suggesting that it may be the system promising the most freedom.

4. *Politics* (U.S. 24, S. 27). This was a small but distinct component; the Americans emphasized *democracy* (14); the Slovenians emphasized *socialism* (18).

5. *Symbols* (U.S. 75, S. 74). The heaviest responses were *flag* (U.S. 18, S. 18), *nature* (U.S. 19, S. 5), *bells* (U.S. 16), and *air* (S. 12). The American group made more references to symbols; the Slovenians gave more responses dealing with diverse natural phenomena.

6. *Restrictions, limitations* (U.S. 78, S. 44). This component was nearly twice as strong for the American group as for the Slovenian. The two highest-scoring responses were *slavery* (U.S. 19, S. 9) and *prison* (U.S. 11, S. 8). Generally there seems to be a stronger American concern with the restrictions of the individual's freedom.

8. *People, country* (U.S. 188, S. 68). This is a strong American meaning component based on such elements as *America* (78), *country* (30), *individual* (29), and *people* (U.S. 19, S. 21). These responses indicate that, for Americans, the United States is the country of freedom. Freedom of the country and freedom of the individual receive about equal attention. Slovenian focus on the *people* and *nation* indicates a stronger emphasis on national freedom, a trend clearly demonstrated by Slovenian history.

9. *Human rights* (U.S. 211, S. 68). This is a weighty, mainly American component. Especially heavy is their emphasis on *speech* (U.S. 53, S. 23), *choice* (U.S. 36), *rights* (U.S. 26), *press* (U.S. 19, S. 14), *life* (U.S. 23, S. 14), and *mind, think* (U.S. 21). This second largest American component suggests that the idea of personal freedom is actively pursued in a variety of contexts of social and political relevance.

EQUALITY (ENAKOST). A comparative analysis of American and Slovenian cultural meaning (Figure 11) involved the following main components:

1. *People, nations* (U.S. 157, S. 132). The American concern with *people* (U.S. 48, S. 64), *everyone* (U.S. 23), and *all* (U.S. 16, S. 29) reflects an egalitarian philosophy. The Slovenian group showed slightly more specific concern with the relationship of *nations* (S. 15) and sexes (*woman*: U.S. 16, S. 29; *male*: S. 14).

2. *Family, friends* (U.S. 0, S. 117). This is a purely Slovenian component, expressing a distinct concern with the relationship within the family: *family* (11), *children* (11), and *brothers* (14). Furthermore, there was a strong emphasis on the ideal of *brotherhood* (S. 67), implying friendship.

3. *Political systems, countries* (U.S. 38, S. 82). This component carries somewhat stronger political connotations for the Slovenians. They stressed *socialism* (U.S. 7, S. 29), *communism* (U.S. 6, S. 17), *state* (S. 14), and *society* (S. 14). The Americans mentioned primarily *America* (U.S. 15) and *democracy* (U.S. 10).

4. *Inequality, confrontation* (U.S. 45, S. 66). There was somewhat more emphasis on inequality by the Slovenians: *inequality* (S. 40), *racism* (U.S. 10), *fight* (U.S. 11, S. 7), and *revolution* (S. 20).

6. *Skepticism* (U.S. 42, S. 34). This is a small but about equally relevant American and Slovenian component. The main expressive responses were *illusion* (S. 12), *future* (S. 12), *needed* (U.S. 15), and *none* (U.S. 15).

7. *Characteristics* (U.S. 68, S. 39). All the reactions were positive; the American responses carry more weight. The three largest responses were *good* (U.S. 21), *fair* (U.S. 14), and *beautiful* (S. 16).

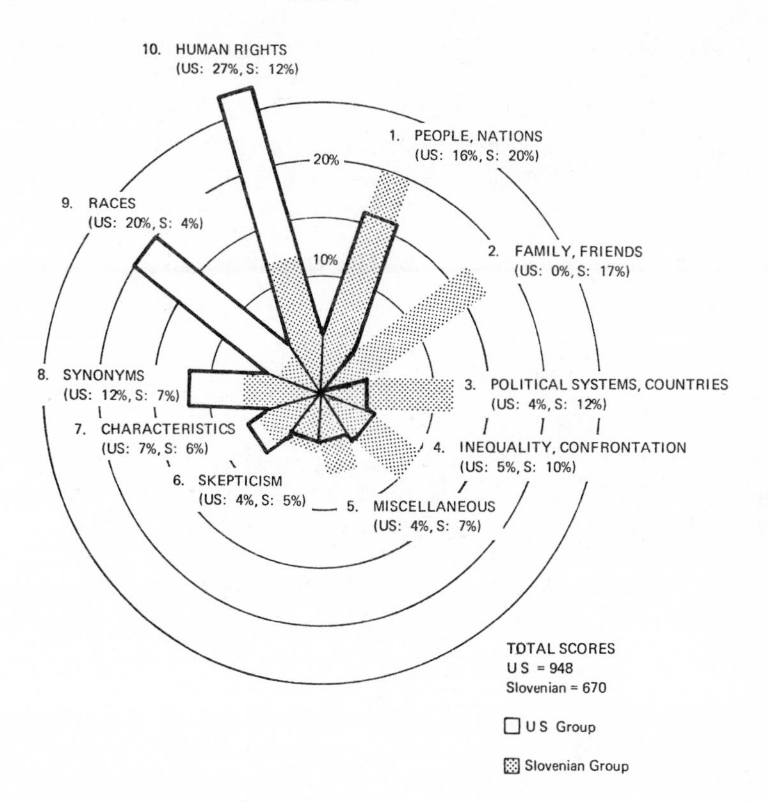

Figure 11. *Equality*: main meaning components

U.S. GROUP: The American interpretation of equality has its strongest emphasis on *human rights*, especially on *freedom*, *law*, and *justice*. In terms of specifics, the most weight is placed on the relationship of *races*. The stress on *people*, *all* and *everyone*, reflects a distinct egalitarian philosophy. The *characteristics* reflect positive attitudes and identification with the idea of equality. Nevertheless, there are some expressions of *skepticism*.

SLOVENIAN GROUP: The Slovenian interpretation of equality shows somewhat less articulation. The strongest single component involves *people*, *nations* with emphasis on the relationship of *nations* and on the relationship of *men* and *women*. *Brotherhood* is a salient ideal and there is a distinct concern with relationships within the *family* and between *friends*. There is some emphasis on *inequality* and some expression of *skepticism*. *Socialism* and *communism* emerge as the representative systems.

8. *Synonyms* (U.S. 110, S. 45). In this mainly American category, the distinct weight placed on *same* (U.S. 47) and *equal* (U.S. 33, S. 8) is perhaps a reflection of the egalitarian principle.

9. *Races* (U.S. 190, S. 28). In this heavy American meaning component, the three most frequent responses were *races* (U.S. 81), *black* (U.S. 50, S. 6), and *Negroes* (U.S. 44, S. 8). These data show that for the

American students the problem of equality has an articulate racial component.

10. *Human rights* (U.S. 256, S. 80). This heaviest American response category included special emphasis on *rights* (U.S. 41), *freedom* (U.S. 40), *law* (U.S. 27), *justice* (U.S. 19, S. 44), and *constitution* (U.S. 18), in general. Furthermore, it included very specific references to particular rights such as *education* (U.S. 16) and *religion* (U.S. 13).

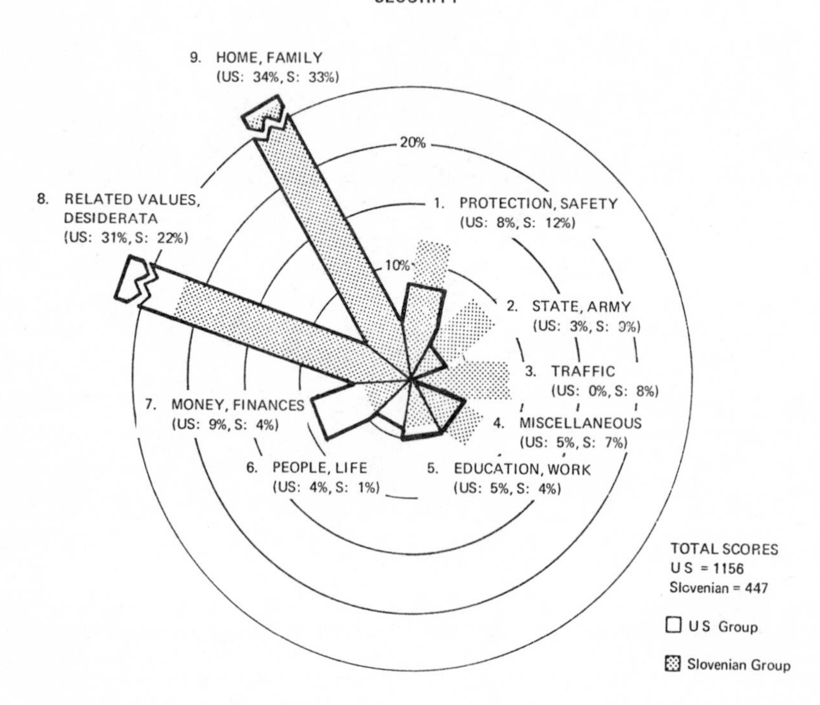

Figure 12. *Security*: main meaning components

U.S. GROUP: The American interpretation of security is richer in meaning, as expressed by the larger total score. *Family*, *home* represents the main context in which this highly popular value is used, suggesting that the focus is on close interpersonal relations. This impression is reinforced by *related values*, *desiderata*, which contains *love* as its central element. The next but much less important component is *money*, *finances*, expressing concern with financial security; *work* and *education* reflect concerns with job security.

SLOVENIAN GROUP: The Slovenian meaning of security is more narrow as expressed by the total response score, which is the lowest of all themes studied. This low score implies low priority and limited meaningfulness. The largest single response, home, indicates that *home*, *family* are especially symbolic of security. *Values*, *desiderata* primarily involve such ideals as *happiness*, *comfort*, and *health*. There are only two components on which the Slovenians actually outscore the Americans: *traffic*, implying traffic safety, and *state*, *army*, used in reference to internal security.

SECURITY (VARNOST). The analysis of this theme (Figure 12) revealed the following components:

1. *Protection, safety* (U.S. 91, S. 55). This primarily American component contains partial synonyms dealing with *protection* (U.S. 12, S. 19) and *safety* (U.S. 40, S. 12) in general as well as items providing protection and safety such as *blanket* (U.S. 35), and *shelter* (S. 64).

2. *State, army* (U.S. 38, S. 41). This represents a component dealing with state or internal security. The three main responses were *state* (S. 17), *army* (S. 16), and *police* (U.S. 14).

3. *Traffic* (U.S. 0, S. 35). This small but distinct Slovenian component indicates that the Slovenian word is used in the sense of safety. The two largest responses are *road* (16) and *car* (13).

5. *Education, work* (U.S. 62, S. 20). The American responses were somewhat weightier and more numerous, emphasizing concern with job security. The three largest responses were *work* (U.S. 9, S. 10), *education* (U.S. 20), and *job* (U.S. 25).

6. *People, life* (U.S. 47, S. 5). This primarily American category was relatively small and was composed of a variety of low-scoring responses indicating that security is viewed as an essential element of human life and human relations.

7. *Money, finances* (U.S. 104, S. 16). This is a mainly American component reflecting the philosophy that wealth implies security. The three largest responses were *money* (U.S. 65, S. 16), *bank* (U.S. 18), and *savings* (U.S. 16).

8. *Related values, desiderata* (U.S. 359, S. 97). This is a heavy American category with diverse responses dealing with values, ideals, and expectations connected with security. Some of the largest responses were *love* (U.S. 95), *social* (U.S. 60), *happiness* (U.S. 38, S. 16), *comfort* (U.S. 16, S. 15), *warmth* (U.S. 33), and *need* (U.S. 22). The category shows that security is a very positive concept for the American group and is closely related to some of the most popular personal values, especially love and happiness.

9. *Home, family* (U.S. 398, S. 148). For both groups, this represents the largest single component of interpretation. The three strongest elements were shared: *home* (U.S. 11, S. 83), *family* (U.S. 73, S. 40), and *parents* (U.S. 21, S. 21). Some additional responses were purely American, such as *friends* (74), *marriage* (31), and *boyfriend* (21).

COMMON TRENDS OF INTERPRETATION: VALUES, IDEALS. A brief summary of the common response trends is presented in Table 3. The following is an analysis of the main points of the Table:

1–2. The consistently higher American references to *people* represent an individualistic orientation, especially when contrasted with the consistently higher Slovenian references to *society* and *nation*, which indicate a distinct collectivistic orientation.

1–3. The Slovenian references to nation may carry some nationalistic undertones. The generally higher American references to *world*, *mankind*, and *United Nations* carry some cosmopolitan, international connotations.

Table 3. Common trends of interpretation in the domain of *values, ideals*

Trend of interpretation	Peace	Freedom	Equality	Security
1. Focus on people, individuals	U.S.	U.S.	U.S.	U.S.
2. Focus on nation, society	S.	S.	S.	S.
3. Focus on country, mankind	U.S.	U.S.	U.S.	—
4. Emphasis on government, constitution	—	U.S.	U.S.	—
5. Emphasis on state	S.	S.	S.	S.
6. More references to American, U.S.	U.S.	U.S.	U.S.	—
7. More references to Yugoslavia	S.	S.	S.	—
8. More weight on safety, security	U.S.	U.S.	U.S.	U.S.
9. More weight on war, revolution	S.	S.	S.	S.
10. More stress on individual rights and race relations	U.S.	U.S.	U.S.	U.S.
11. More stress on social values: brotherhood, friendliness	U.S./S.	S.	S.	S.
12. Greater emphasis on love, happiness	U.S.	U.S.	—	U.S.
13. Greater emphasis on joy, health	S.	S.	—	S.

4–5. Generally the Americans emphasize the *government* while the Slovenians emphasize the *state*. This suggests that the *state* in the sense of *état* represents a more central concept for the Slovenians, while the corresponding American reference is to *government*. The issue is complicated by translation problems. The differences in word use probably reflect the fact that for the Slovenians the state refers to a relatively stable, permanent structure heavily rooted in history and tradition — for example, the concept of national sovereignty; in contrast, Americans conceive of the government as a more contemporary and changing body of elected political-administrative leadership.

6–7. In connection with these ideals, the Americans generally make more references to the United States, while the Slovenians refer more frequently to Yugoslavia as the representative. Similarly, the American references to democracy are numerous and contrast with the constantly higher Slovenian references to socialism and communism. These responses indicate that these ideals are identified with Western democracy by the Americans and with socialism and communism by the Slovenians.

8–9. The American references to safety and security are consistently higher than the corresponding Slovenian responses. A strong concern with these values suggests a distinct security orientation; this may be indicative of insecurity and loneliness in the sense discussed by sociologists such as Riesman (1961) and Fromm (1941). At the same time, the Slovenians generally make more references to *war* and *revolution*, probably because *freedom* and *equality* were popular objectives and therefore frequent slogans in the "War of Liberation" during World War II.

10–11. The American group makes consistently heavier references to individual rights in general and to race relations in particular. This emphasis is not surprising in view of the obviously strong concern of the contemporary American society with racial problems. The generally higher Slovenian references to brotherhood and friendship show the Slovenians' emphasis on collective, social ideals.

12–13. The American group frequently places a heavy emphasis on love and happiness. These values center on the individual, again suggesting the individualistic focus. The Slovenian responses *joy*, *beauty*, and *health* indicate positive reaction and individual approval, elicited by the ideals and values explored in this domain.

Characteristic Priorities

An important characteristic of psychological meanings assessed from free verbal associations is that they can be used to assess how important and dominant a particular theme or domain is for a particular group. This priority, or dominance value, is inferred from the number of associations shown by the group total, or dominance scores. In the comparison of Slovenian and American groups, it must be taken into consideration that over a large number of words ($N = 85$), the American group has produced on the average a higher dominance score (1066) than the Slovenian group (846). The average difference (220) is probably the consequence of the slightly different experimental conditions (instructions and test forms). By using this average difference for adjustment (see relative dominance values in Table 4, column 6), it is possible to make direct comparisons.

The data presented in Table 4 shows that in terms of Slovenian-American differences, *socialism* and *Marxism* score the highest among the themes to which Slovenians assign greater importance than Americans.

On the other hand, Americans assign especially high relative dominance to security and political party as compared to the Slovenians. Considering the data at the level of domains, the *ideology* domain receives above average attention from the Slovenians, while the American dominance average is greater than the Slovenian in the *political institutions* and *political values* domains. Based on the quantity of responses elicited, these findings show closely similar trends as the results derived from the analysis of the American and Slovenian response distributions.

Table 4. Dominance of themes and domains

Domain	Theme	U.S. Score	S. Score	U.S.–S. Score diff.	U.S. (+), S. (−) Relative dominance*
Ideology	Communism	1109	853	256	+ 35
Isms	Marxism	966	924	42	−178
	Socialism	859	827	32	−188
	Capitalism	1138	985	153	− 67
Domain mean		1018	897	121	− 99
Political	President	1066	796	270	+ 50
Institutions	Government	1111	773	338	+118
	Politics	1013	797	216	− 4
	Political party	1090	656	434	+214
Domain mean		1070	755	315	+ 95
Political	Peace	1205	918	287	+ 67
Values	Freedom	1018	695	323	+103
	Equality	948	670	278	+ 58
	Security	1156	447	709	+489
Domain mean		1081	682	399	+179
Mean over 12 themes		1056	778	278	
Mean over 85 themes		1066	846	220	

* The relative dominance values show whether there is more or less difference on a particular theme than the mean difference value (220) found over 85 words. The positive scores show differences greater than the mean, suggesting stronger than average American emphasis compared to the Slovenian; the negative scores show themes and domains on which the Slovenians put more than average emphasis compared to the Americans.

SUMMARY AND GENERAL CONCLUSIONS

The findings on the three domains analyzed indicate some distinct differences between the American and Slovenian groups studied. The main trends differentiating the American and Slovenian groups in the three domains analyzed are summarized in Table 5.

These trends have been briefly elaborated in the context of each particular domain. They reveal some characteristic differences in the political frame of reference of the American and Slovenian students. Some of these differences can be clearly traced to sources in the prevalent ideologies; others appear to be closely related to life experiences specific to the groups studied.

1. *Political process, political ideology.* The American emphasis on political process is a finding which is in fundamental agreement with the

Western principles of participation and choice, as well as the Anglo-Saxon pragmatism emphasizing operational process over principles. The Slovenian trend shows a strong ideological orientation which is in line with the Marxist-Leninist emphasis on theory and ideology. Furthermore, as found in the context of particular themes and domains, the Slovenian belief system also shows considerable amount of similarity with specific themes of Marxism-Leninism (the image of decaying *capitalism*, the ideal image of *communism*, etc.).

Table 5. Main trends of interpretation

U.S. emphasis	Domain	Slovenian emphasis	Domain
1. Political process/ elections/campaign	I, III	1. Politics/ideology/ theory/isms	I, II, III
2. People/government/ country	I, II, III	2. Society/nation/state	I, II, III
3. Civil rights/free speech, etc./individual freedom/individual ideals	I, II, III	3. Social rights/group ideals/brotherhood/ equality/national freedom	I, II, III
4. Identification with Western democracy	I, II, III	4. Identification with socialism, communism	I, II, III
5. Economic references	I, II, III		
6. Love/happiness/safety/ security	III	6. Joy/beauty/health	III

2. *Political institutions*. The American emphasis on the relationship of the people and government is in agreement with the ideology of Western democracies. The evaluation of diverse political systems as representing different patterns of relationships between people and government also reflects the influence of this democratic ideology. The American focus on country is probably a result of the historical tradition of immigration and settlement of peoples from different nations, and of the process of forming a "new country" which for quite a while was not a nation.

3. *Political rights, ideals*. The weight and consistency with which the American group refers to civil rights — the rights of the individual to be free in movement, choice, and expression — reflect the American philosophy of individualism and freedom. The most popular ideals are individualistic; this is consistent with the American focus on the individual and with the democratic focus on the people.

The weight and consistency of Slovenian references to social rights, group ideals, equality, and brotherhood reflect collectivistic orientation; these concepts have their roots in the collectivism of Marxist-Leninist theory as well as in the agricultural ethnic traditions. In the social-political realm, however, the Marxist-Leninist influences are probably the more influential and decisive. This impression is reinforced by the fact that the

Slovenian group uses such ideological terms as *class struggle* and *exploitation*.

4. *Political isms, systems*. The Americans' identification with Western *democracy* is a clear expression of preference, a product of political principles as well as of tradition, culture, and life-style. Similarly, the Slovenian identification with *socialism* and *communism* emerges as a clear case of ideological commitment which stresses solidarity and which is also likely to have historic, traditional foundations as well.

5. *Economy*. The fairly consistent American emphasis on economic variables has an apparently strong foundation in cultural values, in past and present life conditions, and in its relationship to capitalism.

6. *Personal values*. The American emphasis on such personal values as happiness and security probably has its main roots in the cultural milieu and life conditions. Similarly, the Slovenian values of brotherhood, health, joy, and beauty are primarily culturally or traditionally determined; moreover, some of them, like brotherhood, may also be reinforced by the dominant political ideology.

As to the influence of ideology in the various domains, we may conclude that they are in no way limited to the *ideology*, *isms* domain, but are also evident in the other two domains, *political institutions* and *values, ideals*.

Probably because the ideology of Marxism-Leninism (in the sense of a body of political thoughts and principles) is more uniform and distinct than what may be called the "ideology of democracy," the influences of Marxism-Leninism on the actual beliefs are more readily identifiable.

Finally, a certain familiarity with contemporary Slovenians permits recognition of response trends as indicative of characteristically Yugoslav-Slovenian political frames of reference. For instance, references to *socialism*, *communism*, and *Marxism*, and their identification with countries, people, and leaders reflect characteristically Yugoslav priorities. Their ideal image of *communism* and its priority over *socialism* also represents a highly characteristic political identification which is probably fairly unique in central Europe. Moreover, the close connection between Marxist values and national feelings indicates a level of integration which is probably characteristically Yugoslav as well. A more precise identification of such characteristics would naturally require a broader, multinational comparison.

REFERENCES

BAUER R.
1959 *The new man in Soviet psychology.* Cambridge, Mass.: Harvard University Press.
FROMM, ERICH
1941 *Escape from freedom.* New York: Farrar and Rinehart.
INKELES, A., R. BAUER
1961 *The Soviet citizen.* Cambridge, Mass.: Harvard University Press.
KENNAN, GEORGE
1964 *On dealing with the Communist world.* New York: Harper and Row.
NOBLE, C. E.
1952 An analysis of meaning. *Psychological Review* 54:421–430.
OSGOOD, CHARLES E., G. SUCI, P. H. TANNENBAUM
1957 *The measurement of meaning.* Urbana: University of Illinois Press.
RIESMAN, DAVID
1961 *The lonely crowd.* New Haven, Conn.: Yale University Press.
SAPIR, EDWARD
1962 "Culture, genuine and spurious," in *Culture, language and personality.* Edited by David G. Mandelbaum, 78–119. Berkeley: University of California Press.
SZALAY, LORAND B., JACK E. BRENT
1967 The analysis of cultural meanings through free verbal associations. *Journal of Social Psychology* 72:161–187.
SZALAY, LORAND B., J. A. BRYSON
1973 Measurement of psychocultural distance: a comparison of American blacks and whites. *Journal of Personality and Social Psychology* (May).
SZALAY, LORAND B., R. D'ANDRADE
1972 Similarity scaling and content analysis in the interpretation of word associations. *Southwestern Journal of Anthropology* (Spring): 50–68.
SZALAY, LORAND B., DALE A. LYSNE
1970 Attitude research for intercultural communication and interaction. *The Journal of Communication* 20 (June): 180–200.
SZALAY, LORAND B., B. C. MADAY
1973 Verbal associations in the analysis of subjective culture. *Current Anthropology* 14 (February-April): 33–50.
United States Senate, Hearings Before the Committee on Foreign Relations.
1969 *Psychological aspects of foreign policy.* Washington, D.C.: U.S. Government Printing Office.

Application of Anthropological Linguistic Theory to the Study of Autistic Children's Discourse

JOHN REGAN and BEVERLY KILMAN

When dysfunctions are discovered in an individual's behavior, traditional diagnosis focuses on the identification of the deficit. A different approach to which anthropological procedures could contribute is that of identifying the content of the individual's information. This paper will describe one such attempt in a report of a two-year study of the communication of autistic children. This study sought to identify the locus of communication deficit of a sample of autistic children. The principal purpose was to determine whether any direct remedial benefit was likely to derive from the language development aspect of the educational program or whether remedial resources would be more profitably employed along other lines.

The initial impetus came from the presentation of examples of autistic children's conversations to a seminar studying normal language function development. These samples suggested that the program of verbal reinforcement underway for these autistic children was not optimal in that the content of the instruction was already in the children's information system. To examine this initial impression, the investigators drew upon anthropological linguistic definitions and concepts of language as an "instrument" having "functions" and asked two simple questions: Of what is talk made up? What is talk used to do?

Much has been written about autistic children and their unique speech characteristics, but a clear picture of their language and the functional language upon which a remedial educational program can be built remains to be articulated. "Until the structural features of language — in their entirety — are described and understood, I feel we will be unable to advance our understanding of how linguistic systems are acquired or to increase our knowledge of the underlying anatomy of our neurological equipment" (Smith 1976). This study is a contribution toward that goal.

Precise means of identifying communication dysfunctions are preliminary goals in locating the behavioral evidence of inner problems. The discipline of linguistics, specifically anthropological linguistics, can contribute to the understanding of human behavior. Such insights can then be of use to education, medicine, and psychiatry. Knowledge of exactly what is and what is not lacking in, for example, the autistic child's discourse is a precursor of any decision about remedial or compensatory learning experience.

The descriptive method selected in this study was developed from the research assumptions of the anthropological linguistic school of thought associated with G. L. Trager and H. L. Smith, Jr. This method centers on the concept of the value for research of separation of meaning from the mode by which meaning is created. In terms of the analysis system of this paper, this is the separation of instrument and function. Data collected on the nature and function of nonverbal instruments used by these children has been excluded from the scope of this paper.

REVIEW OF LITERATURE LANGUAGE FUNCTIONS

Anthropology, the science of description of human beings and their creations, has been the principal discipline to use the concept of function. Through its exploration, anthropologists have gained insight into culture from the study of the tangible and intangible products of a society. The concepts and method developed are now being applied to language, itself a prime product of the human creative ability to adapt innate capacities in the invention of tools.

Among the precursors of present interest in language function is the philosopher, Wittgenstein (1958), who explored the concept of language games, and the anthropologist, Malinowski (1930), who developed the principle of functionalism in the study of culture. The field is new, and theory and data are still in the process of reshaping one another; hence, there is no fixed set of categories in functional taxonomies. W. P. Robinson (1972), for example, discussed numerous category systems which have been used in classifying the functions.

Vygotsky (1969) identified emotional release and socialization as two functions of speech. Buehler (1934) distinguished three functions: representation, expression, and appeal. Harold Lasswell (1948) arrived at a set of functions for mass communication, an extension of human communication. He found that the communication process serves three functions: surveillance of the environment, correlation of the components of society, transmission of the social inheritance. Langer (1967, 1972) considered the developmental aspect in evolutionary perspective. For her the first function is representation. Hymes (1962), Jakobson (1967), and Ullmann, (1966) elaborate models with functional derivations. Jakobson's functions include process, dynamic, emotive, and aesthetic. Ullmann's functions include expressive, communicative, and effective.

The most exhaustive study of functions has been that of Halliday (1973), who includes seven developmental functions of the child's model of language. These functions are as follows: instrumental, regulatory, interactional, personal, heuristic, imaginative, and informative. Halliday's position is that adult language differs from child language fundamentally in number of functions simultaneously performed, with adult language allowing for a number of them at once. Halliday suggests three macrofunctions of adult language: ideational, interpersonal, and textual.

De Laguna's taxonomy (1963) includes: the social function in relation to the higher forms of intellectual life, and the control of the behavior of others. Dance and Larson (1976) posit three functions: linkings of individual with environment, development of higher mental processes, and regulation of behavior. Leathers (1976) notes that the functional significance of nonverbal communication relates to the purposes, accuracy, and efficiency of the exchange of meanings.

Autism

Autism, perhaps the most serious pathology found in children (Harrison and McDermott 1972) is a relatively rare syndrome that occurs at a rate of approximately three in 10,000 live births (Treffert 1971). It seems to hold special fascination for students of psychopathology (Harrison and McDermott 1972) and has generated extensive research since its first description thirty-five years ago by Kanner (1943).

Language Deficits in Autism

There is general agreement that language delay and deviation represent a cardinal symptom of autism (Kanner 1943; Rimland 1964; Ornitz and Ritvo 1976; Rutter 1975; Ekstein 1964, 1969; Shapiro and Fish 1971; Shapiro 1976). Language differences were first described by Leo Kanner in 1943 who noted delay in speech acquisition plus a general non-communicative use of speech after it developed. Language deficits that have been described include the following:

1. Mutism, the failure to develop speech, is characteristic of a number of autistic children (Creak 1963; Rimland 1964; Rutter 1975; Ornitz and Ritvo 1976). It is generally agreed that the prognosis is unfavorable for children who do not speak before age five or six (Freeman and Ritvo 1976; Fish 1971; Goldfarb 1971; Rutter 1975; Rimland 1964; Eisenberg 1957; Baker et al. 1976; Shapiro 1976). Knobloch and Pasamanick

(1975) found lack of speech was a poor prognostic sign only if mental deficiency was also present. However IQ has been shown to be positively correlated with language level and outcome (Rutter 1975). Rimland (1964) found the nonspeaking autistic children to manifest more autistic behaviors and demonstrate less verbalization in infancy.

2. Diminished vocabulary, frequent use of imperatives and incomplete phrases are characteristics of autistic children (Kanner 1943; Shapiro 1976; Baker et al. 1976). The lower output is seen as part of the general underproductivity in the language of autistic children. Deficits in vocabulary have been found to respond to various remedial measures including behavior modification, teaching machines, and engineered programs (Shapiro 1976; Hewett 1965, 1968; Lovaas, Schreibman, and Koegel 1976).

3. Deviations in articulation, quality, volume, pitch, rhythm, and inflection have been described (Kanner 1943; Shaw and Lucas 1970; Goldfarb 1971; Ekstein 1969). Fish (1971) and Ornitz and Ritvo (1976) have noted the atonal and arrhythmic quality of the speech of autistic children. They find these children lack inflection and note that the atonal quality usually persists into adulthood. Others (Ekstein 1969) have noticed the overformal pedantic quality of the speech of the autistic child. Ekstein (1969) has noted the tendency of these children to reproduce phrases with the intonation of an adult speaker whom they have heard uttering that same phrase in the past. This is probably related to echolalia, the next characteristic to be described.

4. After mutism the most commonly described speech characteristic of autistic children is immediate or delayed echolalia (Rutter 1975). Echolalia is the tendency of the autistic child to repeat what has been heard word for word with exact rhythm and tone of the original speaker. If an autistic child is asked, "Do you want an ice cream cone?", the reply may very well be, "Do you want an ice cream cone?" Delayed echolalia refers to the observation of this repetition after long periods of delay, sometimes of days, weeks, or longer. In general it has been found that the degree of echolalia is negatively correlated with language development (Hingtgen and Bryson 1974), but it is also generally noted that children who demonstrate early echolalic behavior progress to more complex and spontaneous speech more quickly than do children who are mute (Lovaas, Schreibman, and Koegel 1976; Hingtgen and Bryson 1974). Echolalia as a normal prelinguistic vocal behavior emerges at about nine or ten months (Kavanaugh 1971). The emergence of echolalia in the normal child at about nine or ten months may not be evidence or a complete association of the word with a particular item or with a particular part of the environment and may not represent true speech (Kavanaugh 1971). The normal echoing behavior of the nine or ten month child, however, can be differentiated from the echolalia of the

autistic child in terms of the rigidity with which the autistic child echoes the environment (Shapiro 1976). Normal children generate their own grammar to form their own language and apparently do not simply produce carbon copies of adult speech (Kavanaugh 1971); the autistic child, however, tends to produce exact copies of adult speech, often even including intonation and inflection (Ekstein 1969). Shapiro (1976) found that four-year-old autistic children produced significantly more rigid imitations than normally developing two-year-olds. Further, he found that autistics do not make "normal" grammatic errors in tense formation as with –ed endings.

Pronominal reversal in autistic children has been described (Kanner 1943; Kanner and Lesser 1972; Rutter 1975; Wing 1976). Pronominal reversal quickly marks off the autistic's language to the casual listener. The child refers to himself as "you" and occasionally to others as "I," but not to himself as "I." The autistic might say "would you like a cookie" as a request for same. In the past pronominal reversal has been seen in terms of the psychopathology, the lack of personal identity and ego weakness of the autistic (Bettleheim 1967; Hingtgen and Bryson 1974); however, recent authors have viewed pronominal reversal simply as part of the echolalia (Rutter 1975).

5. The limited use and comprehension of paralinguistics, including gestures and kinesics, was noted by Kanner 1943. Goldfarb (1971) has also noticed limited facial and body reinforcements accompanying language.

6. Inadequacies or deviations in comprehension as well as inappropriate and noncommunicative language have been described in autistic children (Kanner 1943; Rutter 1975; Mykelbust and Bannochie 1972). Shapiro (1976) has noted the limitations of communication in autistic children. Analysis of speech samples of an autistic child taken during nursery school revealed functional deficits. Although the child had achieved a language IQ of 99 at four years using the Gesell scale, an analysis based on Shapiro's functional scale showed that less than 50 percent of his speech was communicative. Further, the speech produced revealed major rigidities. Simple naming and frequently heard commands predominated. Goldfarb (1972) relates the communication deficit to a defect in capacity for abstraction and categorization of incoming stimuli; deficits in the integrative process are seen as resulting in a primary cognitive deficit. Goldstein (1948) agrees that an inborn defect in the capacity to abstract is related to communication deficits.

Lack of communicativeness has been related to the autistic youngsters inability to appreciate the possibilities of response from the environment and to delayed auditory processing or "delayed mental audition" (Rimland 1964). Both Kanner (1943) and Rimland (1964) note part/whole confusion in the autistic child which results in use of idiosyncratic and nonspecific phrases. These are often confusing to the auditor in the environment. For example, Rimland reports one three-year-old autistic used the delayed echolalic phrase, "Do you want some catsup honey?" to ask for dinner or for any type of food. Shapiro (1976) and others note the difficulties in remediating in the area of communication. Shapiro found that the children could be taught to increase the number and range of words or phrases and that they could increase intelligibility, but structuralization remained retarded, rigid, and deviant.

7. Kanner (1943) noted the lack of questions in the autistic child; subsequent investigations have found similar lack of questioning and of curiosity or spontaneity in the autistic person (Wolff and Chess 1965; Rutter 1975; Baker et al. 1976; Shapiro 1976). Kozloff (1973), using parents' reports looked at autistics' answers to questions and the use of the word "yes." None of the parents reported ever having heard their child use the word "yes" or ever having answered a question. Baker et al. (1976) reviewed the material on autistics' questions and pointed out discrepancies in the research. They noted that Wolff and Chess (1965) mention lack of questions as a characteristic of the language of autistic children; whereas Cunningham and Dixon (1961) note no difference between autistics and controls in the amount of questioning, and Rutter (1975) pointed out that communication is primarily by a series of obsessive questions in the adolescent.

Comparison of Language Development in Autistic Children and Other Clinical Groups

When autistic children's language development is compared with other clinical groups, findings generally include some overlap but the autistic children are generally characterized by lower developmental level, lack of questions and informative statements, few personal pronouns, greater use of imperatives, limitations in verbal output, and more frequent idiosyncratic uses of words (Hingtgen and Bryson 1974).

Comparison of the language development of autistic children with aphasic children reveals further specific deficits in the developmental pattern of autistic children even as compared to a group of language impaired children (Rutter 1975). Similarities between the aphasic and autistic children were that both groups were reported to have diminished or abnormal babble before speech developed and a large proportion of both groups were suspected of deafness at some time because of their poor or inconsistent response to sound. In both groups abnormalties of syntax, intonation, use, and understanding of words was found. In both groups a family history of speech delay was fairly common. However, the autistic children had more echoing and pronominal reversal. Autistic children did more poorly on WISC IQ tasks of Comprehension, Similarities, and Vocabulary than did the aphasic children. The autistic children were more impaired in their use and/or understanding of gesture. The autistic children produced more stereotypic utterances and showed less imaginative play. The autistic children "chatted" less and generally failed to use their limited language skills for "social communication." The only area in which the aphasic children were more impaired was in their articulation skills. Rutter (1975) finds the autistics' language deficits both "wider and deeper" than other language disorders such as aphasia.

Language as a Tool for Understanding Autism

Recently Baker et al. (1976) have suggested that language studies may provide insights into diagnosis, prognosis, and etiology of autism. Shapiro (1975, 1976) and others have suggested the possibility of developing a system for defining the language of atypical children that will serve as a diagnostic tool. Various authors have used language levels as an index of future outcome (Hingtgen and Bryson 1974; Rutter 1975). Rutter (1975) has suggested that an understanding of the language of the autistic child may be a key to understanding the nature of the disorder.

While the unusual aspects of the language of autistic children has been clearly documented, measurement of these differences has proved more difficult. Several scales have been developed to attempt to systematically measure the language development of the autistic child including scales used by Shapiro (1975) and Richey (1976). However, to date none of these programs has successfully demonstrated its predictive and/or diagnostic power.

Despite a wealth of descriptive work with respect to the language characteristics of autistics, systematic language studies of autistic children are few, and specific and reliable profiles of autistic language do not exist (Baker et al. 1976). A useful scale may well depend on the development of a more reliable and specific profile of autistic language. One of the difficulties is that many of the studies of autistic language are impressionistic and based upon nonsystematic observations. For example, it has been suggested that systematic studies of the number of echolalic utterances will reveal a lower frequency of occurrence than previously reported (Baker et al. 1976). Systematic studies may well clarify other descriptive characteristics as well.

Recent investigations have been moving in the direction of a more systematic approach (Shapiro 1976; Richey 1976). However, careful longitudinal studies are quite rare; one of the very few such attempts has been made by Cunningham (1968). Cunningham also attempted a functional analysis using tape recordings taken during a play session of the autistic child with an adult. He analyzed the first fifty utterances in terms of comprehensibility, length, and function. The functional analysis was limited to a twofold measure; utterances were designated either egocentric or socialized. Cunningham found that the children he recorded produced significantly more incomprehensible and egocentric utterances. Echoes of the interviewer, delayed repetitions, self-repetition, thinking aloud, and inappropriate remarks were frequent in the autistic group. Similar findings relevant to the functional analysis of the language of the autistic are recorded by Shapiro. Functional studies, however, are limited in number; the categories performed also tend to be limited. Analysis has been restricted to dichotomies such as Cunningham's social vs. nonsocial or Shapiro's communicative vs. noncommunicative. Shapiro divides his noncommunicative speech category into four subgroups and communicative speech into two subgroups.

Detailed analysis of the language of autistic children leading to functional language profiles remains to be completed. Nevertheless, work from this point of view may offer an opportunity to describe more precisely the unique characteristic of the language of the autistic child such that characteristics as the failure to ask questions and lack of comprehension, spontaneity, and originality may be clarified. And if the unique aspects of the language of the autistic child can be understood from this fresh point of view and can be mapped in functional analysis, then it may be possible to develop a new developmental remedial program for the language education of these children.

NATURE OF DYSFUNCTION OF THE AUTISTIC CHILD'S VERBAL INSTRUMENT

Attending only to the autistic child's conversation and comparing that with normal children's, what exactly can we identify that he has and does not have? To answer this question we drew upon the Trager and Smith (1951) linguistic structuralist methodology and classified our observations into two classes — instrument and function — the former having four subcategories; the second, ten. The four subcategories or strands that combine into the instrument used in talking are (1) the *phonological*, namely the sounds used in the language to formulate morphemes or words; (2) *morphological*, the basic meaning units created out of the phonological system and combining into words; (3) *syntactical*, the definition of syntax to be used in sequencing rules of each of the strands of the verbal instrument, not only the morphological and the arrangement of words, but also the positioning of phonemes and discourse units as exemplified in the turn-taking; (4) *paralinguistical strands*, i.e., the extralinguistic tone of voice phenomena including kinesics, the structured use of bodily motion.

Whereas Smith and Trager's focus was directly on data concerning the instrument *qua* instrument, S. M. Lamb, whose anthropological linguistic work is also relevant here, regards a speaker's knowledge of these levels as aspects of his "information systems." This term raises the point that human talk is redefinable in terms of information. A component, therefore, of what a child either explicitly or implicitly knows, is specifically about language itself: the systems of sound, morphological reference, syntactics, and paralinguistics. Identifying the child's implicit information about each of these levels was our first task.

As a first step we decided to rely upon anthropological linguistic definitions. We replaced terms customarily used to label the autistic child's communication behavior with more descriptive statements so as to restrict the conclusions to those which derive directly from the method to be used.

The corpus begins with examples initially presented to the investigators for their explanation. These three samples follow:
1.01 with a 4-year old autistic youngster. Names and identifying information have been altered in all of the samples.

Adult: "What is your name, Bill?"
Child: "William Andrew McNeil."
Adult: "Where do you live, Billy?"
Child: "20 Mountain Drive, San Francisco."
Adult: "How old are you, Bill?"
Child: "William Andrew McNeil?"
Adult: "How *old* are you?"
Child: "20 Mountain Drive, San Francisco."
Adult: "No, how many years are you? Show me on your fingers how many years old you are."
Child: "William Andrew McNeil."
Adult: "Do you go to school, Billy?"
Child: "20 Mountain Drive, San Francisco."

1.02.02

Adult: "Why shouldn't you play with matches?"
Don: "You might get sick."
Adult: "Can you tell me more about that?"
Don: "Tell *me*."
Adult: "Can you tell me what you mean?"
Don: "You tell me the words and then I'll tell you."

Adult: "Let me say the question again: Why shouldn't you play with matches?"
Don: "You might get burned."

The second group is of children in remedial classroom interacting with the teacher over a longer period. A selection from this corpus follows:
2.01.001

Teacher: "What color is that candle, Matt?"
"What color is the candle?"
Matt: "Pink."
Teacher: "Can you say it in a sentence? What color is the candle? The — "
Matt: "One — two.:
Teacher: "What color is the candle? The —"
Matt: "The cake — the — the candle is pink."
Teacher: "Very good Matt. Very good Matt."

2.01.002

Teacher: "What color is the girl's dress?"
Matt: "Orange — red."
Teacher: "Can you talk in a sentence?"
Matt: "Red."
Teacher: "The. . ."
Matt: "The girl is red."
Teacher: "The girl's dress is red."
Matt: "Dress is red."
Teacher: "Can you say the whole thing — say the whole thing Matt."
Matt: "Dress is red."
Teacher: "Good for you."

2.01.003

Teacher: "Today is a special day — it is Kay's birthday. Kay is four years old today. Kay is four years old — how old is she Matt?"
Matt: "Four years old."
Teacher: "Good — can you say the whole thing? Kay — Kay."
Matt: "Four years old."
Teacher: "Who is four years old? Kay is four years old. Can you say the whole thing?"
Matt: "Four years old."
Teacher: "Kay is four years old. Matt — look at me."
Matt: "Four years old."
Teacher: "Kay is four years old."
Matt: "Kay is four years old."

Teacher: "Good for you — good boy, Matt — That was nice talking. Good."

2.01.004

Teacher: "What color is the frosting Matt? Get ready to work"
Matt: "Frosting — cake."
Teacher: "What color is the frosting?"
Matt: "Blue."
Teacher: "Say — the —"
Matt: "is blue."
Teacher: "The — the — what is it?"
Matt: "Blue."
Teacher: "The"
Matt: "The frosting is blue."
Teacher: "Good for you."

2.01.005

Teacher: "What kind of cake does Karen want Matt?"
Matt: "Cake is Karen."
Teacher: "What kind of cake does she want?"
Matt: "Karen."
Teacher: "What kind of cake — she wants a chocolate cake."
Matt: "A chocolate cake" (at same time with teacher).
Teacher: "What kind of a cake does she want? No."
Matt: "She wants a birthday cake."
Teacher: "She wants a chocolate cake." (Matt says at same time). "Good, get ready to work Matt. Good."

A third phase of data gathering was begun in an effort to find samples of the children's talk in less formal situations, two selections from this portion of the corpus follow.

This first is of seven children and their two teachers in a special class for autistic children. Ages range from seven years to eleven years. The children move outside to the lunch table, carrying their lunches and not speaking. At the table the children begin eating without verbalizing. The first words are from the teacher:

T1: "Little bites." (period of no verbalization)
T1: "Where's your sandwich?" (period of no verbalization)
T1: "No, Roger. Little bites."
Roger: "Little bites." (hits head) (period of no verbalization)
Art: "School bus stop." (looking at a lunch pail with school bus lettered over a picture of a yellow bus) (period of no verbalization)
Jim: "Coffee drink."

T1: "Not coffee. What's that?"
Jim: "Juice, juice." (period of no verbalization)
Mac: "Walk, walk." (some children not from the group come by chattering)
T1: "It's okay, they can walk through." (long period of no verbalization)
T2: (brings a tray with milk cartons) "Donna, Donna, Donna" — (does not respond).
T2: "Art, get her to look at me." (Art is next to Donna who is about two feet from the end of the table where the teacher is standing.)
Art: (does not respond for several seconds, then) "No."
T2: "Donna, Donna, what do you want?" "Donna, Donna, tell me what you want."
Donna: "Milk."
Jim: "Milk, milk." (pause for several minutes)
Jim: "Chips!"
Mary: "Lunch."
T1: "No lunch for me today."
Jim: "Chips."
T1: "What do you eat first?"
Jim: "No chips."
T1: "You eat all your sandwich first, then your chips." (long pause)
T1: "Look, you have something more in your lunch."

The children continue eating without verbalizing. One by one they leave the table as they finish putting their lunch things away and move about the yard. One child is finally left alone as the teachers also leave the table; he continues eating for another five or more minutes, then leaves also. In the play yard the children move about with wagons, scooter board, in the sand box, jumping on large inner tubes, etc. Young children who help are with them. No verbal exchanges were noted between the children in the program. A few interchanges between child helpers occurred.

Helper: "Do you want a ride?"
Norma: (sits down in the wagon) (time delay)
Helper: "I'll push you." (time delay)
T1: (observing Norma ride by in the wagon) "What are you doing Norma?"
Norma: "Wa."
T1: (in aside to the observer) "At least she responded." (galloping along behind Jim with book in his hand comes very close to him, Jim moves off quickly, Art turns, gallops in the opposite direction, *humming*)

Don: (walks up to child in the sandbox, hits him, walks off)
Mac: (pulls back, but does not respond)
Helper: (showing Donna how to pull the wagon around the tile), "You back up and then go this way."
Donna: (no response)
Cross-age girl 1: (follows Donna across the yard with the wagon) "Donna shall I pull you?"
Donna: (gets in wagon)

Bell rings; teacher 2 opens the door, the children begin to move inside. After a few minutes, all, but one are in.

T1: "Time to go in Dan."
Dan: (does not respond)
T1: (moves over, takes his hand) "Time to go in."
Dan: (goes limp, lays on the ground)
T1: Waits a few moments; then "let's go." (attempting to pull him up by the hand) Child continues to resist; teacher finally picks child up and carries him into the school.

The second example is selected from observations of children working in a group program. When the children were first observed they were engaged in a reading activity on the blackboard. The children were observed to pick up errors purposefully made by the teacher and to use the written sentences as prompts for verbal replies. The teacher announced that it was play period. The children began moving about the room in different directions. One child picked up a string from a pull-toy and began pulling it around the room. Another child moved to the headphones and record player. Another child picked up a deck of cards which were moved from hand-to-hand and the remaining four children wandered aimlessly about the room, stopping to rock.

Tim: "I want to go to the bathroom."
Carl: (pulling the pull-toy back and forth) "Tucka, tucka, tucka. . .oh."
Ron: (putting down the cards, moving about the room and rocking from foot to foot with hand to face) "Ow." (after hitting his own face)

For several minutes the children move about without comment.

Ron: (picks up a mirror and begins dancing about the room, looking into the mirror)
Ricky: (moves to a large beachball, throws himself on it and begins rolling back and forth)

Don: (listening to the earphones), says "Diane." (name of teacher and then begins dancing about)

Teacher: "Don is dancing."

Don: (after a pause) "Diane, Diane. . . look it." (the teacher does not respond this time and Don goes on with no apparent concern)

Ricky: (moves from the rolling ball to pick up a stuffed dog, moving it back and forth, making small sounds)

Carl: (continues to move the pull-toy up and down and is screaming) "Don, no Wa-Wa Conkite."

Ben: (stops rocking back and forth and moves to the sink for a drink of water and then begins roaming about the room again)

Mark: (comes back, runs over, grabs Don who runs away)

Mark: (to teacher) "Mark got hurt."

Don: (grabs Carl to take the pull-toy, this is followed by screaming and disengagement follows)

Mark: (picks up the large truck and begins pushing it back and forth across the room)

Don: (picks up his foot) "A toe."

Ricky: (moves over to the large ball, rolls on it), says "Ball."

Mark: (runs across the room, grabs Don by the shoulders, begins to scramble)

The teacher moves toward them.

Mark: "I love you Don, I love you Don, I love you Don." (he looks at his fingers and moves away)

Ben: (standing in the center of the room sucking his thumb)

Don: (standing by the record player), "Wide World of Sports." (jumps repeatedly in the air)

Ricky: (rocking on one foot, says nothing)

Don: (after a pause), says "Skateboard."

Ben: (rocking back and forth on one foot, looking at his hand), says "Go around in a circle."

Carl: (continues running back and forth across the room with the pull-toy)

Mark: (pushing a large truck is following roughly a similar pattern, but the two do not look at each other)

Ricky: (standing in the middle of the room), says "Peep, peep, peep." (in a small voice as if making the sound of a chicken)

Ricky: (alternately is rocking, looking at a bulletin board, rocking and looking)

Mark: (grabs Don, hits him on the head), says "Oh, Don, oh Don, I love you Don, I love you Don."

Mark: (leaves Don, waits, then jumps, twirls), says "Hey."
Don: (with the cards again, runs, hits the ball, looks back at the cards)
Ben: (goes for another drink)
Carl: (moving around the room) "Lazy Lion, Lazy Lion." (repeating this in an increasingly louder voice, followed by a scream and) "Ah da."
Don, Ben: (are both twirling around)
Carl: (hits Don and runs)
Don: "No, no, no, no, don't touch, no, no, no, no, Diane." (the last is in a scream)

The teacher moves toward Don, he quiets and then begins jumping in the air.

Ben: (in another part of the room, is jumping), saying "Oh, oh, dance."
Ron: (hides in the closet, returns walks out)
Mark: (continues to move up and down with the truck)
Carl: (continues to move about the room, pulling the pull-toy)
Ben: (is still in the center of the room sucking his thumb)
Ricky: (is now moving to the ball to lie and rock on it)
Ron: (is in another corner, standing intermittently, jumping and rocking or standing silently)
Ricky: (leaves the ball and picks up the dog again)
Don: (continues to stand by the record player, apparently engaged in listening to the music)
Ricky: (moves over to Don as he changes the record, then again moves away to his spot in the middle of the room where he is rocking)

After approximately twenty minutes, the teacher notifies the children it is time to work (with verbal and sign communication). The children in a reasonably orderly manner move to where their folders are kept, select the appropriate folder with their name on it, and move to their spot to begin the structured work activities.

FINDINGS

Working with the corpus from which these examples were taken, we first asked ourselves if it appeared that the phonemic and morphemic constitutes and rules of English were intact.

With respect to the phonological system the children's productions have been shown to be more like than unlike other youngsters. None of the responses were outside normal limits for preschool age children with respect to the phonological system.

The autistic children's language did not demonstrate irregularities in using morphemes. In this regard the children are more like than unlike other children.

Analysis of the syntactical system defined here as to include sequencing of morphological segments revealed some variations from the normal expected syntax of this age level. Omissions of articles or parts of sentences found in general usage were not considered syntactically different.

Analysis of Child 1's fifty-eight responses resulted in two utterances being classified as unusual. For Child 2, two of the ninety-three responses were possibly syntactically different. Both of these instances involved omission of the verb, for example, "ring in the box," "cat in the box," "fish in the box," "gas car," "toilet in the bathroom." In addition to verbal omissions, for Child 3, there were three responses which were considered unusual, perhaps peculiar, syntactically: "snow is sled," "feet rollerskate," "drink bottle." For the three children, sixteen of the 345 responses (or 4.6 percent) were syntactically questionable or different.

The conclusion was that these autistic children were in full control of the syntax of the phonological and morphological strands. However, on the level of discourse organization, the first evidence of some pathology was detected. Despite this conclusion, the quantity and quality of information related to sequence of discourse units far outweighs any evidence of dysfunction. The echo loop-track-like repeat of the name/address sequence (1.01) was recorded as a malfunction on the discourse syntax level beginning after the end of the first run-through sequence. Such designation, we concluded, provided a practical designation and an opportunity to pinpoint the dysfunction more precisely.

We speculated as to what explanations would be proposed if this example were found by regular teachers in the conversation of a nonautistic child. The transcript therefore was given to ten regular classroom teachers with the request to make comments on the child who had made these statements. The judgments were that either the child was stubborn or not attending. Other evaluations that could be hypothesized were (a) that such a child was especially sensitive to a critical questioning of his skill such that continued questioning led to progressively poorer replies; (b) that the child might not recognize a management question as such, deeming them to be requests for the essence of information as Sample No. 2.01, for example, "four years." Subsequent replies could be then interpreted as a search for an alternative interpretation that might suggest the child's having a limited view of the functions of language.

Considering the tragic dimensions of the autistic child's interactional problems, such commonplace explanations may not seem to hold much promise of insight into the general problem. However, they should not be dismissed, for such explanations when combined with an exhaustive profile of the child's communication instruments do serve as a heuristic classifying device.

Many of the teachers who examined the transcripts but were unfamiliar with autism were surprised at the emphasis on complete sentences. The uniform consensus among them was that the partial represented a more normal response than would a full sentence. The incidents of full and partial sentences for the children were analyzed. Of the 156 responses for the three children of the original sample, 122 utterances were partial sentences and 34 full sentences.

Although reports of the language of autistic children often refer to the noncommunicative aspects of their speech, the teachers failed to be concerned with this aspect of the sample. Noncommunicative speech has been defined by Shapiro (1975: 308) as including "isolated expressive, imitative and context disturbed speech." Shapiro defines isolated expressive utterances as "those in which a child seems to be talking to no one, vocalizing either with gross motion or with music or play or perseverating in a way that has not essential communicative value." Shapiro continues, ". . .context disturbed speech includes utterances that show poor or tangential relationship between what the child has to say and the immediate surroundings or utterances that confuse the examiner by the essential irrelevancy of the remarks. In such speech even if one could pick out a referent, it would be very difficult to pinpoint how a particular sequence of events in the play led to the child's calling up this referent. Imitative speech includes immediate delayed echoing within the examination" (Shapiro 1975: 308).

The incidence of noncommunicative speech was examined for the three children in the original sample. Child 1 produced fifty-eight responses; five were categorized as noncommunicative. Child 2 produced only one noncommunicative response in ninety-three utterances, and Child 3 produced eleven out of 194 which were noncommunicative. It is important to note that these samples were taken in a classroom setting where the demands of the situation, the repetition of tasks and the effects of training to criteria might well be expected to decrease the number of noncommunicative utterances.

An analysis of the noncommunicative utterances of the children in informal situations reveal a higher percentage of noncommunicative remarks outside the formal classroom situation. In the lunch time and informal discourse samples the number of noncommunicative responses is sixteen of a total of thirty-seven responses.

We also looked at the prevalence of echoing over originating (echolalic speech) in a sample of the total. Of fifty-four utterances analyzed, ten were excluded because of insufficient clues being available from the data to determine whether they represented original productions or not. Of the remaining forty-four, twenty-seven were analyzed as nonecholalic, fourteen echolalic, and three as containing echolalic elements. Malfunction echolalic speech on the discourse syntax level presented approximately 35 percent of the responses.

We closely analyzed each of the utterances which were deemed unclear. Other examples were also analyzed individually in terms of possible difference. For example, in the exchange 2.01.005 (cake is Karen) an initial evaluation can be made. The phonemic and morphemic constituents and rules of English are intact in this statement. However, questions can be raised on the level of syntax. In example 2.01.003 (four years old) the length of time the child did not give the complete sentence is obviously exceptional. In the example 2.01.004 (frosting is blue), the child again holds back for an exceptionally long time before repeating the sentence, but since the teacher does not model the response, the possibility that some gap in the linguistic information system prevented the child from forming a complete sentence indicated he had an originating capacity, not merely an echo ability. Indeed, other samples show him capable of organizing the morphemes into a complete sentence. Was the lack that of not knowing what parts of a sentence were crucial to ongoing discourse? Did he, for example, lack the ability to discriminate relevant from irrelevant consequent discourse items of such a statement as "The boy is four years old?", hence selecting the words randomly each time. This possibility was also rejected. Such morphemes as "the" or "is" would as easily have been linked were the dysfunction in the level of arrangement of words. Indeed, rather than give nonsense morpheme combinations Matt three times gave close approximation of the essence of a normal answer. On the more general level of his implicit information about the syntax of discourse, the boy understood the role of turn-taking. He waited for and filled his lexical discourse slot as would be expected from a normal child. He neither overlapped nor unduly paused, but systematically spoke and ceased speaking, as would be expected of normal conversants. However, it is in the extension, duration, or tenacity of repeating the answer that the appearance of abnormality appeared more dramatically. This is the feature attracting concern, providing the common basis of judgments that such children "do not have language" and, therefore, remedial work on language is necessary. Thus, we turned our attention to this feature and compared the material with examples of similar situations with normal children in teacher/pupil situations, for example, the following extract from a classroom (Sinclair and Coulthard 1975).

Following further examination of delayed and undelayed discourse units in classroom interaction, we concluded that the rule that had been violated concerned length-duration. Did other samples, for example, show a distortion of a discourse duration, consistently and similarly sustaining the sequence in such question/answer situations? Did abnormal shortening occur as well as lengthening in the length-duration dimension of information? To answer these questions we compared the samples from the autistic children with samples from ordinary classroom situations. The following example comes from a regular classroom.

1. Now tell me: why do you eat all that food? Can you tell me why do you eat all that food?
   Yes.
2. To keep you strong.
3. To keep you strong. *Yes*. To keep you strong. Why do you want to be strong?
4. Sir — muscles.
5. To make muscles. *Yes.* Well what would you want to use — what would you want to do with your muscles.
6. Sir, use them.
7. You'd want to —
8. Use them.
9. You'd want to use them. Well how do do you use your muscles?
10. By working.
11. By working. Yes. And when you're working, what are you using apart from your muscles? What does that food give you? What does the food give you?
12. Strength.
13. Not only strength; we have another word for it.
    Yes.
14. Energy.
15. Good girl. Yes. Energy. You can have a team point. That's a very good word. We use . . . we're using energy. When a car goes into the garage, what do you put in it?
16. Petrol.
17. You put petrol in. Why do you put petrol in?
18. To keep it going.
19. To keep it going; so that it will go on the road. The car uses the petrol, but the petrol changes to something, in the same way that your food changes to something. What does the petrol change to?
20. Smoke.
21. Water.
22. Fire.
23. Smoke.
24. You told me before.
25. Inaudible.
26. Again.
27. Inaudible (Energy)
28. Energy. Tell everybody.
29. Energy.

30. Energy. Yes. When you put petrol in the car, you're putting another kind of energy in the car for the petrol. So we get energy from petrol and we get energy from food. Two kinds of energy. Now then, I want you to take your pen and I want you to rub it as hard as you can on something woolen, if you've got something woolen. Just rub it. There. And then — you're using a lot of energy now, aren't you? — and then take your bit of paper and tear it into tiny little bits, little bits smaller than that; put them on the desk; and then, just put the end of the pen that you've rubbed near the paper, and see what happens. — Activity —	
	31. Sir, Sir. 32. Static electricity. 33. Sir, electricity. 34. Sir. 35. Sir.
36. Not in your hair. On your jumper. And then see what happens.	37. Sir. 38. Sir.
39. When you've done it, tell me what happens.	40. Sir. 41. Inaudible. 42. Pen, Sir.
43. You haven't got what? 44. Tiny bits, please. Much smaller than that. Much, much smaller than that. Now. Put it near your tissue paper. Tell me what happens when you put your pen near your tissue paper.	45. Sir; the pen, er, picks it up.
46. Yes. Would you say the pen is doing some work?	47. Yes, Sir. (chorus)
48. Yes. Would you say that the pen was —	49. Energy.
50. using something?	51. Yes, Sir. (chorus) 52. Energy. (chorus)
54. Yes. It's using energy. Yes. Where did you get the energy from?	53. Energy, Sir. 55. Sir.

56. Sir.
57. From your arm.

58. From the rubbing. Yes. Right. Put your pens down. All eyes on me. Now what we've just done, what we've just done (is) given some energy from? Just think where we've got the energy from. This is what we did. Where did the energy come from?
Keith.

That children give approximations of teacher's desired answers is not uncommon. Nor is it uncommon for teachers to allow approximations and deviations for durations of the interaction sequence. Unusual in the instances we studied is the continuation of such a reply for longer than the time expected. The children do know when and when not to speak, do comprehend that another approximation of the previous answer is required, and do select a close approximation of the teacher's desired answer.

The teacher, beginning formal pedagogical questioning, has a sense of how long the question-answer phase will continue before she brings the procedure to a halt, either by openly extracting or giving the answer herself. This sense of discourse duration is one of the components of information shared by participants in a conversation, and apparently this is also an awareness that autistic children share. In the name-address sequence and the "cake is Karen's" examples above, the answer sought by the teacher is finally given by the pupil in the form sought. After close clue laden coaching by the teacher this occurs after a longer period of extracting than is typically expected. Nevertheless, the children do come upon the correct answer within a relatively short time. Hence, again by looking at the discourse in terms of its constituents and separately considering the exact nature of the interactants' match or mismatch, a clearer picture of what the autistic child has within his linguistics information system can be obtained.

FUNCTIONAL ANALYSIS

The children demonstrate that they possess the tools of communication, but what they do with these tools is perhaps where the essence of the difference lies. To explore this possibility we completed a functional analysis of the utterances. We employed a system developed during a ten-year Claremont study of children's language. The method employs a ten-slot category system. These functions slots can be understood as a spectrum rather than definitely separate units.

The first function of taxonomy is *expressing* reaction with words, not with vocalization such as shouts and grunts. The second function is that of *play*. Each strand of the instrument is material with which a speaker can play. For the young child the play combines with practice by which he increases his skills with the instrument. The third function, *contact*, is that of making or maintaining distance, not with any sound or even direct action but with language. The fourth function involves the characteristic capacity of language of *naming*, of referring, categorizing. The fifth, is the *recognizing* function operating both on a level of reception and production. The child recognizes others in part by virtue of their language and is recognized by virtue of his use of it. The sixth, *thinking* with the linguistic categories, occurs as soon as the child has names to give phenomena; he also has an instrument with which to do his thinking about them. In the seventh, the *affiliating* function, as the child begins using the categories of his language, he enters a way of seeing the work of which he/she is now a part. The eighth function, *explaining*, is laying out a picture of events for others. The ninth, *social control*, is attempting to influence others in terms of their resources, to acquire or defend resources. The tenth is the *heuristic* function, the use of words to find out information. The characteristics and proportions of these last three educative functions were discovered to constitute the essential differences between autistic and other children.

A functional analysis of each utterance was completed. Each utterance was chartered separately and on a discourse analysis chart. Any utterance which could not be mapped with certainty because of the lack of scope of the context was coded as uncertain, and each alternate possibility was specifically coded. This discourse was compared to a nonselected sample from peers in terms of the degree of certainty in mapping.

Functional analysis of the classroom sample resulted in the following findings:

1. It is more difficult to attribute function with certainty to the autistic sample than to normal samples in the ten-year Claremont study corpus. Twenty-three of Child A's fifty-eight responses required more information for classification than was available; twenty-six of Child B's ninety-three responses were uncertain as to function, and forty-four of Child C's 194 responses were classified as uncertain. Uncertainty as to the function of an utterance is an unusual finding in normal populations with the ten-year study.

2. Those utterances which could be classified fell into three function groups: (a) play (second function); (b) explain or answer question (eighth function); (c) influence others (ninth function).

No utterance could be labeled clearly as heuristic (tenth function), however, there was one possible example of it. In the samples of play and lunchtime "conversation," Mary said the word "lunch"; she may have been asking the teacher if she was going to have lunch. This finding is relevant to the data regarding question asking in autistics. A total of thirty-seven responses were obtained for a total of 110 minutes of observation. Of these the functions of play (function 2) and to influence others (function 9) are about evenly represented and account for 65 percent of the responses. Adding function 8 (to explain) account for 81 percent of the responses. In addition, there was one possible occurrence of each of functions 1, 2, 3. There was one function 5; three function 6; one possible 10. As in the classroom sample, functions 2, 8, and 9 predominate.

Further analysis of the data from the expanded sample of nonclassroom settings reveals the limited productivity of autistic children in a nondemand setting. Other investigators have pointed to the fact that autistic children tend not to "chat" or talk to one another or as Rutter (1975) notes, tend not to use the limited language skills they possess for functional communication. This finding is borne out by the present study. Language data from nonclassroom settings tend to be sparse; few utterances are produced during periods of play observation. During the average ten-minute sample of lunchtime or play conversation, 3.36 utterances are produced.

DISCUSSION

It is at the point of interface between the individual linguistic and cultural information system that the autistic's deficits are most demonstrable. Autistic language may be seen as a failure to develop experiential forms, to appreciate or use gestures, to use or answer questions, to engage in imaginative play. The prolonged exclusive "attention" to repetitive stereotypic tasks, the simultaneous apparent deafness and hypersensitivity to sound, the obliviousness to certain visual stimuli with hyperalertness at subsequent moments, the tactile defensiveness and simultaneous insensitivity to pain, can be seen in terms of integrative failure. Sydney Lamb (1974) has posed this question: "Can the relation between language and culture be considered as a relation between two possibly intertwined semiotic systems, the linguistic and the cultural?" Accordingly, the linguistic dysfunctions of the speaking autistic child could be considered as an inability to structure experience forms — a failure at the cultural linguistic interface that disallows an integrated perspective.

Normal language development is dependent upon both accommodation (i.e., the learning of culturally agreed upon phonological, morphological, lexical systems) and assimilation (i.e., the structuration and reorganization of the material for use in original reformulation).

The autistic child experiences some success in accommodation as reflected in the imitation in autistic echolalia, but, unlike the normally developing child who uses fewer congruent echoes, the autistic demonstrates almost no newly created imitative forms. In assimilation, the autistic child fails; integrative failure in Piagetian terms is assimilative failure. The greater use of creative imitative forms by the normally developing child suggests assimilation as well as accommodation in equilibration, in reaching a new developmental level. It is at the point of reformulation, the structuration, that the autistic is found limited. The cultural forms are swallowed whole, undigested; and the child is left dependent upon a repertoire of imitative signals rather than symbols with which to develop an appreciation of future and past.

The autistic child's failure to structure the world with language severely limits the capacity to learn through the linguistic mode. It has, however, been proposed that the language failure itself is part of an earlier more general deficit in the structuralizing capacity of the organism. General cognitive processes which are dependent upon organization and structure of experience are delayed in the autistic. Language, as more than a rote repetition of sounds, is dependent upon prior organizations and integration and upon transformations into symbolic function. This capacity for transformation which is apparently a part of the biological equipment of normally developing infants (Lenneberg 1970) appears deficient in the autistic (Shapiro 1976).

The failure of attachment and the isolation of these children may be considered to be one of integrative failure at the most elementary level. Some of the evolutionary, bioadaptive mechanisms upon which the mother-infant attachment depends are apparently deficient in many autistic infants: normal gaze; clear preferences for the human face over other stimuli and the human voice over other auditory stimuli; the assumption of an anticipatory posture to being picked up; and the adjustment of posture when held. Without this basic bioadaptive equipment, the failure of mother-infant attachment is less surprising and the subsequent "inability to relate themselves to people in situations' (Kanner 1943) may relate to this initial biophysiological failure.

Experience forms develop mainly around significant persons or significant and important events and mainly when circumstances, locales, or love objects are habitual or stable. Without the early ability to differentiate human from nonhuman stimuli, the autistic child cannot accumulate specific experience entities around certain individuals. Without the capability to differentiate, to single out the human figure from the object ground, the organization of attributes of a single figure is delayed and no single individual can obtain a level of salience and stability. In this case the process of accumulating experience entities around a single individual and the binding of these intercoordinated experiences and coordinated configurations which are the experience forms, does not develop normally. The autistic delays or fails to differentiate experiential self-experiences from others; hence, behavioral differences and social failures are understandable in terms of the failure at structuration of the cultural/linguistic world.

The uneven abilities of the autistic person ("islands of precocity") also can be understood by reference to an inability to structure experience. The "experience entities" of the autistic's world remains as separate, discrete, unorganized data which are put forward idiosyncratically and in isolation, and on occasion with a startlingly precocious appearance, but without the utility achieved only through some location of the part into the larger organization of behavior. Autistic children at preschool age and early elementary years have on occasion been observed to demonstrate surprising academic feats. For example, one child could add columns of figures of any length at nearly the speed of a computer but could not answer, "Which ball is bigger?"

Autistic children are often easily taught to "read" the sentence, some regardless of difficulty, but are unable to demonstrate any comprehension skills (Kanner 1943). Isolated talents in other autistics have been limited to more practical matters as illustrated in comments, such as "He always knows where the peanut butter is hidden" or "He can undo any lock or security system ever developed regardless of its difficulty." Not all autistic children demonstrate such dramatic "splinter skills," and while many present more or less uniformly subnormal functioning (Rutter 1975), even these children demonstrate relatively greater educational strength in reproducing isolated tasks. The failure to structure these fragments into large cultural forms limits the utility of tasks learned.

Parents frequently complain of the child's inability to make use of past learning. The children are said to seem to forget tasks that they have learned or "not remember them" in new situations. One mother reported that it seemed to her as if her child were always surprised; as if he were always seeing the world "for the first time" and had no way of carrying over events from previous situations. This was an example of the child's failure to develop integrative patterns or experience forms which would allow recurring but modifiable patterns and the use of the personal history necessary for the development of self.

The failure of integration of experiential structures which normally give values of things, persons, and relations, allowing for continuity of means of responding to similar environmental circumstances, results in the failure to transfer and generalize learned behaviors. Most individuals recall abstractions from the global structures of their experience forms. If the child has failed to develop integrated experience forms but has available only fragments of nonintegrated stimuli, then one would predict memory functions much like those demonstrated by the autistic child and would expect language development to proceed in terms of rote or, as it were, "undigested fragments." The failure of assimilation of the language and cultural experience, i.e., of the intertwined, linguistic, and cultural semiotic systems, may be seen as the underlying deficit in autism.

EDUCATIONAL IMPLICATIONS

In recent years there has been an increase in the development of treatment programs and approaches for the autistic child. The increasing demands of parents for public school education for their autistic children has intensified the need for development of coherent educational plans. Despite the increased attention to educational and treatment programs, the results have not been particularly encouraging (Miller 1975). Miller notes there is considerable question still as to the ultimate effectiveness of any of the therapeutic modalities. "Initial optimistic reports by a few investigators . . .have not been duplicated, and the accumulative experience of most treatment centers seem to be that the great majority of children (2/3 or more) fall far short of normality and continue to manifest bizarre and stereotyped behavior and concrete thinking" (Miller 1975: 391–392).

The type of intervention advocated depends upon the investigators theoretical orientation. Methods selected for speech training are related to the therapist's concept of causation. Those who regard lack of communicative speech as a function of withdrawal from social contact tend to stress the development of a relationship as the primary treatment approach and report that language improvement is concomitant with improved relationships if it ever occurs. Those who look upon the lack of development of speech as a type of behavior deficit tend to use behavior modification techniques as the primary means of increasing the speech (Hingtgen and Bryson 1974). It is generally agreed that behavior modification techniques increase verbal productivity and vocabulary; however, it is becoming increasingly apparent that there is little generalization from behavior modification training and that there is little increase in spontaneous and creative speech and very little improvement in functional language use (Hingtgen and Bryson 1974; Shapiro 1975, 1976).

Rutter (1975) notes that even with other than behavior modification techniques the outlook for improvement may not be optimistic and speech may continue to be mechanical and concrete and Shapiro (1976) believes that cognitive psychologists and linguists may have a crucial experiment in these children. "These are groups of children who appear to defy the premise that reinforcement alone yields 'natural speech.' It yields rigid echoing and contextual irrelevance along with some natural speech" (Shapiro 1976:238).

Graham (1976), using behavior modification techniques for language training as developed by Lovaas, Hewett and others, found that the program resulted in an increase of specific skills but a failure to "learn to learn."

The teacher of the autistic child is well aware of the student's characteristic pattern of "learning" a task but subsequently being able to demonstrate that particular task only in a certain context. For example, many teachers of children diagnosed as autistic complain that their students are able to demonstrate a response to criteria nearly 100 percent of the time when seated in a particular chair in a particular part of the classroom, but are unable to demonstrate the behavior when at home or on the playground or even in a different part of the classroom. Accumulating new behaviors appears to be easily facilitated through behavioral approaches, but it seems to continue to be very difficult to demonstrate generalization of specific tasks learned. Learning for other children ordinarily involves a process of developing new connections to a particular aspect of a much larger structure.

The current educational methods which lean heavily upon behavior modification techniques are initially dramatically effective. Their positive value in behavior management and self-help skills have been demonstrated. However, in so far as these techniques provide a means of acquiring more, still structured, behavior bits to the aggregate, it is suggested here that these techniques will be of limited value to linguistic development at the more abstract levels. Behavioral repertoires, even quite complex ones, can be taught with this method, but integrations cannot. While it is clear that these children are desperately in need of an expanded behavior repertoire to operate in their world and to make life bearable and predictable for those persons closest to them, more is needed.

While it is recognized that the functional linguistic training of autistic children is an extremely difficult task, it is important at this time to attempt to design a functional-developmental program for language education of autistic youngsters. Graham (1976) came to a similar conclusion based on empirical experience. She reports an expanded program of language development based upon an experimental model. Teachers are given the task of helping children learn to play appropriately, using a plan based upon the play attainment developmental level of each child. The teachers are expected to respond to the child's play level irrespective of chronological age.

A similar developmental play and language approach to language stimulation has been instituted as part of the Claremont study with two children. The children are at a stage of language development where they are just beginning to respond with some recognition and pleasure to a repetition of their own vocalizations. The two adults working with these children have been assigned the task of responding with accurate repetitions of their vocalization during a prescribed period of time each day. No other intervention is attempted during this time aside from the usual attempts to initiate contact and to provide tactile stimulation. This experiment has only been underway for a short time but some improvement has been noted.

Piaget (1973, 1976), Bruner and Sherwood (1976) and others have emphasized the importance of play for cognitive and language development, but play as an educational or treatment tool has been generally overlooked with the autistic persons except in psychotherapeutic approaches.

We observed limited play as we looked over language samples; this is consistent with reports of other investigators (Rimland 1964). Developmental reports from parents of autistics suggests severe limitations, or no occurrence of play interactions as infants. The work of Bruner (1975) and Bruner and Sherwood (1976) and others suggest that play is implicated in early language acquisition. "Part of the child's first mastery of language is, in exchange games, in peek-a-boo and in other structured interactions, young children learn to signal and to recognize signals and expectancies . . . play serves as a vehicle for language acquisition" (Bruner and Sherwood 1976:19).

In our view the developmental deviations in the play and in the functional language development of autistic children do not represent separate symptoms of the syndrome, but rather two aspects of the more fundamental failure of the semiotic. Experience with play may be a necessary condition for the full assimilation of the tool of language. Play at the symbolic level, fails to develop in the usual manner with the autistic (Hingtgen and Bryson 1974); symbolic function in both language and play is deficient in the autistic child. Play, language and culture are interdependent semiotic systems. Early failures in reciprocal play have implications for language development; limitations in the development of linguistic tools have implications for the development of the more advanced forms of symbolic and social play. Limitations in linguistic and play functions have implications for knowledge of the culture.

If one bears in mind that the environmental milieu of the autistic child does not differ from that in which other, nonautistic children have developed, then the educational task becomes one of supplying a developmentally appropriate play and language milieu at sufficiently concentrated and high intensity levels to impact the autistic child's development.

CONCLUSION

> *Think of the tools in a box; there is a hammer, a saw, a screwdriver, a ruler, a glue pot, glue, nails, and screws. The functions of words are as diverse as the functions of these objects and in both cases there are similarities. Of course, what confuses us is the uniform appearance of words when we hear them spoken . . .*
> L. WITTGENSTEIN, *Philosophical investigations*, p. 6

This has been a report of a study initiated by a group of educator-therapists whose analysis of children's conversation had raised questions about the remedial outcomes from behavioral shaping approaches to language. The conclusion reached was that behavior shaping is insufficient.

The basis of this conclusion was that the children studied had control of each of the structured levels of the verbal instrument which in combination constitutes the normal speaking child's linguistic information system. Specifically, they had control of the constituent parts and rules of the phonological, morphological, syntactical, and paralinguistic strands. By expanding the definition of syntax to include that of conversational units, the first evidence of a deficiency was identified. Despite the obvious peculiarities of many responses, there was evidence that the children indeed did know the rules of discourse structure. Therefore, the findings did not support the judgment that they were not in control of their linguistic instrument as such, despite their problems in interacting with others.

Hence, little could be found that practice with the phonological, morphological, lexical, or syntactical components could hope to remedy. The principal lacunae were uncovered in the analysis of the data into function categories. Remediation programs, such as sentence completion, rephrasing, and repeating of the morphological/syntactical level, therefore, were deemed not to be of direct value. Rather, it is suggested that remedial work relating to experience forms held greater potential for change.

There is an organizational focusing effect of intense question and answer teacher-pupil verbal interaction. Studies of normal classrooms around the world indicate that the question/answer ("answer in a complete statement") approach, provides an efficient means of managing and focusing attention. This unnoticed value of the teacher/child interaction provides a productive outcome which can be erroneously interpreted as the result of the linguistic intervention rather than in the organizing time/attention management advantages. The question may be asked: What greater returns could be anticipated from a time/attention management approach if it contained information the child did not have?

Linguistic instruction for the purpose of attention management seems to be an unfortunate waste of an opportunity and could have other than positive effects.

REFERENCES

BAKER, L., D. CANTWELL, M. RUTTER, L. BARTAK

1976 "Language and autism," in *Autism: diagnosis, current research and management*.Edited by E. R. Ritvo. New York: Spectrum.

1970 "The life course of children with autism and mental retardation," in *Psychiatric approaches to mental retardation*. New York: Basic Books.

1971 "The nature of childhood psychosis," in *Modern perspectives in international child psychiatry*. Edited by John Howells, 649–678. New York: Brunner/Mazel.

BETTLEHEIM, BRUNO

1967 *The empty fortress: infantile autism and the birth of the self*. New York: The Free Press.

BRUNER, J.

1975 Language and cognitive development. Paper presented at the third Biennial meeting of the International Society for Behavioral Development, University of Surrey, Surrey, England, July 12–15.

BRUNER, J., V. SHERWOOD

1976 "Peekaboo and the learning of rule structure," in *Play: its role in development and evolution*. Edited by J. Bruner, A. Joley, C. Silva. New York: Basic Books.

BUEHLER K.

1934 *Sprach theorie*. Jena: Fischer.

CREAK, M.

1963 Childhood psychosis, a review of 100 cases. *British Journal of Psychiatry* 109–184.

CUNNINGHAM, M. A., C. DIXON

1961 A study of language of an autistic child. *Journal of Child Psychology and Psychiatry* 2:193.

CUNNINGHAM, M. A.

1968 "Comparison of the language of psychotic and nonpsychotic children who are mentally retarded." *Journal of Child Psychology and Psychiatry* 9:229–244.

DANCE, FRANK, CARL LARSON

1976 *The functions of human communication.* New York: Holt, Rinehart, and Winston.

DE LAGUNA, G. A.

1963 *Speech: its functions and development*. Bloomington: Indiana University Press.

EISENBERG, L.

1957 The course of childhood schizophrenia. *AMA arch Neuro Psychiatry* 78:69–83.

EKSTEIN, R., E. CARUTH

1964 On the acquisition of speech in the autistic child. Reiss-Davis *Clinic Bulletin* 1:63–79.

1969 Levels of verbal communication in the schizophrenic child's struggle against, for, and with the world of objects. *Psychological Study of the Child* 24:115–137.

FISH, B.

1971 "Contributions of developmental research to a theory of schizophrenia," in *Exceptional infant*, volume 2. Edited by Jerome Hellmuth, 73–482. New York: Brunner/Mazel.

FREEMAN, B. J., E. R. RITVO

1976 "Cognitive assessment" in *Autism diagnosis, current research and management*. Edited by E. R. Ritvo. New York: Spectrum.

GOLDFARB, WILLIAM

1971 "Therapeutic management of schizophrenic child," in *Modern perspectives in international child psychiatry*. Edited by John Howells, 685–705. New York: Brunner/Mazel.

GOLDSTEIN, K.

1948 Language and language disturbances. New York: Grune and Stratton.

GRAHAM, V. L.

1976 "Educational approaches at the N.P.I. School: the general program," in *Autism: diagnosis, current research and management*. Edited by E. R. Ritvo. New York: Spectrum.

HALLIDAY, M. A. K.

1973 *Explorations in the functions of language*. London: Edward Arnold.

HARRISON, S. I., S. F. MCDERMOTT

1972 "Introduction" to Part VII: Childhood Psychosis in *Childhood psychopathology*. New York: International University Press.

HEWETT, F. M.

1965 Teaching speech to an autistic child through operant conditioning. *American Journal of Orthopsychiatry* 35:927–936.

1968 *The emotionally disturbed child in the classroom*. Boston: Allyn and Bacon.

HINGTGEN, J. N., C. Q. BRYSON

1974 "Recent developments in the study of early childhood psychoses: infantile autism, childhood schizophrenia and related disorders" in *Annual progress in child psychiatry and child development*. Edited by S. Chess and A. Thomas. New York: Brunner/Mazel.

HYMES, DELL H.

1962 "The ethnography of speaking," in *Anthropology and human behavior*. Edited by T. Gladwin and W. C. Sturtevant. Washington, D. C.: Anthropological Society of Washington.

JAKOBSON, ROMAN

1967 "About the relation between visual and auditory signs," in *Models for the perception of speech and visual form*. Edited by Weiant Wathen-Dunn. Proceedings of a symposium, 1964. Cambridge, Mass.: M.I.T. Press.

KANNER, L.

1943 Infantile autism. *Nervous child* 2:212–256.

KANNER, L., L. LESSER

1972 "Early infantile autism" in *Childhood psychopathology*. Edited by S. I. Harrison and S. F. McDermott, 647–699. New York: International University Press.

KAVANAUGH, J.

1971 "The genesis and pathogenesis of speech and language," in *Experi-*

mental infant, volume two. Edited by Jerome Hellmuth, 211–247. New York: Brunner/Mazel.

KNOBLOCH, H., B. PASAMANICK
1975 Some etiologic and prognostic factors in early infantile autism and psychosis. *Pediatrics* 55 (2):182–191.

KOZLOFF, M. A.
1973 *Reaching the autistic child in a parent training program*. Chicago, Ill.: Research Press.

LAMB, S. M.
1974 "Discussing language," in *Discussing language*. Edited by Parrett Herran. The Hague: Mouton.

LANGER, S.
1967 *Mind: an essay on human feeling*, volume one. Baltimore: Johns Hopkins University Press.
1972 *Mind: and essay on human feeling*, volume two. Baltimore: Johns Hopkins University Press.

LASSWELL, H.
1948 "The structure and function of communication in society," in *The communication of ideas*. Edited by L. Bryson. New York: Harper and Row.

LEATHERS, D.
1976 *Nonverbal communication systems*. Boston: Allyn and Bacon.

LENNEBERG, ERIC
1970 "On explaining language," in *Annual progress in child psychiatry and child development*. Edited by S. Chess and A. Thomas, 104–125. New York: Brunner/Mazel.

LOVAAS, O. I., L. SCHREIBMAN, R. L. KOEGEL
1976 "A behavior modification approach to the treatment of autistic children," in *Psychopathology and child development*. Edited by E. Schoplen and R. J. Reichler, 291–310. New York: Plenum Press.

MALINOWSKI, BRONISLAW
1930 "The problem of meaning in primitive languages," in *The meaning of meaning*. Edited by C. Ogden and I. A. Richards. New York: Harcourt, Brace and Co.

MILLER, ROBERT T.
1975 "Childhood schizophrenia: a review of selected literature," in *Annual progress in child psychiatry and child development*. Edited by S. Chess and A. Thomas, 357–401. New York: Brunner/Mazel.

MYKLEBUST, A. BANNOCHIE
1972 Emotional characteristics of learning disability, in *Journal of Autism and Childhood Schizophrenia*, 2 (2):151–159.

ORNITZ, E. M., E. R. RITVO
1976 "Medical assessment," in *Autism: diagnosis, current research and management*. Edited by E. R. Ritvo. New York: Spectrum.

PIAGET, J.
1973 *The child and reality: problems of genetic psychology*. New York: Grossman.
1976 "Symbolic play," in *Play*. Edited by J. S. Bruner, 555–569. New York: Basic Books.

RICHEY, ELLEN
1976 "The language program," in *Autism: diagnosis, current research and management.* Edited by E. R. Ritvo. New York: Spectrum.

RIMLAND, BERNARD
1964 *Infantile autism: the syndrome and its implications for neural theory of behavior*. Englewood Cliffs, New York: Prentice Hall.
ROBINSON, W. P.
1972 *Language and social behavior*. Baltimore: Penguin.
RUTTER, M.
1973 "Relationships between child and adult psychiatric disorders," in *Annual progress in child psychiatry and child development*. Edited by S. Chess and A. Thomas, 669–688. New York: Bruner/Mazel.
1975 "The development of infantile autism," in *Annual progress in child psychiatry and child development*. Edited by S. Chess and A. Thomas. New York: Brunner/Mazel.
SHAPIRO, THEODORE
1975 "Language and ego function of young psychotic children," in *Explorations in child psychiatry*, 303–319. New York: Plenum
1976 "Language behavior as a prognostic indicator in schizophrenic children under 42 months," in *Infant psychiatry: a new synthesis*. Edited by Rexford, Sander and Shapiro, 227–338. New Haven: Yale University Press.
SHAPIRO, T., A. ROBERTS, B. FISH
1971 "Imitation and echoing in young schizophrenic children," in *Annual progress in child psychiatry and child development.* Edited by S. Chess and A. Thomas, 421–440. New York: Brunner/Mazel.
SHAW, C. R., A. LUCAS
1970 *Psychiatric disorders of childhood*. Englewood Cliffs, N. J.: Appleton-Century-Crofts.
SINCLAIR, J. MC H., R. M. COULTHARD
1975 *Towards an analysis of discourse*. London: Oxford University Press.
SMITH, H. L. JR.
1976 Linguistics as a behavioral science. *Forum Linguisticum* 1 (2).
TRAGER, G. L., H. L. SMITH JR.
1951 "An outline of English structure," in *Studies in linguistics: occasional papers*, number three. Norman, Oklahoma.
TREFFERT, D. A.
1971 "Epidemiology of infantile autism," in *Annual progress in child psychiatry and child development*. Edited by Chess and Thomas, 441–454. New York: Brunner/Mazel.
ULLMANN, S.
1966 *Words and their use*. New York: Hawthorne.
VYGOTSKY, L. S.
1969 *Thought and language*. Edited and translated by Eugenia Hanfmann and Gertrude Vakar. Cambridge, Mass.: M.I.T. Press
WING, L., *editor*
1976 *Early childhood autism*. London: Pergam International Library of Science, Technology, Engineering and Social Studies.
WITTGENSTEIN, L.
1958 *Philosophical investigations*, third edition. Translated by G. E. M. Anscombe. New York: Macmillan.
WOLFF, S., S. CHESS
1965 An analysis of the language of 14 schizophrenic children. *Journal of child psychology and psychiatry*, 6:29.

Meaning and Memory

DWIGHT BOLINGER

To start this discourse I cite three texts, the first from Freeman Twaddell, the second from Raimo Anttila, and third from Peter Ladefoged. But first a word about the topic.

For a long time now linguists have been reveling in Theory with a capital T. If you assume that language is a system *où tout se tient* — where everything hangs together — then it follows that a connecting principle is at work, and the linguist's job is to construct a one-piece model to account for everything. It can be a piece with many parts and subparts, but everything has to mesh. That has been the overriding aim for the past fifteen years. But more and more evidence is turning up that this view of language cannot be maintained without excluding altogether too much of what language is supposed to be about. In place of a monolithic homogeneity, we are finding homogeneity within heterogeneity. Language may be an edifice where everything hangs together, but it has more patching and gluing about it than architectonics. Not every monad carries a microcosm of the universe inside; a brick can crumble here and a termite can nibble there without setting off tremors from cellar to attic. I want to suggest that language is a structure, but in some ways a jerry-built structure. That it can be described not just as homogeneous *and* tightly organized, but in certain of its aspects as heterogeneous *but* tightly organized.

Specifically what I want to challenge is the prevailing reductionism — the analysis of syntax and phonology into determinate rules, of words into determinate morphemes, and of meanings into determinate features. I

This paper was originally delivered as a lecture at Brown University, April 29, 1974. It has been revised slightly from the previously published version in *Forum Linguisticum*, 1976, 1: 1–14. See also "Idioms have relations," *Forum Linguisticum* (forthcoming) for further discussion.

want to take an idiomatic rather than an analytic view and argue that analyzability always goes along with its opposite at whatever level, and that our language does not expect us to build everything starting with lumber, nails, and blueprint. Instead it provides us with an incredibly large number of prefabs, which have the magical property of persisting even when we knock some of them apart and put them together in unpredictable ways.

My three texts are not a proof of this, but they are persuasive because if what they say is true, then what I am going to say is at least possible. All of them point to the same kind of enveloping memory that has to be assumed, if the human brain is to have the power that I want to impute to it.

Twaddell is speaking about syntactic rules. He admits that a lot can be done with them, but he cautions that "there is also much in linguistic activity which seems to be more plausibly described as the recall of quite specific memories." And he adds that "it is uneconomical to invent a rule to account for behavior which can be accounted for as an autonomous communicative signal, without any necessary systematic relation to the rest of syntax" (Twaddell 1972:26).

Anttila looks at morphology rather than syntax. He feels that just about as many mistakes are being made now in the positing of underlying forms as used to be made in the chopping of words into morphemes. "The errors are chiefly seen in the general tendency to write rules for fossilized connections of the type *drink/drench*, *bake/batch*, *hallow/whole*, and so on" (Anttila 1972: 130–131). Speakers are being endowed with productive mechanisms that no longer produce anything. The truth is that we have the words, but they are stored as independent units; they are not connected in our subconscious. The absurdities here almost exactly repeat those of the 1950's, when speakers were endowed with morphemes that had long since been swallowed up in words, like the *dox* in *doxology* or the *ham* in hamlet.

Ladefoged writes:

> The indications from neurophysiology and psychology are that, instead of storing a small number of primitives and organizing them in terms of a [relatively] large number of rules, we store a large number of complex items which we manipulate with comparatively simple operations. The central nervous system is like a special kind of computer which has rapid access to the items in a very large memory, but comparatively little ability to process these items when they have been taken out of memory. There is a great deal of evidence that muscular movements are organized in terms of complex, unalterable chunks of at least a quarter of a second in duration (and often much longer) and nothing to indicate organization in terms of short simultaneous segments which require processing with context-restricted rules (Ladefoged 1972:282).

If I may play St. Augustine to St. Peter and exaggerate a bit, I would say that the human mind is less remarkable for its creativity than for the fact that it remembers everything. In another place Anttila says that "Memory or brain storage is on a much more extravagant scale than we would like to think; even the most 'obvious' cases can be stored separately" (Anttila 1972:349).

The picture we have always had drawn for us is a layercake, with syntax on top of morphology on top of phonology. In spite of some interpenetration, especially in morphology, the three layers have been regarded as fairly distinct. The separation between morphology and syntax is seen as quite sharp, with bound forms on one side of the line and free on the other. The implications of my revised view should be apparent. They are that speakers do at least as much remembering as they do putting together, and a great deal of what we have been regarding as syntactic will have to be put down as morphological. The picture that emerges is a vast continuum between morphology and syntax, with perhaps a slight crease where the two domains come together but nothing like the abrupt edge that we are accustomed to putting there. The relationships between form and meaning become identical from the top to the bottom of the scale, the only real difference being that structure gets more rigid the closer you are to the bottom — as if solidified by pressure from above. Everyone knows that words are comparatively rigid in the way they condition morphemes. (*Anxiety* not only diverges from *anxious* but conditions *-ty* in a way that *sensitivity*, for example, does not: it is more "bound" in the first noun than in the second.) What is not appreciated widely enough is that sentences can also be rigid in the way they condition words. Much less rigid, but we have to see the phenomenon as a pervasive one if we are to understand language as an organism and not an Erector set.

To restate Anttila: There is no sense in taking two such forms as *book* and *beech* and trying to derive them from a single underlying form. It does make sense to do this with two such forms as *bring* and *brought*, because they are remembered together — in whatever filing system we carry in our heads, there has to be a folder somewhere in which *bring* and *brought* are side by side. Anttila's example is an extreme one. Nobody but an etymologist knows that *book* ever had anything more to do with *beech* than it has to do with *back* or *buck*; in fact the back of a book bound in buckram has a better chance of being associated than *beech* does. If we wanted a more debatable case we could take something like *renege* and *renegade* or *disease* and *ease*.

Twaddell's point is that the situation is the same in syntax. Suppose we took the phrase *out of patience* and looked for an underlying representation. It would have to contain the same *out of* that is found in *out of money*, *out of time*, *out of ice cream*, out of anything that one formerly had a supply of but has no longer. It had never dawned on me, until January 23, 1973, that the *out of* in *out of patience* had any connection with that other *out of*. Penetrating the source of the expression was one of those odd strokes of illumination that we get every so often with something we have been using all our lives. It is clear enough that most people do not associate the two, because if they did they would not turn *out of patience* into a command, or use intensifiers with it (*Don't be so out of patience*!; **Don't be so out of money*!)[1]. When we say *out of patience* we are not pulling *out of* and *patience* separately from storage and putting them together but retrieving the whole thing at once.

Ladefoged's point is that precisely this is what we have the psychological capacity and tendency to do.

So far my arguments have amounted to no more than a claim, along with some examples that nobody would be much disposed to argue about. Every linguist recognizes that there are limits to how far one can go in saying that two forms are connected in some kind of paradigm in our minds, rather than being simply independent representations of whatever their meaning is. And every linguist makes room in his scheme of things for lexical units larger than words. He calls them *idioms*, and it was from the stock of idioms that I got my *out of patience* example. To make a case on a higher level, it is necessary to show that idiomlike connections extend far up and beyond what are ordinarily regarded as idioms. Many scholars (e.g., Bugarski 1968; Chafe 1968; especially Makkai 1972) have pointed out that idioms are where reductionist theories of language break down. But what we are now in a position to recognize is that idiomaticity is a vastly more pervasive phenomenon than we ever imagined and vastly harder to separate from the pure freedom of syntax, if indeed any such fiery zone as pure syntax exists.

If morphology and syntax have the overlap that I claim, we should be able to see it by standing above it and looking down, as well as by standing below it and looking up. So before taking the upward tack it will do no harm to go in the opposite direction for a moment to see what degree of syntacticity may be found dipping downward into morphology. As a first example, take the word *ago*. It has several interesting characteristics. First, it is always suffixed — we say *a year ago*, not **an ago year*; and we do not use it independently — there is no **He got there ago* meaning "He got there at some time past." Second, it is always suffixed to expressions of a particular kind, namely those referring to time. In this respect it differs from its synonym *back*: either *ten years back* or *ten miles back* is normal, but not **ten miles ago* (unless we intend to metaphorize the noun as a noun of time: *a grief ago*). Third, it is unstressed: *It happened a year báck*, but *It happened a yéar ago*. Fourth, it is context-restricted. With definite temporal quantities, there is almost complete freedom, but the indefinites carry some peculiar restrictions. *Long ago* is fine, but not **briefly ago*,

[1] The asterisk means "unacceptable;" a question mark preceding a cited form means "doubtful."

though either *a long time ago* or a *brief time ago* is normal. *A short while ago* is heard, but less likely **a short spell ago*. As for *much* and *little*, they are barred completely, even with the word *time* added: **much ago*, **much time ago*, **little time ago*, and so on. The restriction is not because *much* and *little* refer to quantity whereas *long* refers to extent, because *long* cannot be replaced by other adjectives of extent either: *a long time ago* but not **an extended time ago*. With indefinite plurals, *centuries ago* is all right but ?*decades ago* and ?*eons ago* are doubtful and **millennia ago* I think is impossible. So *ago* behaves in most important respects like an unusually productive affix. The only real difference I can detect is the word's high degree of semantic integrity. *Ago* stays the same in meaning. Few if any lexical suffixes in English do that (though there may be some prefixes that do).

Another example will dispel the idea that *ago* is unique. Take the word *else*. In modern English it is always suffixed. It is also context-restricted — we say *somebody else* but not **some person else* — the restriction is to indefinite pronouns. And within that restriction there is a further one excluding or at least limiting indefinites referring to time: *something else*, *someone else*, *someplace else*, *who else*, *where else*, *how else*, *what else*, *nobody else*, etc., but not **sometime else* and for at least some speakers not **when else*. (*None else* is literary, and the borderline status of **one else* and ?*which else* suggests a gradient in definiteness — they seem not to be indefinite enough.) The only mutant in the species is *or else*, as in *You'll do that or else*, but even there *else* is still suffixed. It has only one nonsuffixal trait, which is its stress.

These two examples show that if criteria were set up to determine what elements belong on which side of the line between syntax and morphology, a wide middle ground would probably manifest itself. At least this appears to be true looking down from syntax toward morphology to see what penetration there may be in that direction. What of the worm's-eye view, looking up, to see how far morphologicity or idiomaticity extends into syntax?

The examples *ago* and *else* can be used again for a preliminary look. The question is, why do we not generate **an extended time ago* if we generate *a lifetime ago*, and why do we not generate **sometime else* if we generate *somewhere else*? It is not because the generative mechanism is lacking. I suggest that at least in part we do not do it because we have not heard it done. We have no memory of it. The *else* combinations are learned as combinations. Even a rule that specifies "temporal" as well as "indefinite" will not suffice to generate correctly, because "manner" is also partially restricted: *We'll do it somewhere* and *We'll do it somewhere else* are both normal, and so are *where else* and *how else*, but whereas *We'll do it somehow* is normal, **We'll do it somehow else* is not — it is necessary to paraphrase the latter as *We'll do it in some other way*.

Looking now for more substantial examples, we come first to idioms, which are well known for differing among themselves in what and how many syntactic processes they allow to go on inside them. Consider two idioms with the word *easy*, *go easy*, and *take it easy*. Neither one allows *easy* to be replaced by *easily*, and this is idiomatic by comparison with *to go softly* or *to take one's responsibilities lightly*. But *take it easy* is open to variations on the verb, whereas *go easy* is not: *I take it easy*, *He takes it easy, They took it easy, We're taking it easy*; but *If they go easy*, *everything will be OK*; *If you go easy*, *everything will be OK*; *??If he goes easy*, *everything will be OK*; **If you just went easy*, *everything would be OK*; **I'm going easy from here on*. The bare form *go* gets by regardless of its subject, but the more it is inflected the worse it sounds. There are idioms even more iron-clad than this. The expression *Take that*, on striking a blow, I think would never appear in any other form, even with the subject: **You take that*. It has to be laconic, like its companion idioms *So there* and *That for you*!

Idioms as tight as these have a lexical status close to that of individual words. As they loosen up, they gradually fade into the background of phrases that can be generated by rule. So the question arises whether even those expressions that we take to be freely generable may be infected with the idiomatic virus. How free are they, anyway?

This is a question that British linguists have been asking themselves for a long time. They recognize another level above that of idioms, which they call *collocations*. Mitchell (1971) defines a collocation as "an abstract composite element . . . which can exhibit its own distribution qua compositum." Idioms are different in that they have meanings that cannot be predicted from the meanings of the parts. The meaning of a collocation *can* be predicted, but is nevertheless particularized. Mitchell gives this example: "Men — specifically cement workers — work *in* cement works; others of different occupation work *on* works of art; others again, or both, *perform* good works. Not only are good works *performed* but cement works are *built* and works of art *produced*" (1971: 50). It is fair to ask why builders do not *produce* a building or authors *invent* a novel, since they do produce blueprints and invent stories and plots. It is, of course, a matter of terminology whether *collocations* should be classed separately from idioms or as a major subclass. Makkai (1972) calls them "idioms of encoding."

Mitchell's examples remind us of other more striking ones, the prefabrications that are generally called clichés. Some are so thoroughly amalgamated that when we hear them we no longer suspect that the speaker or writer is striving for effect, and so we do not object to them. They are not examples of fads but of folkways. One instance is the highly specific intensifiers that the common adjectives in English have acquired: *hot as hell*; *sharp as a razor*; *thin as a rail*; *hard as nails*; neat as a pin; *crafty as a fox*; *wide as a door*; *white as a sheet*. Verbs have them too: *beaten to a pulp*; *drenched to the skin*; *dressed fit to kill*; *armed to the teeth*; *spoiled rotten*; *blown sky-high*. Where many other languages have numeral classifiers based on size and shape, English has minimal quantifiers like *flurry of snow, sprinkle of rain, smattering of knowledge, glimmer of hope, bite of food*, *dash of salt*. Peter Maher has a phenomenon that he calls "salient feature copying," where adjective and noun are matched according to a feature that the noun typically has: *high mountain*, *sharp knife*, *heavy load*, *strong ox*. With other clichés we may be conscious of the striving for effect and then we put the expression down as trite: *inclement weather*; *signal honor*; *unconscionable liar*; *patently absurd*; *to harbor a grudge*; *to cherish a hope*; *a modicum of good sense*; *steeped in tradition*; *We have been treated to the spectacle of . . .*; *It smacks of . . .*; *I discovered to my dismay*.

These examples are so close to idioms that the two categories merge imperceptibly. We can deduce the meaning of *blown sky-high* from the meanings of the parts. The similar expression *knocked galley-west* does not yield so easily. We need some examples at this point that are less stereotyped, to get farther above the level of idioms. I offer an indecorous one. Here are two sentences that I find normal: *I wet my shoes so badly cycling in the rain that I had to dry them overnight*; *I got my pants so wet cycling in the rain that I had to dry them overnight*. Now try switching the verbs. It is obvious that there are syntagmatic associations for the verb *to wet* that force us to steer a pretty narrow course when we try to use it.

The test for a collocation, like the test for an idiom, is that given the meanings of the elements and of the rules, there will be something highly unpredictable about the derivational output. With an idiom there will be something quite unpredictable. With a collocation it may cover any degree of improbability, but we are always to some extent wise after the fact. For example, given the means to generate the sentence *That story is likely*, we ought to be able to generate *That's a likely story* from the same source, but we cannot unless we mark *likely* as [+Ironic]. But since *That's a likely story* is probably the only place where a [+Ironic] feature on *likely* is obligatory, the procedure is ad hoc, and would not in any case guarantee that we would actually get *That's a likely story* because *That story is likely* could be ironic too. It is not only such extraneous things as irony that disrupt a normal derivation. Some collocations are semantically incoherent. Take the expression *I have no choice except to* . . . . This refers to a situation where there is only one course of action — that is its collocational meaning. But if there is only one course, there is no choice, and consequently the word *except* is nonsense. We do not put an interpretation on the parts of the expression but on the whole of it, perhaps by way of a blend between *I have no alternative to* and *I can't do anything except*.

A collocation may involve normal senses of all the words in the string, but without the easy possibility of substituting some other word with the same meaning. Most of the time we can substitute a synonym — that is what synonyms are for. But here it is relatively difficult. Take the words that mean sleeping and waking. Nearly all of them can be used to refer to attention or inattention. *Wake up* is commonplace to tell someone to pay attention. The verbs *to sleep*, *to doze*, *to nod*, *to nap*, *to snooze*, *to drowse* can all be used to mean "inattentive." In spite of this, if we were to say that someone missed the point because he was *nodding as usual*, we would find it probably a little less appropriate to say that he was *napping as usual* — this tends to suggest that he was really asleep. But in the expression *it sort of caught you napping*, *didn't it*?, nodding is not very appropriate. Much of the information on usage that we get from the examples cited by dictionaries is actually collocational material. In this particular instance the Merriam-Webster *Third new international dictionary* remarks that *nap* is often used with the verb *to catch*, and cites *He was caught napping*. Adjectives stereotyped with nouns are a common occurrence in this type of collocation. Thus *mutual* and *common* are synonyms, and you and I may be *mutual enemies* or *mutual friends*; we may also be *common enemies*, but we are not apt to be **common friends*, although we may have friends *in common*.

Of course there are collocations that involve more than just one element of stereotyping. Here is a snatch from an anecdote in a popular magazine: "I asked him where he had been. He sort of smiled. '*Just meditatin' with the Lord*, *son* — *just meditatin' with the Lord*.' " This formula has four set features: (1) the word *just*; (2) repetition of the whole phrase, a sort of extra long reduplication; (3) insertion of a vocative, usually a patronizing one (*son*, *dear*, *old boy*, *friend*); (4) inconclusive terminal intonation.

There are collocations that confer not a restriction but some last vestige of freedom on some element within them. We can sometimes look back in history and see how this has come about, just as we so often can with the more fossilized morphemes that we call affixes. For example the noun *drink* is no longer freely used in a sentence such as **They brought him drink*. But we are just as free to say *They brought him food and drink* as to say *They brought him food*. The collocation *food and drink* shelters the noun *drink* just as the word *cobweb* shelters the now unused noun *cop* for "spider." The only difference is that nobody would misunderstand a person who said, *They brought us drink*, but it might affect our views somewhat to hear that Miss Muffet was frightened by a big cop that came and sat down beside her.

There are collocations that do not merely exist independently but are found in clusters. Among the illocutionary expressions that are getting attention these days — those formulas that put the hearer on notice as to what stance he is to adopt toward what he is about to hear — there is a group involving the word *what*, all equally stereotyped but related in meaning. The following list may or may not be complete:

Know what?
Guess what!
Tell you what:
Tell you what you do:
I'll tell you what I'll do:

The last is familiar as a favorite of the carnival confidence man. What it says is, "OK, let's make a fresh start, I'll meet you more than half-way," etc. All five expressions have the meaning "something new coming." In addition the last two have a suggestion of "disarmingness" about them.

Finally there are collocations that are all but indistinguishable from freely generated phrases, and here of course is where we must ask, if such things can be, may there not be some degree of unfreedom in every syntactic combination that is not random? A useful example is the verb *to hurt*. It shows the way in which specialized meanings come attached to perfectly ordinary combinations that look as if we ought to be able to shake them loose and reassort them, but we cannot. Take the sentence *They've hurt her*. It is a reasonable guess that the hearer will infer that she has been hurt physically, or in her career, or in some other material way. The same is true with *They've hurt her badly*. But with a stronger intensifier than *badly*, say *cruelly* or *terribly*, as in *They've hurt her terribly*, one is more apt to infer that they hurt her feelings. A touch of exaggeration pushes *hurt* out of things and into sentiments. The same is not necessarily true of other verbs — *They've wounded her terribly* could be either physical or moral. However we may have learned the expression, *to hurt terribly* is in the language, stored as a unit. It might even be that we regenerate a phrase like this every time we use it, but its having been used before is a spur to its regeneration, from some trace in our minds.[2] There are many possibilities of degree of interplay between remembering and remaking; but memory is not to be denied its effect.

The selectional restrictions of generative grammar are one kind of collocational restriction, a special kind that can be described in relation to a category of some type, which Mitchell and others call a *colligation*. For example the verbs *to look at* and *to regard* are synonyms, and we can say either *He looked at me strangely* or *He regarded me strangely*. But whereas we can say simply *He looked at me* and let it go at that, or say *He looked at me for five minutes*, we cannot say either **He regarded me* or

[2] Compare the use of the term "trace" as used by Lamb, Reich, Lockwood, and others (Makkai and Lockwood 1973).

**He regarded me for five minutes*. The verb *to regard* in this sense has to collocate — or colligate — with a manner adverb. We find the opposite too — cases where in spite of its meaning a word is unable to form a colligation. The verb *to publish* is defined as a synonym of *declare*, *disclose*, and *proclaim*, any one of which can be used with a *that* clause — but *publish* cannot: *He proclaimed that he had been involved*, **He published that he had been involved*. One can even say *He made public that*, but the nearest to this with *publish* is *He published the fact that*. *Publish* is in the grip of constructions like *publish the news*, *publish a paper*, and syntactic lag keeps it from moving in the direction of *reveal*, *announce*, *assert*, *confirm*, etc.

The difficulty with trying to identify all colligations with some sort of selectional restriction is both the potentially vast number of categories that would have to be recognized and the vagueness of their borders. How for instance would one characterize the adverb *nothing*, which can occur in the archaic *nothing loth* and the still used but highly literary *nothing daunted*, *nothing perplexed*, *nothing perturbed*, *nothing dismayed*, but not in **nothing worried*, *concerned*, or *puzzled*? The participles that can be used with *nothing* do have something in common, but it reflects both meaning and register.[3]

There is no guarantee that the reader will have the same intuitions as mine regarding the acceptability judgments I have been using, but a reasonably good match is enough to make the point. If the examples I have cited were really generated from scratch, it is hard to see how they would run into those peculiar restrictions. But then how do we account for the fact that we can manipulate them to a certain degree? If they were learned as indivisible chunks they should remain just that, like idioms, though as we have seen the same problem comes up even there with idioms that allow some syntactic operations but not others.

I suggest that the reason for it is that learning goes on constantly — but especially with young children — in segments of collocation size as much as it does in segments of word size, and that much, if not most, of our later manipulative grasp of words is by way of analysis of collocations. This does no more than put words on the same footing as morphemes. It does not seem odd to us to suppose that our understanding of the prefix *un-* comes by way of some analytical processing of the words that we learn in which *un-* occurs. Since it never occurs alone, there is no other way. And it would be absurd to imagine that once we have learned the prefix *un-* we will proceed to forget all the words from which we got it, and then coin them anew whenever we need them. Clearly we possess *un-* as a unit and

[3] The survival of *nothing* also probably reflects field relationships with synonyms and antonyms. The adverb *something* is no longer used, regardless of register. **Something loth* is obsolete, being replaced by *somewhat loth*. But the negative term *nowhat* never really competed with *nothing* and no longer survives even in set phrases.

we also possess *unwise*, *ungracious*, *undo*, and *unwind*. By the same token we must retain the collocations even after the individual words have become entities in their own right, at least a large number of them; it would be interesting to know what the psychological concomitants of *forgetting* are. This explains the failure of words to combine freely: the analytical separation of many words from their contexts has simply not gone all the way. A theory that endows all human beings with limited memory but unlimited generative power cannot account for this. The raw material is there, but much of it remains in the rough and is not turned into any smaller finished products.

The best example that I can think of to illustrate a struggle toward freedom that never quite makes it is the verb *to bear*. It is a veritable jungle of idioms and collocational restrictions but with no semantic center, like an animal with no heart to pump blood through it but with peripheral organs that manage to keep functioning anyway. Idioms include *bear up*, *bear down*, *bear the brunt*, *bear right*, *bear left*. Collocations include *bear a resemblance*, *bear a grudge*, *bear a child*, *bear a burden* — we are less likely to *carry a burden* and quite unlikely to *bear a load*. But the puzzling part is the syntactic possibilities of transforming which are simply not carried out. For example, it is normal to have *bear* in a relative clause in an expression like *the love that I bear them* (notice that this is affirmative). And a declarative or interrogative is all right with a negation or implied negation: *I bear them no love*. But a simple straightforward affirmative declaration is out of the question: **I bear them love*. How do we generate *The love that I bear them* if we cannot say *I bear them love*? The same goes for infinitives and gerunds: we can say *She bore his neglect bravely* and either *She can't bear to be neglected* or *She can't bear being neglected*, but we do not say **She bore to be neglected* nor **She bore being neglected*, though *She endured being neglected* is normal enough. The only easy way to explain these phenomena is that a child first learns the verb *to carry* and later learns *to bear*, but failing actually to *hear* the verb *bear* in contexts where *carry* is normal and from which metaphorical extensions can be made if one wants to, he never fully dissociates *bear* from its collocations, even though all the transformational possibilities are there for him to do it. This argues for the kind of memory storage and retrieval I have postulated. The fact that it has to be invoked for *bear* suggests that it is potentially present throughout the system.

When conditions *are* right and the child does accomplish the analysis, I believe that stored units are to be found in three overlapping levels, corresponding to morphemes, words, and collocations. For any given collocation I would maintain that all three may coexist. The phrase *indelible ink* makes a good illustration. It is not an idiom by the classical test — the meaning is clearly deducible from the separate meanings of *indelible* and *ink* — and yet an association test quickly shows how tight the internal connection is. I gave a group of twenty-seven adults the word *indelible* and asked them to write the first word that came to mind; eleven responded *pencil*, seven, *ink*, eight, *unerasable* or a synonym, one, *tracing*, and one — the usual lunatic fringe — wrote *individual*. If one tries to use *indelible* in other combinations it is possible to go only so far: *indelible impression* is figuratively all right and can perhaps be stretched to *indelible memory*, *indelible recollection*. But perfectly valid uses in the literal sense, for example *An embosser leaves an indelible mark*, I think would momentarily puzzle a hearer or reader. The primary connections are with *pencil* and *ink*, and they have led us to expect a writing *material* of some kind for this word.

Figure 1 is intended to show the coexistence of the three levels. The vertical broken line signifies that morphemes as morphemes may or may not be stored; a lot depends on the perceptions of individuals and of course on how productive the morphemes are. Not many speakers in a whole lifetime will ever make some new combination using the prefix *in*-, say **inaustere* or **inobdurate*, and some will not even give it passive recognition; hence, the description "low yield" in putting morphemes together to make words. In learning our language we read the diagram down; it has the advantage of showing the degree of penetration into ever tighter structure that different individuals are able to manage. Linguists tend to read the diagram up, and there I think is the basis of misconcep-

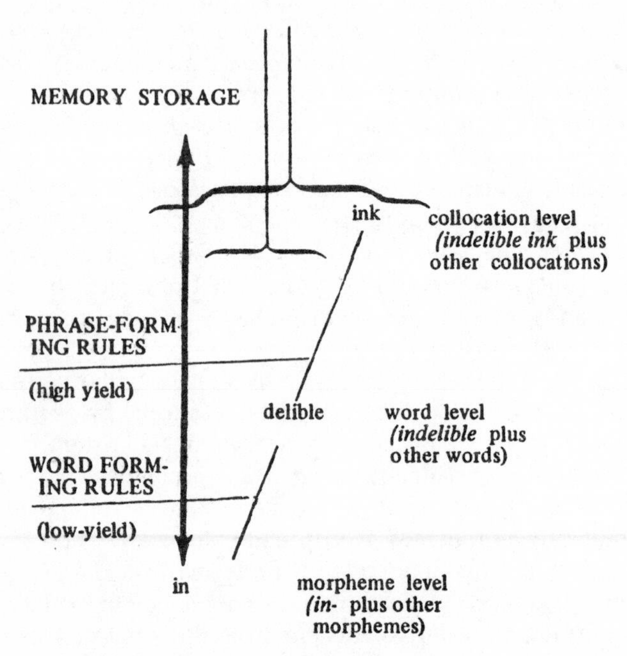

Figure 1. The three levels in which stored units are found

tions in trying to fit linguistic descriptions to psychological reality. There is a sort of rough parallel to certain old-fashioned theories of the origin of language. Instead of seeing the psychological capacity to use language as growing and developing along with and out of the evolution of language, it was thought that human beings came into existence with all those capacities and at some point passed a law that there was going to be language. Similarly, linguists start with the idea that all those nice morphemes were there waiting for you. You didn't have to prick your fingers tearing them out of the brambles; all you have to do is put them side by side and they turn into words.

Of course the big question is still how the extrication from the brambles is done. All I can do is speculate, but I suspect that psychologists have not advanced beyond the speculative stage either. Taking another metaphor, one might compare it to cloud-seeding — the early provision of a few word units around which others are precipitated. I believe that Charles Osgood (1971:18) is right when he says that the child begins to acquire words and meanings "from actual behavior toward things signified." This includes early situations in which individual words are highlighted in concrete situations, which in turn sets the stage for analyzing collocates as soon as they begin to come along. And it is significant that the collocate is what the young child produces if you ask him for a definition. *A hole* is defined as *a hole in the ground*, to use an example from Cazden (1972: 129). The verb *throw* does not elicit a synonym such as *toss* but a collocated noun such as *ball*. But if a child knows *ball* and also knows *sand*, and furthermore knows *throw-ball* and *throw-sand* and has behavioral associations into the bargain, he can arrive at *throw* by the ancient device of proportional analogy.

The picture that I have proposed I think fits better than a highly structured reductionist model with the kinds of indeterminacy and heterogeneity that are found everywhere in language. I will not say that it accounts for them, because more has to be filled in, only that it accords better with them.

Take indeterminacy. When meanings are built up from below with determinate features, there is no way to get the elasticity that one always finds with meaning. William Dwight Whitney, whose credentials are said to include the fact that he read all 21,000 columns of proof for the *Century dictionary*, declared that "Hardly a term that we employ is not partially ambiguous, covering, not a point, but a somewhat extended and irregular territory of significance" (Silverstein 1971:43). This is to be expected if individual words and meanings are arrived at by abstraction from collocations. If a child defines a hole as a hole in the ground, he is giving an example of the only kind of concrete existence a word has, which is in its remembered associations. The anthropologist who first started linguists playing with kinship terminology as a model for semantics should have remained in his wickiup. Nothing could have been less typical. For an example of the real problems in abstracting a meaning, take the word *fruit*. The child learns it in collocation with such words as *eat*, *taste*, *dinner*, *table*, given of course in contexts of situation that then become the appropriate settings for those collocations. (Unlike such "seed" words as *apple*, *pear*, *banana*, it is not apt to be heard in direct association with the object: *Do you want this apple*? but not *Do you want this fruit*?) Out of them he abstracts a meaning that includes such foregrounded components as "sweet" and "edible" along with some secondary ones such as "round" and "seed-bearing" — qualities that we notice about something when we see it or handle it or, in the case of a fruit, eat it. Later the child encounters some rather disturbing new collocations that force him to overlay the homelier features with other ones, which eventually can be summed up as "ripened ovary of a seed-bearing plant." The category of "fruit" covers a range of gradations, with apples, pears, and oranges in the center, melons a little farther out, tomatoes and avocados still farther, and gourds and hedge apples on the fringes. But there is no such indeterminacy about the collocations. *Fruit on the table* is edible fruit. It is not that *on the table* "disambiguates" *fruit*; there was no ambiguity to begin with. The reductionist theories of language would have us believe that ambiguity is present in practically everything we say and disambiguation is a continuous process. Instead, most ambiguity is as artificial and contrived as a psychologist's reversible figure and ground. That sort of phenomenon is aptly called an optical illusion, and I suggest that ambiguities be regarded as semantic illusions. In both cases the effect is achieved by depriving the subject of certain locational or collocational cues.

The indeterminacy of such words as *fruit* when they are encountered separated from their collocations is the subject of a series of ingenious experiments by a group of psychologists at Stanford University (Smith, Rips, and Shoben 1974). Their prime example is the word *bird*, which they find that subjects are willing to use as a defining category only under certain conditions. A subject given two sentences, say *A robin is a bird* and *A chicken is a bird*, and told to judge the statement true or false, will take more time to react to the *chicken* sentence. The experimenters conclude that there are degrees of birdiness, depending on what they call defining features and characteristic features. My interpretation of their results is that the word or category *bird* represents a coordinate through all the collocations that we have stored up and from which bird was abstracted in the first place, with the help of an obliging parent pointing at an object on a telephone wire (but not at one in a hencoop) and saying *Look at the birdie*.

I mentioned heterogeneity as another trait language that a rich memory capacity makes possible. Heterogeneity is of course heterodoxy, as far as linguistics is concerned. Formal deductive grammar since its beginnings has striven to eliminate variety, not only in the idealization of the system away from the realities of "performance" but in the inner structure as well. This of course is the basis for reductionism. If there is the barest possibility of accounting for something in terms of elements at a lower level, that possibility is adopted, often in the teeth of two kinds of negative evidence: first, that the procedure may give inaccurate results; second, that it may be psychologically unnecessary. All the debate up to the present has been on the first point — that of accuracy. But now it is time to take account of the psychological side and give the greater weight to Anttila's remark that "even the most 'obvious' cases can be stored separately," which is to say that even when a linguist *can* analyze, it does not necessarily follow that he ought to. If grammars merely aimed at descriptive adequacy this argument could not be advanced; but when grammarians begin to claim psychological reality for their models, they are presuming on explanatory adequacy of a deeper sort than they had bargained for.

If the psychological debate has hardly begun, the one on accuracy is still unsettled. A key motive for reductionism is not having to say something more than once. This leads to assuming the underlying identity of different structures. But more and more of the paraphrases that were assumed turn out to be false equations. The "transformationally introduced particles" are not as empty as they were supposed to be. Active-passive, particle insertion, particle movement, gerund vs. infinitive complementation, *that* complementation, indirect objects — all of them involve differences in meaning. What we find is not identities but similarities and differences, both of which have to be remembered. This adds up to heterogeneity and a much heavier demand on memory.

Accepting a principle of heterogeneity means a different attitude, I think, toward the object of our study. Prime attention to the overarching system and more or less incidental attention to the local detail follows naturally from the assumption that grammar is homogeneous and self-confirming. In a formal deductive system everything is explained when everything fits. There is no need to worry about missing a detail or two because sooner or later any mistake or oversight will show up in a grinding of the gears. It even seems rather petty to insist on full and precise documentation in each separate province of the realm. The power of the system as a whole to reveal a flaw is so great that mastering the grand design outweighs any small help that can be picked up from one of its mere neighborhoods. But once homogeneity is denied, this security is gone. Separate — and, yes, ad hoc — explanations for each part become essential. It is no longer safe to assume that the system will correct itself. Unquestionably it will in many ways; certainly there are local regularities as well as universal ones; but the self-corrective remedies must be sought *after* one has canvassed as thoroughly as possible in each neighborhood, approaching its indwellers on their own terms.

Recent work in neurolinguistics points to heterogeneity not only as a possibility but as a reality. There is for example the research of Diana Van Lancker and Victoria Fromkin on dichotic listening. It appears that there are two kinds of language, automatic and propositional, related to the lateralization of functions in the cortex. The separation points to a side that files things and a side that puts them together — a scheme that could readily accommodate itself to the storing of vast quantities of remembered stuff. Collocations would be the automatic or semi-automatic syntagms that continue to be more or less automatic even when passed through the analytical sieve that separates them into their parts and makes propositional language and elaborated codes possible. Figure 2 shows Van Lancker's gradient of propositional-to-automatic language. It is significant that she titles her study *Heterogeneity in language and speech*.

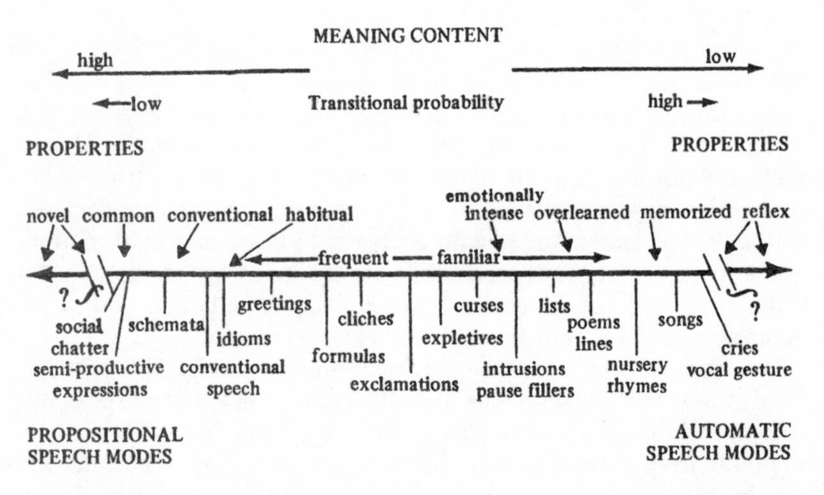

Figure 2. A hypothetical continuum of propositional and automatic speech modes and their properties. *Source*: Van Lancker (1975:173)

I conclude on a mathematical note. J. D. Ratcliff in an article in a popular magazine, which shall be nameless, since it is the *Reader's Digest*, points out that Joe's brain contains 30 billion nerve cells each of which connects with others, some as many as 60,000 times. The response of the reductio-linguist to this could only be "Who needs 'em?" My answer would be, if we didn't need 'em, they wouldn't be there. The human brain is not a vestigial organ.

REFERENCES

ANTTILA, RAIMO

1972 *An introduction to historical and comparative linguistics*. New York: Macmillan.

BUGARSKI, RANKO

1968 On the interrelatedness of grammar and lexis in the structure of English. *Lingua* 19:233–263.

CAZDEN, COURTNEY

1972 *Child language and education*. New York: Holt, Rinehart and Winston.

CHAFE, WALLACE

1968 Idiomaticity as an anomaly in the Chomskyan paradigm. *Foundations of Language* 4:109–125.

LADEFOGED, PETER

1972 "Phonetic prerequisites for a distinctive feature theory," in *Papers in linguistics and phonetics in memory of Pierre Delattre*. Edited by Albert Valdman. The Hague: Mouton.

MAKKAI, ADAM

1972 *Idiom structure in English*. The Hague: Mouton.

MAKKAI, ADAM, DAVID G. LOCKWOOD

1973 *Readings in stratificational linguistics*. University, Ala.: University of Alabama Press.

MITCHELL, T. F.

1971 Linguistic "goings on"; collocations and other lexical matters arising on the syntactic record. *Archivum Linguisticum* n.s. 2:35–69.

OSGOOD, CHARLES

1971 Explorations in semantic space: a personal diary. *Journal of Social Issues* 27: 5–64.

SILVERSTEIN, MICHAEL, *editor*

1971 *Whitney on language*. Cambridge, Mass.: MIT Press.

SMITH, EDWARD E., LANCE J. RIPS, EDWARD J. SHOBEN

1974 "Semantic memory and psychological semantics," in *The psychology of learning and motivation*, volume eight. Edited by G. H. Bower. New York: Academic Press.

TWADDELL, W. F.

1972 Syntax: past, present, and future. Paper presented at the Linguistics Meeting of the English Language Institute, University of Michigan, Ann Arbor, Michigan.

VAN LANCKER, DIANA

1975 Heterogeneity in language and speech: neurolinguistic studies. Working Papers in Linguistics 29. Los Angeles: University of California.

Comparative Characteristics of Experience Forms

GEORGE G. HAYDU

Experience is not a continuous stream. It is a sequence of events, and each event is an experience entity. I have described its structure previously (Haydu 1972). Briefly summarized, the experience entity was seen as an integration of many components, chiefly an aroused need that seeks and finds its instrumentalities, concepts, images, and action. Action is then fed back to indicate results and consequences. This multicomponential integration, the experience entity or psycheme, is the smallest unity that is still a full-fledged psychological happening. It has its particular shape or form quality. Experience entities cluster and combine mostly according to their most significant features, mainly around significant persons or powerful events. Such structures are the experience forms. They can be expressed and they can be communicated. They are basic to interpersonal relations and depend on and can be encoded in the semiotic ecology of the culture of which the person is a member. Relations, objects, and intentions are viewed and weighed through the outcome of these events: the outcome-oriented aspect of experience forms constitute axiological value (Haydu 1977). Values can be divided in two rather artificial categories according to outcome: drive object outcome and drive integration outcome.

To speak of experience forms may bring to mind Plato's or Whitehead's forms — configurations from another world — but these are not implied here. When components are thoroughly integrated in a functional connectedness, then form arises. It is the expression of something tightly, economically, even elegantly structured. When amino acids are lined up in protein production and the various forces of the constituent chain become engaged, then they assume a condition where the least amount of energy is spent for maintenance: the string of amino acids folds into a stable conformation. Anything that is so structured and aligned is a form (Christian 1973). It is determined by the component forces. To discover that similar forms have similar characteristics and similar functions is a great help and insight. But to abstract the forms and infer that these grant the whole structure its qualities is a prideful imputation. The generalizeable effects and functions of forms merely assert that similar relationships have kindred characteristics. Life and its functional entities and their interactions are reliable, not capricious.

Here I will attempt to show how experience forms — their characteristics as to further development and transformations — could be compared, what principles such comparison involves, what standards might be applied to judge experience forms, good or evil, high or low. We pass such judgment every day, every moment of our existence; and a certain degree of clarity, if possible, may not be superfluous at this cultural juncture of our time.

PATTERN DISCOVERY

To see experience form in behavior, or rather, to see experience form as a particular instance of behavior — the objects expressing it — to give resonance to it, to judge its appropriateness or excellence, brings this activity into the realm of ethics and aesthetics. In both fields we give resonance to patterns of experience forms. Only in moral judgment we consider these patterns as to their consequences to others, or as to consequences to the progression of the life of our culture. In aesthetic participation we let the pattern resonance stir or confirm or even develop our own personal system. Basically these activities — in ethics and aesthetics — are intimately related. Pattern discovery and pattern resonance are basic to both what we find good and what we find beautiful. To appreciate or even to recognize an aesthetic or ethical existing structure we must have the skill of form resonance. All of us have this to a considerable degree. But some are keener at it and can see and appreciate a configuration better than others. Parenthetically, we must not make the mistake of thinking of the realm of desirable activities as something settled by a code or a prescriptive standard. You can measure the height of a child by a measuring rod, but his height is not achieved by it. Good and evil activity originate from experience forms that constitute the person who performs them. These structures do include rational views, concepts of origins, and consequences that are implicated by the particular behavior judged. But the recognition and appreciation are, first and foremost, a function of resonance we can give to the experience forms.

Value, as the outcome-oriented aspect of experience forms, is in a sense subjective and culture-bound. But that does not make it arbitrary or chimerical. It has its own dynamics of genesis, consistency, even constancy. It also has its own dynamics of transformability and competence. When people say that all this is a matter of "taste," it is often implied that this ability is a superficial acquisition. The word taste is used in too many ways. At times it suggests something casual, a finicky sophistication easily satirized. Such people of "taste" may have little palate for substantial things (and usually, in addition, have a bad nose). At other times, taste suggests great discrimination, a capacity to be moved in one's fundamental and substantial being, a capacity to feel and see qualities in human behavior, figurations in interpersonal relations. This latter requires cultivation, wide experience, in giving resonance to a great many experience forms as met in human communion, in suffering, in great art, in the basic poetry of the self. By dint of such cultivation, one may become rich and free-resonating, full yet receptive.

In a similar way, the word aesthetic is used in a very confusing way. It may refer to the appreciation of a Brahms symphony, at other times to pretty food coloring. To pretty up something may be said to give it the aesthetic touch. To express a total creative experiential state is aesthetic activity, too. Yet, the difference is enormous.

Also, pointing out a basic feature of a semiotic creation is very far from describing its fundamental quality. We can talk about the lilt and sprightliness of the waltz. But just think of the um-pa-pa of the village waltz and then think of other varieties: the waltzes of Chopin, Glazunov, Lehár, the mariacchies of Mexico, the waltzes of North Carolina, the waltzes of Brahms, those of the Rosenkavalier. To judge even a waltz, where it fits, and what it expresses, requires a great deal of experience.

Anthropological fieldwork taught us this old lesson with new vigor. There was a time when understanding a faraway people or a society known to us from artifacts required only that we see the activities and the employment of skills and instrumentalities in real life. Fieldworkers discovered that right in the middle of a strange culture you are almost as incomprehending as when you read old documents. In both instances the giving of proper resonance and complemental analysis are essentially the same hermeneutic activities that are needed (Gadamer 1965).

Giving proper resonance requires a very substantial activity. The discovery, for example, of something bad, something unjust, or something obscene must reach those structures of one's being where it matters, where the irking, gnawing, painful responses dwell, where one's desires, as acquired in living, reside. And so the other way, if something does not exhilarate you, if it does not send you, then there is no discovery of it as a beautiful thing. For the creator of beauty made it for his own whole-making, and that always goes along with intense feelings of longing and delight. Discovering certain features of a work is not the same as giving resonance to its entirety with our substantial self.

One can communicate discovery in two basic ways. Expressive activity is not descriptive activity. One can communicate the savoring of a delightful soup by just saying "Ah." It is expressive. Or "Ah — great!" This, in addition, uses a concept of excellence. The Mock Turtle who said: "Soup of the evening, beautiful Soup," had made the bittersweet realization that his fellow creatures were part of this repast. The communication is still very expressive. But he could say: "And that evening the soup was truly magnificent." A lot of expressivity is still with us. One can further push the statement to more descriptive modes: "Everyone commented that the soup was very tasty," or "This soup, which was generally liked, was made of vegetables and fresh turtles." From full expressivity to scientific description there are innumerable gradations.

To have good judgment, a man of wisdom must have the feel of the dynamics of the human world. It begins close to home. He must learn first what hurts others and what pleases them; what is momentary, what is more lasting; what helps and what injures others and one's self. He must learn this through living and through having given resonance to others, to other people's experience forms. In our present world, in order to become properly perceptive and wise, one must give resonance also to experience forms of faraway places and of many other historic times in order to acquire the capacity to understand and feel not what man is, but the many diverse beings that he had been and the great variety of beings he still could be.

MULTICOMPONENTIAL INTEGRATION

Experience forms arise in living. Their sequence constitutes the trajectory of individual lives. They also can be encoded, through interpersonal and other semiotic interchange, into the semiotic ecology that is one's culture. In their final tastes and terms, experience forms are incommensurable. They have a shape that is savored or felt as a simple quality and cannot be compared on that basis with another. Yet the components which an experience form unites in its configuration can be studied, can be listed, can be used for comparison with another. These two ways — the configurational and the analytic — must be carefully distinguished when the comparative characteristics of experience forms are considered.

Recent neurobiologic discoveries demonstrate to us that many different and disparate elements and components can be kept together in one harmonious pattern (Katchalsky, Rowland, and Blumenthal 1974). That is a very general aspect of life. The most individual and crystalline configurations can be viewed (actually seen by now) that arise out of a great many diverse forces. So we do not have to despair when contrary appearing and contrary working forces are at play. There can be a resolution, a new pattern of integration, a new accommodating structural idiom.

Experience forms are conative structures. Through them an aroused need or desire finds its scope of action. As such, a transactive experience form has consequences. There are many components to every experience form and a great many consequent effects. To discover a few of them may be helpful but not sufficient for a satisfactory understanding. We must try to sense and find what else, beyond the few factors that interest us, is contained in a particular experience form. In comparative studies we must even sense and find what it does not have. As we shall see, an experience form can be often bad or evil, not so much by the components it contains, but by those it lacks.

It is important to state that experience forms originated in a world where people lived a nomadic or agricultural life. The forms undoubtedly exhibit characteristics that reflect such important components. But to use a few such standards in comparative studies will give us only typology: types of cultures, types of experience forms. Typology is helpful but completely insufficient to understand life forms, their comprehensive functionings, their individuality, and inability to switch from one kind to another in a carefree way (we all find out sooner or later that this is not a pick-and-choose universe). The change from agricultural existence to industrial ways was accomplished many times, but the transformations are not understandable on the basis of typological examples. Transformations involve all other components as well. And because comparative studies of characteristics must see what consequences experience forms have in the continuous transformations of cultures, the components which allow or inhibit such transformations must be carefully considered.

The science of the modern age (1600–1900) has been too static. The post-modern view begins with Nietzsche's anthropology and linguistics: the gay science. We are now steeped in this. We see growth and transformation everywhere, but not by enlargement or thickening or substitutions or by any unnatural miracle. We see everything as developmental, as structure goes through sequential changes and enjoys integrative transformations. As far as we have knowledge, this organic world is one by virtue of its commmon origins and by the necessity of mutual accommodations and not by repetition of dead forms. We see in it necessity; we see in it response; and we see in it delight. We recognize how much the new owes to its past, how much of it is potent past; yet the new is totally so in its individual quality, in its final tastes and terms. This view may be lyrical but it is not romantic: The enthusiasm of transfiguration is tempered by the sense of indebtedness.

In the study of comparative characteristics of experience forms we see many components. When a component is invalidated by a new discovery, the experience form is obliged to undergo transformation. This is the locus of a great many difficulties. For invalidation of a component is often construed as invalidation of many others which are integral constituents of that particular form. For example an experience form may have as its integral component the conviction (a certainty, a piece of knowledge) that one must care for one's fellow human being. It may also contain the component that the world is supervised by a benign king who resides in the sun. It may come to pass that it can be demonstrated that the sun is just one star among many and is unconcerned with human affairs. New experience forms will arise and compete with the old one claiming that there is no benign king in the sun and, therefore, there is no need to care for one's fellows. We may find the new knowledge that abolishes the notion of the sun king quite convincing. Yet to judge the superiority or inferiority of the experience forms cannot be done on that aspect alone. The truth of a piece of knowledge, as a component, is important to the degree it tells of what is effective and operative in the human world. By and large, whether the sun is just another star or whether the earth moves or not is external to the main function of experience forms. An experience form may contain superior knowledge of the planets, on the astral or atomic universe, but if its interpersonal action and consequences are evil or destructive, then it is inferior. The Confucian conviction that man and his interpersonal relations are the measure of things means just that: seeing how an instance of behavior affects the human world, recognizing the quality of give-and-take, the communion of people, and judging by such standards.

We discover experience forms from behavior. Instances of behavior originate in experience forms which tell the person how a desire fared in the past. They indicate to him how a particular constellation of needs or desire forms pursued their goal of completion and what consequences resulted. It indicates a kind of lesson that modifies other similarly toned experience forms. Since a person's experiential self is his person proper (his "identity"), when we judge a person's behavior, we judge his experiential self, this internal and integrated system of understandings that mark out his purposes. Many confuse this with a teleology that speaks of purposes which are placed into people. The purpose, the intention, seen here is an organic aspect of the event which resulted in a particular experience form. It is a variety of completion. Intention is an intrinsic purpose — or entelechy, if you please.

PSYCHOGENESIS AND ENCULTURATION

Psychogenesis and enculturation are coterminous. Experience forms are culture dependent. And, of course, their outcome-oriented aspects (axiological value) are also culture-bound, i.e., exhibit the forces and convictions of a particular culture at a particular juncture. If happiness (or even a modicum of contentment) consists of desire pursuing and finding fulfillment, then happiness and the good life is also thoroughly culture-bound. Any actual desire is some basic human need which finds its definition in a particular culture and in a particular individual development. Only through the study of evolutionary development of the human condition can we hope to find what kind of desire and happiness is fundamentally superior to another.

In any society, certain varieties of human beings are discoverable. There are some strong and full-blooded persons whose balance is not easily thwarted by some fussy injunctions, and there are sensitive people whose fragile system is very much affected by contradictory demands of a culture. There are always the hopeful young ones, the flirtatious wife, the weak and fearful pantaloons (even if they wear loin cloths), and so on. To describe some of these perennial types does not in the least describe the culture or even the individual. There is no doubt that the tough outgoing man in one culture will resemble the tough outgoing man of another. But the solidity and tenacity seen in these individuals are very much altered by the culture where they grew into full individuality. A tough outgoing man of the Sung Dynasty will be very different from the tough outgoing man of nineteenth century Montana. And a sensitive "poetic" man of the former country is very different from his counterpart in the Rockies. One culture accommodates certain types better, gives them different scope, while another may heap contempt on them.

The deflections and extensions of basic human needs used to be called the spirit of man. Spirit as the system of semiotically gained modifications can be contrasted to the more basic and general human needs. By now we have experimental evidence that any modification, and thus the realm of the spirit, is a qualifying metarchic operation (Haydu 1972). Any deflection or modification is structurally recorded. Spirit is corporeal. Yet there is some sense of seeing it as a somewhat separate dimension. The modifications can be inhibited, enhanced, and indeed, pharmacologically influenced differently than the simpler basic drives upon which they are grafted. In this sense, spirit is different. And many for such reason call it superior to the most basic drives. Yet a blanket judgment of this sort is not tenable. Just as the spirit can (if we should reify this dimension) make the simple needs richer, larger, more potent, more at home in the world at large, more ingenious, more compassionate, more autonomous — so it can make simple needs more devious, destructive, obfuscating, cruel, and disintegrating. Particular semiotic forms can be enabling or ennobling, others can be enfeebling or evil. But to have some principle of judgment of what is good and what is bad is still the essential requirement.

Each historic "time" is understandable only by its own activities, beliefs, economies, arrangements, and instrumentalities. Each is a transformation of a preceding one. History is a continuum, a sequence of self-contained integrations. As such, the basic values of a historic time are, if not absolute, then certainly normative. The values and norms of a particular historic time are enforced, both by law, and by the weight and power of interpersonal pressures. There is no escaping from that. But one thing we know for certain: Every integration is superseded by a new one. No historical time stays. The new is a transformation of the old and retains much of the constituting forces and components. But in its active functioning surfaces it has its own physiognomy and its own incomparable quality. Since values are the outcome-oriented aspects of experience forms, they cannot simply be adopted. They are a developmental result of experience. The old forms undergo transmutations when the old ways are obstructed, either by some external force, new opportunities, or by an increasing disharmony within. That is where the vital predicament lies. The potent past in a creative interchange with an obstruction emerges as a new configuration of expectations and desires. The old structure of skills, instrumentalities, and certainties through which the older members formed and pursued their needs (i.e., the old historic time, the old diachronic juncture) transforms into a new one.

There has been a great deal of discussion on "advanced culture" superseding bourgeois sensibilities. Such view takes a special case for a more general regularity: the revolt of the new against the established ways. Bourgeois, or even the Western Romantic tradition itself supplies here the word: avant-garde. For the revolt and the new view is not considered here as a rediscovery of an ancient and wiser outlook, as the early Renaissance thought had it, or as the Chinese thinkers imagined it. The avant-garde were supposed to be the advanced troops of change against the bourgeois. But we have no doubt by now that every establishment is superseded, willingly or unwillingly, peacefully or violently. It is not just a post-bourgeois stance that opposes its predecessor; every new integration opposes its predecessor.

Any new structure, even the most serviceable and functional creation, has great difficulty in establishing itself or even in becoming recognized at all. A new idiom is bound to stay separate and unappreciated by the majority for a considerable time. The distinction between a newly created, competent, and authentic experience form, on the one hand, and the more or less deficient appreciators, the better or worse recreators, on the other, will always stay with us. Often on such grounds, judgment of higher or lower quality is made. But it is too much to ask of second or third line talents to come up to the quality of the creator. This is never the main issue. The most functional creation has great difficulty in convincing others of its worthwhileness. The more perceptive the majority is as to qualities of experience forms, as to the dynamics of whole-making, as to the service and function of experience forms, the less deformation, the less damage, the less incomprehension the newly created and communicated idiom will suffer.

TRANSFORMATIONS

Dewey (1934) suggested that anything good, beautiful, and convincingly true is in its final quality as valid as another. Only analysis of components and consequences can tell us what is better or superior. He and many other reasonable men made little fuss and little inquiry into how a possibly better idiom can become desirable to people. A new point of view, a new life stance (which experience forms represent), cannot be easily picked up just because somehow it could be shown to have more benign or efficatious consequences. Dewey and like-minded people have not tackled the second — and the more difficult — half of the problem, i.e., how an old beautiful thing may be abandoned and a new one appropriated. Of course, the word appropriated is not sufficiently descriptive of the process. Spinoza devoted the third chapter of his *Ethics* to showing how a desire is modified by greater knowledge. After that he felt he could demonstrate that passion, tutored by a knowledge that is the largest and most comprehensive, will be also the most puissant of passions. To the post-modern view, the acquisition of a new desire form is not quite that easy. To realize that in some ways a particular life stance is not conducive to good consequences, gives rise most often to evasive action, complex justification, external rationalization of the old. The "acquisition" of a new object of desire or a new idiom that we enjoy as beautiful — these are not easily come by. It takes a fundamental, radical transformation. It may be an outcome of patient cultivation, or sudden conversion, but, in all events, it is a fundamental process and not an easy annexation. Here post-modern psychology, the psychology of drive modifications *and* their integrations, can show us what the process consists of, where it occurs, and what are its limitations. What happens is not appropriation or acquisition. It is the internal work of the artist, it is the repair work of mourning, it is the creative resolution of some vital instability (disharmony): It is the creative process. Creative transformation (a redundant term) is the process whereby one acquires new desire forms.

An instance of behavior is mainly a feature of a much larger structure, an aspect of an activated experience form of a person. Yet, regardless of origins, a deed has consequences on others, on one's self, and on the semiotic ecology in which it participates. In truth, one can never tell the extent of the vibrations and the uses of which any action becomes part. As some issues and predicaments predominate at one cultural juncture and not at another, and as disharmonies and their resolutions form an ever novel continuum, we can speak of the present concerns of a cultural world. To judge the competence and, indeed, the basic role or function of experience forms, one must have some feel and understanding for such concerns at the time of judgment.

An instance of behavior has effect not only on present attitudes and on present responses of others. It has a reorganizing effect on our whole system of experiential structures, on our experiential self. If it makes it more competent, fuller, more harmonious, then the action is anagogic, life-furthering. In a strict sense, every action is a modifying or confirming process. Whatever we do has consequence, not only on some immediate human relationship, but on that system of encoded ways which guide and monitor our future desires. Superior or inferior behavior must be judged by this aspect, too.

An instance of behavior, a deed, has its roots in a very deep history. It has a vast and unexplorable base. For this reason some advance the claim that they are not responsible for the consequences they create. While it is true that each person receives in interchange a personal activation and effect, most of the time the basic quality of an action is not only present in the act, but is amenable to prior regulation. We can express the same experience form in a great many ways, sometimes harshly, sometimes tactfully, partially at times, tangentially or comprehensively, and so on. Neither everyday practice, nor scientific study supports the view that an experiential structure can be expressed in only one way or not at all. There are infinite gradations. Frankness was said to be the stepsister of truth. It is often used as an excuse to express in a deed a personal truth hurtfully.

It is also rather fashionable to state that a helpful or supportive act is done for the sake of the self of the doer, for selfish reasons and that no gratitude is due. For after all, the person who performs it does it for his own pleasure. This cool view wants it both ways. It wants a natural, authentic, even spontaneous action of help and support, and then it maintains that, because of it, no appreciation is due. The actual situation is different. The person who carefully cultivates his desires and the habitual transactions of his needs in such a way that his instances of behavior become spontaneous responses accomplishes a substantial piece of work. The result is a system which by dint of training and discipline can act at ease, naturally and spontaneously, so as to do the right thing and find in it satisfaction and self-fulfillment. To know what is right is one thing, but Confucius knew that the superior man not only does it, but enjoys doing it so. His self becomes such that it desires what is right and abhors what is evil, loves what is higher, grieves over what is low.

It is a kind of Puritanical view to think that something is superior when it is done against one's true motives. Yes, that is a feat. But to hold against the doer his enjoyment when he does something helpful or gracious is not tenable. It should be even more welcome when a particular instance of behavior originates from a willing person enjoying the larger context of his spontaneous responses. The question of higher and lower qualities require constant care and development of the person in light of events, outcomes, consequences, new knowledge, new obstructions, new predicaments. It is not an affair of a sudden stopping, weighing, and deciding, according to some simple rule or forensic code. It is a matter of development of an ever-modifiable, competent, experiential self that has its responses built in-depth. That is why exercise, good use, extension, and function are *sine qua non* of superiority.

CULTIVATION AND EXPERIENCE

What is this cultivation, operationally, and the development of new habitual ways? From Plato to William James, the mark of a superior person is that he does what is right by habit, the right ways are in his case embodied. The new objects of desire (new reinforcers, to use a Skinnerian term, in the system of which very few Skinnerians are interested) must become intrinsic components in a new developmental phase of an experience form. To use a somewhat trivial example, let us see what happens when a person develops eating habits that are considered pleasing to others, or at least not offensive, and that, in proper development, are pleasing to the individual as well. How can we cultivate proper table manners? It is certainly not easy when hunger is great. The force of another desire, i.e., to be agreeable and inviting to some interpersonal give-and-take, will not matter much when hunger is terribly pressing. We can discover that when hunger is not so imperious, the desire to be agreeable is reinforced by certain behavior we call good table manners. The lessening of alimentary pressures, the trial and error of engrafting the new ways and seeing their consequences, the production of smooth integration within these instances of behavior — in one word, cultivation — becomes possible. Such is cultivation of the experiential self in the less trivial, but in just as common an arena of everyday living. It needs foresight; one must go beyond the celebrated gut reactions. It requires the understanding that the new habit is important in the total picture of life. It needs time, repetition, patience. To force violent connections has very strong counterproductive effects. One must let the harmonious new developmental integration happen, one cannot blueprint it.

In any culture, at any time, a certain kind of view, mood, conduct, one might say, a certain kind of personality, is desirable. A more or less complete acquisition of these features is the aim. These may not be authentic in many lives. They may not be more than a persona, a kind of action pattern we inhabit while our fundamental self does not correspond at all. Yet most people try to acquire the courtesies and the rudeness of the day in such a superficial manner. That is why manners and style have been equated by many with foolish frills. One whose experiential self and this persona are almost identical is "adjusted." He has no qualms. He is also likely to be a commonplace person, not burdened by basic questions nor challenged to creative activity.

Whether repetitive or creative, any experience is a point of view or, rather, an angle of view and action. A particular perspective that constitutes the climate of a period can be discovered in so many of its activities. Take existential philosophy. It emphasized the now, the here-in-the-world, authenticity, and so on. And then we had the now generation, also an effort toward something genuine, the abhorrence of the phony, etc. Did the popular angle of view come from the former? It may have happened that the philosophic persuasion trickled down from the philosopher to the humanist witness, then to the popularizer of ideas, then to the talker and prattler of fashionable notions, and finally to the gullible and impressionable nonthinker. But that is not likely. The predicament, the pain, the frustrations that resulted in a particular philosophical perspective eventuated in a similar view and action in others quite independently. It often happens that one kind of activity and the resulting new dimension of human understanding gives rise to an angle of vision that is worked out later by entirely different persons. It can be a synergistic give-and-take. It usually is. That is why careful study of what goes on in many dimensions of a culture — in the popular arts, in the changes of the idioms of feelings — can tell us what agitates and preoccupies a particular period.

Whatever pressures or causes require change in a culture (or in a person), it must reach the experiential system, it must become part of a new configuration, must be expressed and received by others before it can become a way or form of life. All else is external or instrumental to this final event. It is very much like creative transformations in nonhuman life: The new constraints and the old system meet and all potent forces become components of a new configuration. This latter cannot be contrived by even the most insightful of men. This is the locus of the creative happening and the process of art-creation. In order to have anything new as a way of life, (in fact, even to maintain the old properly) art-creation cannot be bypassed.

Once a new angle of view and action is achieved, new cognitive structures (concepts) can be derived through it. When that is not left alone as a conviction, but patiently translated into operational and structural terms, then we have a new scientific view that can be validated. It appears awfully objective, for tests and measurements prove it and logic supports it. Yet its origin is in a new way of being.

Many say that behavior forms — a style, a musical piece, a painting — are an affair of emotions and basically have nothing to do with concepts and, therefore, with intellect. This is a natural reaction to the exaggerations of the detached intellectualism of our recent past. All the above ways are experiential forms expressed and put in action through some created object. As experience, they all have a sizeable emotional aspect. For emotion is that abstracted feeling response of experience that signals how things go for our purposes, well or ill, or any thousand modalities in between. But intellect, pieces of knowledge about regularities and reliable connections of our world do go into experience, into this vital contexture, just as powerfully. As Dewey says, poetry (and art) absorbs the intellectual into immediate qualities. Yes, these immediate qualities are singular. Moore (1903) was right. But this does not, in the least, suggest that the final form is something out of this world. That claim was *his* supernatural fallacy. Final forms are represented by all organisms throughout the world.

Some also speak of eternal verities that originate apart from this world of transformations. It is true that certain regularities, certain dynamic features can be abstracted from the fullness of concrete functioning structures. In this sense, certain lawful connections are reliable and look permanent. But these verities are feature abstractions. Each historical station, each personal juncture is unique and perishable as to its final effective idiom. On the other hand, any member of these developmental sequences continues in the very life of the progression: The later stations could not be without the developmental previous ones. In this sense, the utilized previous idioms are preserved.

The concept of eternal verities also relates to the observation that certain relations can be expressed very succinctly in mathematics. They can be further developed, and new relations can be described which find, at the time of such description, no empirical counterpart in the world. It happens that only decades or even centuries later will observations in the concrete world find situations where these discoveries apply. There is no need to be mystified by this. There ought to be only pleasure in seeing how the dynamics of concrete transformations of the world are not different from the dynamics of transformations that happen in our own psyche and that we call creative thought.

SUPERIOR AND INFERIOR EXPERIENCE FORMS

Looking at experience forms in cultural and individual development as we did above, can we judge one superior to another? Can we deem one good, another evil? If we had a reliable end point to the development of the life of mankind, our task would be easier. If there should be a fixed eternal standard to the conduct of life, we could judge with greater certainty; it was done in the past. But we have no such criteria. Ample anthropological observations show that the life of mankind is a succession of cultural junctures which cannot be blueprinted or foretold. Yet one could ask: Does the life of mankind so regarded offer principles for judgment?

It seems to me that anything that promotes, enhances, empowers, safeguards, or confirms life must be deemed good in comparison to forces which inhibit, reduce, or extinguish life. This much we must grant without additional justification. If we should say and believe that death as nothingness is preferable to life, then our discussion would end right here. But if we are thoroughly enthralled by the incomparable power, creativity, continuity, and basic kinship of all that arose in the world, then life-maintaining and life-furthering events and actions will be the measure which we must use and understand in order to judge superior and inferior. We must know what this ever ongoing process is, what are the dynamics of these sequential transformations, and what characteristics of the constituting events are in its service.

To say that experience forms have superior characteristics when they are life-enhancing is not an explanation in itself. It merely equates the desirable good and beautiful structures with life-promoting aspects. But it enables one to proceed with the inquiry in a more pragmatic and empirical fashion. We can ask as we did before: What are the features of the events which are life-promoting in the human world? The characteristics discoverable in that long sequence of integrative transformations which further it, support it, those will be the characteristics of what is superior.

We can ask, in what way has man's estate changed in the last million years, or even in the last 100,000 years. We must ask such blunt questions. (Common parlance elects to contrast day with night when large distinctions are made. And yet, there are times, at dusk, when the judgment between day and night is not easy.) To compare at the start closely related experience forms makes judgment very difficult. First we must see what happened to man in the main. The fact is that he or she was able to form larger and larger cultural systems with more power, productivity, population, more knowledge, foresight, and understanding of causes and consequences. The length of individual life, good health, and opportunities for individual growth have increased tremendously.

The life of mankind, like that of the whole living world, is a progression that is perfectly opposed to the one contemplated by the second law of thermodynamics. This hypothesis requires a running down into random movement and dissolution. Yet the whole course of evolution of life forms is an increase of information and autonomy embodied in them. And the sequential trajectory of cultural transformations, in a large overview, is no exception. To see everything in thermodynamic terms is strange right from the start. To see cultures as dominant powers merely because they show large energy production per capita (Sahlins and Service 1960) is taking this ability as a sort of basic given without the primary functions that are obligatory for its presence, i.e., cultural competence that keeps the social world functioning and lasting. Competence of culture has conditions which grant it the generation and accommodation of superior power. Anything that enables the participants of a culture to work together, to coordinate, to know what leads where will be critical to the establishment of power. A culture may be "thermodynamically superior" yet at the threshold of disintegration. The dynamics of competence involve those encoded semiotic structures which grant integration among members in their beings, in the transacting of the business of life, and in common action. To try to discount this view by calling it anthropocentric is rather odd. It should be quite in order to object to human views and motivations (the anthropocentric stance) when such is attributed to plants or planets. But to object to impelling human motives when man's society and its transformations are studied is strange. One must be resolved to find nonhuman causes at all cost to exclude those impelling forces which operate in the everyday life of mankind.

Of course, to observe that a high energy yielding method will predominate is an entirely different matter. That is safe to say. A technique or system of ways that provides easier livelihood than those employed hitherto will be adopted for exactly anthropocentric reasons (Coon 1955). But, again, to equate a culture with a particular method or technique flies in the face of solid anthropological observations. It is true that important components can be discovered in any important behavioral activity of a culture. All pastoral people, all agricultural, all industrial people carry the features of the particular method of their livelihood; there will be a common feature to each variety. The disappearance of a previous technique of livelihood does not mean the disappearance of that particular culture. It only goes with a substantial transformation. Many are the components to any integrated conformation. Medieval England produced and employed energy very differently than eighteenth century England. Yet it was not one culture supplanting another. It was a new integration achieved in one culture's developmental sequence.

The sequence of cultural integrations which is called cultural evolution is what we must regard. Any particular station of it is a transformation of that integration which preceded it. The potent past and the obstruction constitutes the predicament. The creative transformation of individuals and subsequent encoding is the creative process by which the predicament is composed. By virtue of integrating transformations, the life of man was able to become more powerful, more autonomous, more spontaneous. This is not to say that we can observe inexorable progress. Many a retrogression can be cited. But the facts are such that the estate of man in the last 100,000 years shows a fundamental change toward the establishment of larger, more powerful, more insightful (as to causes and consequences) cultures, with members able to look forward to and enjoy a longer life. Any species that has gone in such directions could be considered a thriving life form. We found that integrative transformation is the basic dynamism that accomplishes this (Haydu 1958). Any instance of behavior, any object of creation, any innovation that represents a full and competent version of the process can be judged superior. It is here that we must look for our criteria of high or low. (In fact, the very maintenance of a culture's existence requires similar dynamics.) We must ask: What are the components and consequences of a new integration in relation to the overall thriving features of the life of mankind? The new idiom is superior when it includes more of what proved itself life-promoting in the past and when it includes the present obstruction in such a way that a substantial resolution is possible through it. A structure of such components constitutes a new life-enhancing creation. Of course, giving resonance to it is one thing, analysis is another. Any form can be enjoyed as much as another by a close devotee. But a proper critique must do something additional. It must find the components and consequences in relation to man's overall trajectory: how they fit, how they promote the emergence of a competent next integration that will not, at the same time, bind the process from further development.

In some periods exceptional creative and innovative activity can be noted, so much so, that we often neglect the presence of such activity at every juncture. These exceptionally fruitful loci in the trajectory of sequential transformations, like fourteenth century Florence for example, are of great interest. For they can point up certain important features. It seems that compared to the preceding evolutionary locus, the apparent efflorescence of creativity coincides with a *relative* liberation of the individual and an indigenous view held as certainty, that the individual's basic insights and innovations, and indeed his variety of existence, matter. This should not be a great surprise when we are convinced that any creative and innovative construction can happen only in the person, in the basic dealings of the individual undergoing creative transformations. Society encodes collectively as a sort of natural selection for what proves, or what seems at the time, to be functional. When the individual is safeguarded and nourished, when individual growth is not stunted or overly inhibited, then we have a cultural juncture that appears like a sudden spurt toward greater insights, spontaneity, strength, and autonomy. This is the nature of greater cultural competence. To see easier, more adequate, integrative transformations in a culture and observe that they correlate with an increase of individual growth is really seeing two sides of the same process.

THE POWER OF COMPETENCE

A new individual structure, if it is a successful resolution of a contemporary cultural predicament, is a powerful thing. In its architectural unity, one might say, architectural simplicity, it has the power to fashion others of kindred predicaments according to its own configuration. The new structure may not be popular in a sense that it is known by name, or even how or where it came to be. By its competence as a sense-making integration, it may become the next important experiential idiom of a whole culture. Beethoven set the tone of the next hundred years in musical feeling in a large part of the Western world. Yet the earlier creations of Cherubini, whom he so admired, are safely contained in Beethoven's crowning achievements. An individual may not be remembered as such, but his individuality can be preserved in the flow of sequential transformations or in the more sovereign constructions of a great genius. Individual qualities are formative in the lives of many others. A great many anonymous experience forms abide in this sort of organic incorporation; they are actually partners of the creative sweep of man's development.

Brute force and violence can be impressive. It appears powerful. But here we are considering the power of cultural competence that grants not only some short-term domination but achieves a more lasting adequacy of standing and withstanding. Language and language changes show a good deal of these processes. For language gives you more the conceptual specificity of experience forms of a culture than any other semiotic system. The Mongol invasion did not abolish the Chinese language or the Chinese point of view. The Norman conquest produced a new language that retained the original Saxon base. The language of the Romans remained basically intact in Gaul and Spain. One cannot make a blanket rule, but the examples we know show that even in conquered places the competent integrative forces can prevail.

Anything that inhibits form creation is bad or inferior. Anything that maims is harmful, evil. Yet it has been said that trials or injuries can be overcome by people, and suffering will produce a deeper, broader result. There is a good deal to this view. But this does not suggest that debilitation is desirable. It only indicates that even severe mutilation can be overcome. Evil can be overcome. It can be transmuted. Anything that promotes form-creation is higher than its opposite, all else being equal. Evil conditions become transfigured. Art-creation is the development of competent, work-doing, original experience forms realized in pigment, stone, and other media, or in the words of beautiful truths. The conditions which we may deem ugly are often part of this liberating structure. But it takes a large soul to transfigure deficient and even obscene realities. The prostitutes of Roualt (even the bland prostitute of Vermeer) have nothing to do with pornography. They are as far from calendar "art" as one can go.

To encourage form creation is not an invitation to fantasy-land. Fantasy, if we should analyze it, would prove to be the creation of a construction that contains elements and components not available in the cultural ambience and usually a fusion of them that is altogether lopsided. The creation of a *competent*, new idiom, on the contrary, is exceedingly observant and careful about what actually is, what forces and obstructions actually are at hand. So a creative new experience form, in this sense, is just the opposite of a new fantasy. Fantasy often actually denies or obfuscates a present predicament and its generative components. Such denials and invalidations appear not only through discursive means but through the expressive forms of the person deficient in the appreciation of what truly is around him.

It was said that fantasies of hell and damnation were enacted and exacted on innocent human beings in concentration camps. Perhaps so. But we need not propose an inherited special faculty of man's wicked nature for this reason. When enculturation is done by punishment, threat of hell fire and damnation, and by the implantation of guilt and vindictiveness — then persons in charge of punitive environment will construct an Auschwitz.

There is a great variety of bad forms, inferior instances of behavior. To lump together all those that offend our "taste" with those that exert calculated cruelty or deprive others of life is not only inexact but misleading. Evil is a disintegrating force. The model of such is dismemberment. That is horrifying even to experimental animals. In man's development, evil effects can often prevent the creation of a competent new integration. Any old structure, no matter how harmonious it may be in itself, once it is functionless becomes hateful and ugly: It is bent on defeating something that answers the present concerns of the life of a culture.

Just as the bad and ugly is a very wide palette, so is the good and the beautiful, even in the same culture. From evil deeds to tactless gesture there are infinite gradations. So is there an immense range for whatever creates and expresses competent experience forms. A work of art, a piece of music may have plenty of charm, grace, or dash. (And here we must realize that humanistic, literary, or critical concepts like those just mentioned are the only terms we can use for describing certain qualities of an instance of behavior even if we later define them operationally.) Yet the work may not be overly substantial or profoundly moving — usually it is not. As long as its expressive means are sufficiently clear and appropriately structured, as long as it has inner consistency, and does not cut a larger and more elaborate garment than is needed — in a word, when it is authentic and sincere — then it has superiority. One cannot enjoy the highest expressions of sublime experiences all the time, not even most of the time. But the variety of superior forms, the dimension of lesser highs, is far from that of the disintegrating, life-inhibiting, katagogic forms. A critic often makes distinctions that are too fine. We must always keep in mind that something not quite so high is still very far from what is noxious or an invitation to disintegration.

THE OBSCENE AS KATAGOGIC

We cannot equate high versus low with high versus "mass culture." Some experience forms that contain and use for expression very complex and sophisticated means may be quite low. And some very simple forms may be the beginning of a highest expression of a people. It should be a mistake to identify kitsch with mass or popular "culture." Negro spirituals were as far from kitsch as you can go. Yet we could not deny that they were folk or mass products. The separation of mass and high domains happens when a specialized separated group of people can produce a highly complex realm of expression forms. That takes leisure and concentration, a special, concentrated tradition. But authentic, original, and important experience forms do develop in plain living. Kitsch and camp are hollow derivations from genuine experience forms into something flimsy and external to the basic rhythm of experience. They are related to inflated and hollow sentiments. They are outcomes of tinkering and puttering for some specious purpose by people who do not feel the basic sources and significance of the originals.

Looking at these distinctions, the nature of the obscene can be understood more easily. When a thorough-paced Brahmin sees meat eaten, he becomes nauseated. When a Spanish philosopher is offered standing roast beef that is red inside, he will pale and lose his appetite. Yet, to us, it is mouth watering. And so with pork for Hassidic Jews, or a mixture of fish and meat for the Desana. An attack on the person as an experiential system is at the bottom of the obscene: It offends. To be a competent experiential system, at ease with one's self and one's particular world requires that the experience forms respond without too much thinking, i.e., without the fresh necessity to weigh every time and decide what a thing or an event is. It is a resonantial affair. It is quite natural that a well-encultured person will feel uneasy and uncomfortable when the bad and wicked is offered. That response is the mark that his engraftings are authentic and organic. Whatever revolts such a person is obscene. And as culture changes, so do the concrete examples of obscenity. There is a broad range to what is obscene, as there is to what is evil. The obscene offers pleasure in evil, it commends the dissolute (in both senses). To restrict obscenity to sexual activity is a common preoccupation. The maiming and violating of people shown for entertainment is highly obscene for many, I hope. To formulate this as an offense of ordinary good taste is a kind of bland euphemism. Good taste can be roiled by inappropriate or gauche actions. Instead of the expression "man of good taste," one should rather think of persons of great sensibilities, persons of discernment, and insights.

All accomplished selves have a certain seriousness, even solemnity, and they often like to deny that they come from simpler elements which are essential parts of obscene events, too. And we know in sexuality what subtle and potent effect play and references to the crude have, at least in our own culture. An allusion to wicked relationships can be a powerful aphrodisiac. It can liberate many from that doleful seriousness that carries with it the feeling of its struggles. Play may let the basic forces have enough play. It may relieve one from sanctimoniousness. It is an acknowledgement of the common in human dynamics and of the fact that we all have the flaws of incongruities. This concept is expressed by the word ribald. The acknowledgement of the simpler, coarser base does not negate great and glorious creative attainments. It only concedes and takes pleasure in the simple origins of anything great. When one can distinguish resonantially (by feeling) between the obscene and the ribald, then one's experiential world is on firm footing.

Pornography is an attempt to produce a desire for katagogic sexual behavior. Sexual relations are just as varied, just as comprehensive and specific as any other human relationship. A Hindu sculpture in the Metropolitan Museum of a nude standing man and woman in the act of loving sexual union is as far from pornography as one can go. The nude body itself, its expressive stances, is as varied and as semiotic form-conveying as is the body's movement in dance. The nude body can exhibit and convey the highest form of being: a human condition incorporating the most complex and ripe integration of a culture's experience forms (in common parlance, that particular culture's highest values). And as such, it may even be erotic, but certainly not pornographic. When a picture expresses an approval of and invitation to short-circuiting the desirable ways of a full person (a full grown member of a culture) then it becomes pornographic. Any short-circuiting has its attractiveness, a kind of prurience we all harbor, especially when sexuality is a highly tortuous affair. In this sense, pornography can be exciting and erotic. But when a pornographic work concentrates on techniques divorced from the wholeness of the person, then it becomes dehumanizing. (Yet the most unusual techniques can be nonpornographic.) Without the knowledge, or at least the feel, of what constitutes a full and desirable communion, one cannot properly discuss pornography. The intention and conviction that what it exhibits is dirty, humiliating, or violating the wholeness of the person is a necessary part of pornographic portrayals and discoverable in the very configurations themselves. Another idiom showing perhaps the same subjects but in a context of understanding and compassion for the weakness in all of us can be a tragic, ennobling stance.

In Gorky's lower world we see degradation. Yet it is not obscene or pornographic. We see it in the total scheme of things. We are confirmed and moved to compassion. That some may find a scene, so full of pathos and poignancy, obscene does not alter the fact. Judgment is difficult and personal, but is not to say that we equate all judgments as we surely cannot equate all persons. The point is that corruption cannot be banned from these discussions. Obscenity and pornography and the urge and temptation to make one unfit for a complete, superior, full existence are coterminous. In this basic sense, pornography and obscenity are harmful. They may be crippling to some.

Closely allied to obscenity is the nature of dirty words. There are dirty words in all cultures. They are not just the designation of something considered disgusting. In our own language, excrement, feces, do-do is not shit. The attitude is very different in these four words, they convey very different particular experiences and relations. The different experiential stances are encoded in the words; they signify different meaning complexes. Encoding of words cannot be arbitrarily, all at once at will, altered. The word must be used, bounced, batted back and forth in interpersonal give-and-take. Only so can it change, become dirty or clean.

The dirty word is like a blasphemous expression: It is a form of defiance that wishes to shock others or to shock one's self. If a curse word does not shock the user, it loses its proper function. Why can a blaspheming expletive cleanse away a lot of bile? Because the anger goes against something very real and limiting. If you can successfully swear at God, you believe in him. An unbeliever finds such blasphemies insipid. So is the case with dirty words. Anyone who finds sexuality a robust and envigorating activity which can be playfully transacted without guilt and furtiveness will not emphasize the use of words that convey sexuality as guilt-ridden, assaultive, and illicit. That is why dirty words are used to affront others, to convey the feeling that a person called a shit or fuck is a despicable character.

The feeling of harmfulness or horror is not the feeling of dirty, and the signifying words are not the same either. When something is enjoyed furtively with the titillation of a guilt-ridden person whose development was arrested at puberty, then we come to attitudes of dirtiness. The hanging of holy pictures in New Orleans bordellos is a perfect example of this attitude: It represents the titillation of people who believe in some external puerile purity. When a person's development occurs mainly by threat and external injunctions, there forms what Freud called the superego: an external structure to the system of adient drives. In a cultural colony where sexuality is considered illicit except for very circumscribed practices, the attitude encoded in dirty words arises easily.

For the post-modern persuasion, it is not the dirty words as such that are irritating, for we are fairly free concerning excrement and sexual play. But we are very much involved and concerned about cruelty and the violation of others and the source of these in the people of dirty words.

Every culture keeps certain physiological activities somewhat discreet. Yes, discreet: kept apart. To see someone else's lovemaking may be a beautiful experience, it may also be a disgusting experience. These activities are highly individual. To see someone eat and drool and make loud chomping sounds can be disgusting. Someone else's eating may be wholly appetizing. That is one reason why table manners are standardized in every culture. Society makes the generally acceptable ways routine; one is used to them, one is not scandalized by them. But to make lovemaking conform to social bed manners like table manners do with food, that was seldom tried, and hardly ever successful. A measure of privacy is the alternative. (It also creates an environment in which two persons can follow the bent of their highly individual desires.)

One man's way of eating may be another's poison. And surely, one man's lovemaking can be another's turn off. That is why such experiential events on the screen are so difficult to portray. It can be done, but only by a great artist, not by Marlboro Country or Madison Avenue "slicksters." One must realize that discrimination subsumed under "good and bad taste," "good and bad manners" forms a steady continuum with "the right and the wrong way of living" (that is extant morality) and finally with "the required way of living" as expressed in laws that punish or reward what is minimally demanded by a particular culture. In this sense, good taste and good manners are training grounds, or at least they can be.

We can see how close "good taste" is to what used to be called "consideration." A considerate person was one who was concerned about what hurt or pleased others. The old adage, "You don't mention rope in a hanged man's house" seems good advice if you want to avoid hurting his family. You are considerate if, in your conversation, you can insert the filter: "Please, no rope in this house." But this filtering ability must come without effort, through earlier cultivation. To start speaking of rope and then stop and say "excuse me" will hardly do. The same way, to act happy in a sad house is inferior behavior. To speak of delectable dining in a house of a man who just underwent a gastrectomy constitutes inferior manners and behavior. To exhibit much exuberance where people are old and decrepit is bad form. Yes, a great deal must be learned and must become part of one's experiential system to make a person of good sense. Yet we must realize that ways of consideration, good form, and taste change and that the rigid insistence on them produces rebellion and rightful defiance. That is why laws ought to be restricted to basic "bad" behavior that is to be avoided; and consideration restricted to people whom you must tolerate in a tolerant world; and good taste accepted and required from people you hope to claim as friends.

The greatest issues of high and low insensibly blend with questions of style, manners, tastes. They can be easily ridiculed. To show that a man who takes his hat off to ladies in the elevator, thereby behaving as a gentleman and then point out how such a person is a sham is the legitimate approach of the satirist. In the same way, external qualities which have little to do with basic attitudes of the person can be easily lampooned. To some degree, this is not an outcome of malice on the part of the scoffers of excellence. The fact is that small signs, preferences, and style often reflect deep structures and practiced habits. And these external signs can easily be acquired by selves that have little else. Little signs of courtesy are not kindliness of the person. Borrowed features of love are not the reality of love; in fact, love may not be present at all. But to deny, therefore, the characteristics of the basic forms does not lead to a better understanding.

CONSEQUENCES OF BEHAVIOR

Every instance of behavior has consequences for one's self or for others. Experience forms tell us in advance what such consequences were in the past. Renner (1972) could show this effect even in experimental rodents. An enlightened person is just that: Many of the consequences of his behavior are not only known to him but they are reckoned with in the act itself. The consequences are embodied in the action, not as a carefully analyzed set of considerations, but within the contexture of his experience forms.

The experiential self is central to the transacting of values. It is instrumental in this regard. The much maligned Dewey was never so silly, like many of his followers, as to think that all selves are equipotential or equivalent. Expressivity may be equally sweet to each, but its quality and worth are not. Experience forms and an experiential self can include many far-reaching and distant truths (relations held true). Nearby effects and faraway consequences can become "the spread-out environment of a single situation." And thus highs and lows develop how much and what kind of consequences are embodied in the experience event. The knowledge of consequences is a powerful component.

To know and be mindful of other people's needs, to know what hurts or pleases them, requires a substantial breadth, familiarity with many varieties of the human condition, with a great variety of experiential selves and their expressions. It was at a particular time in the course of man's culture changes, at the time of city formation, that the conditions for such breadth and diversity came to be. To be civilized meant to be citified. To consider others, what offends them or what upsets them (long before they wince or strike back), and then act spontaneously without laborious calculations—that used to be good upbringing. It cannot be acquired all at once, nor realized by a sudden conversion to "goodness." It involves a condition of life that offers enough time for communion and choice.

Let us note that choice, in this respect, is not a free-floating arbitrariness. Individual development and contingent habit formation deal with large scale necessities and continuities. The incorporation of the potent past into the next state of being finds its reason at this locus. Any transformation in this regard could be said to be determined; but only in a sense that the new structure includes the major forces *as* components. They are discoverable as essential features. But the new form is a brand new configuration. Looking at choice with these limitations, we can see that a society where this is present constitutes a radically distinct variety of culture compared to those that do not have it. The point is not whether there are varieties of occupations. The point is whether a person is fixed or not in one place more or less permanently by race, family, or even by choice. Where such fixity is absent, life is "civilized"; where it is present, life is quasi-slavery.

The comparative characteristics of experience forms were based on their life-affirming function. The development, growth, and thriving of the life of man was based on the competence of integrating transformations. And competence, in this regard, was seen as the creation of a new integration which included in its incomparable idiom components of the present predicament. This latter consists of the forces of the potent past and the forces of the unsettling obstruction (which may even be a beneficial new technique or energy source). Anything that supports and enables the emergence of a new competent form of existence is superior to events that do not. Because such emergence cannot be blueprinted, can only arise in individual human beings, can never be ordered by any kind of authority—for all these reasons, a new insight in the seventeenth century Western world became a singular beneficial contribution. It followed the hounding and oppression of opponents for "truth's" sake. The religious wars and atrocities of seventeenth century Europe, the religious persecutions of England made John Locke a refugee in Holland. He became convinced that if various understandings could be given a fair hearing, the ones with greater function and competence would find their scope of action without the need to burn the books and bodies of their competing antagonists. Toleration, without the regard for truth, became a very novel, fruitful, and inspiriting notion.

Holland's ruling oligarchy had a sort of magnanimity towards those who dissented. It came after a cruel struggle against tyranny which enforced conformity by death and torture. Give the individual some security, some domestic solidity, let him make a fool of himself, he will also make a great soul of himself. That is what began in America 200 years ago. And that is what happened in Florence when its people transformed and transfigured Europe. But even in Athens, the flowering came after a hard struggle against an inflexible empire, after which the city state could pursue commerce in relative toleration. It is not that toleration is the necessary consequence of political liberation. But when it follows a difficult struggle, then we have this happy explosion of functioning, competent creativity.

To point out the place occupied by an instance of behavior is one thing; to repress it is another. Criticism is important service. It can save a lot of floundering by others who could never come to any real judgment. The flowering of toleration, as it has become an important component of the culture of the United States, permits opinions and judgments to make their way if they have the power and competence. It may lead to the neglect of the substantive or the concerns for a deeper understanding; it may lead to simplemindedness. But once the Jeffersonian play of what is offered is done, the competent views and modes can emerge more easily and even more fully. The tradition of toleration also defuses much of the harmful intensity that is generated when expressions held untrue or judged harmful are oppressed by the power of state or other corporate power. In this respect, the concentration of power over outlets of mass media is a great peril for the way of life of a country of toleration.

Many who inhibit discrimination and claim that judgment is impossible do so in the fear that acknowledging superior or inferior forms equals their persecution and oppression. That is not the case with people who are truly democratically tuned, i.e., who are convinced that when all voices are heard the more competent and comprehensive is also the more powerful. That is the Jeffersonian conviction. The study of integrative transformations supports his view that ripeness, decency, continence, a large participation in the life of the understanding will empower you with functioning, work-doings, mighty abilities. As Emerson put it, the true democratic temper is convinced that learning, discipline, and the lofty pursuits of the present concerns of the culture confer upon people powerful virtues. In this sense, the denial of high and low is no service to democracy; it fosters the ascendancy of the mob and mob leaders. A full democracy supports the emergence and use of the excellent.

As long as we judge behavior as examples of types we shall be summary, statistical. Individuality will escape us, we shall see very rough approximations only. That is why statements about types of people and behavior, when that is all we deal with, becomes so external and irritating. What we need, and are in the process of acquiring, is the capacity, training, and attention to individuality, the recognition of patterns that are unique and thus truly expressive. When a person is understood in true mutuality as the only such individual in the world (as it is the case), then only will he or she be confirmed and nourished. This does not imply that certain general features or regularities cannot be seen or reckoned with. In any society there will be strong ones, weak ones, combative persons, submissive characters, dull ones, intelligent persons, protective creatures, and a thousand more. That is why the portrayal and empathic expression of such can find resonance in so many cultures and ages. The persons of the *commedia dell'arte* may not be universal, but certainly fairly common in most complex societies.

In a society of go-getting toughs, it is natural that "nice guys finish last." Considering other people's thriving, not to speak of their sentiments and what makes them feel pleased or hurt, becomes a foolish burden. A new standard of valuation develops. It scoffs at the gentility of the gentleman, at the courtesy of a "well-bred" person, at the keenness of a well-honed intellect. Yet such stretches in cultural continuance cannot last too long. Power, function, the ability to act in a concerted way, the offering of opportunities to individuals to become more potent and harmonious structures — all these are essential to maintain a relatively thriving culture. Relative in the sense that any culture can be seen only in its own time and place as it relates to others and to its own previous stations. Only thus can we attempt to find clues to its power, to the wisdom of its collective action and ultimately to the survival of a people as tone-making, pace-setting, and even united system.

We found that for a competent resolution of a cultural predicament, and even for maintaining a culture's way of life, the participation of a large variety of individuals is optimal. Anything that inhibits, sequesters, oppresses the genesis and care of a new creative transformation of persons runs counter to the emergence of a new competent life-stance, and indeed, life-structure. That is why individuality and its care requires the corollary activity and feeling of compassion. For without the nurturing and supporting effect of others, the individuality of a person cannot function and even has difficulty emerging. From this view, compassion is power: It gives rise to the individual, to variety, and so to the capacity of the culture to thrive.

FACTORS IN JUDGMENT

Even if we should agree that principles can be winnowed from integrative transformations of experience forms, the fact remains that judgment is difficult and personal. One can show so many instances where a highly trained and knowledgeable critic proves to be mistaken. Yet through individual preferences and judgments on experience forms, large agreements have been validated. There are contrary opinions, but the superiority of experience forms conveyed by Shakespeare, or Socrates, or Jesus, or Rembrandt, or Bach, or even Stravinsky, and innumerable lesser ones will not be easily shaken.

But what is the process that constitutes judgment? To begin with, the judgment of any experience form is first the discovery of it. If it is embodied in some object, we have to comprehend it. At this point, the dimension of appreciation revolves around issues that tell us whether the experience form is expressed well or not, faithfully or not, and whether the giving of resonance on our part is comprehensive and appropriate, or not. To be able to give resonance to experience forms of others is a skill and ability that can be cultivated or stultified. At any rate, experience or value forms at their functional fullness are not commensurable. From this primary point of view, there is no high or low. Only when we can analyze what components went into the form, where it functions, what its relations are to the past and present concerns of life — only then can we find ways to speak of wider, richer, fuller, more ingenious, more spontaneous, more nurturing functions and their opposites. But even then, we must locate judgments in the trajectories of our evolutionary world. We must remember that any experience form has its meaning only in the contexture of particular transformations of a particular human world.

Judgment comes first as a response to our experiential system. We find reasons afterwards. Rationalization is finding specious reasons, an excuse. That is not what we are discussing. But even discoveries of true components and their place comes rather late. First we feel the thing, the action, and the persons. Aroused experiential structures of our own are matched to the perception of the outside object. Our first response tells us what kind of object or event we are beholding or are confronted with. It is an immediate categorizing or typological assignment. The matching of the object or event to one's experiential structures is already a kind of judgment, for our experience forms tell us where the object or event belongs in preference or avoidance. The more basic the aspects of our experiential self that are activated, the greater the personal interest that will appear — a greater personal involvement. In a situation of danger, a sound is met by very different experiential structures than the same sound when we are at leisure. So judgment is first an assignment of the object by our experiential structures to its place in our particular world. Just as any experience entity is sorted and, by resonance, stored with other similar structures, so by this same dynamic of resonance, the present object is colored and indeed marked by the arousal of particular concordant psychemes. When they are energized and enter awareness, then we get the feeling of the particular significance and meaning of what we behold. We can only later express this quality in an image or concept and words. This last stage is judgment proper. Emotional judgment is a poor expression. Yet it indicates that our judgment first is an experiential assignment by the work of resonance with our *available* experiential structures.

So a lot depends on what experience forms are available. We do not simply own them. Our selves are made up of them. To understand an object that conveys an experience form or even an instance of axiological value (the outcome-oriented aspect of experience forms), one must give this resonance and delight. As such one man's pleasure is as good as another's. You can be in ecstacy with one value as much as with another. It is, in a way, a reanimation of our experience forms. So much depends on what is available in the responder. A good deal depends on what ingredients went into the final configurations, what earlier experience entities are involved, how much training, breadth, and ripeness has gone into it. A person who never savored a painting by Mantegna, or La Tour, or Chardin, or never heard of Manet or Matisse can have a perfectly strong and candid opinion on Norman Rockwell. Yet his judgment, or rather, the chance that he will truly discover the content conveyed by any of the above painters is very slim. His range and appreciation is limited. Value judgment depends on the range and quality of the judging person's experiential life. Some persons are large, rich, harmonious; others are not. In this sense, judgment is personal, but a person's competence can be observed. A sort of consensual validation goes on all the time; in fact, that is the life-blood of our everyday existence. We compare our own understanding of others with the understanding our friends have of them. This goes on all the time. And value judgment, i.e., giving resonance to what meaning and significance things, people, and relations represent, is constantly trimmed and scrutinized by the person as to its validity, reliability, and scope. That is why judgment, although perfectly subjective, is not an arbitrary, chance, or erratic process. It can be studied objectively and profitably so.

Again, just because experience forms are singularly individual, does not mean that they are not communicable, or that there is no kinship between such structures. It does not suggest that there can be no communion between individuals. Kindred forms respond to one another incredibly sensitively with very low intensity thresholds. It requires no crude or massive onslaught which used to be called sensitivity training (but turned out to be insensitivity training).

The more, and the more perfectly, an idiom includes basic human experiences, the longer it stays relevant and felt to be kindred by other cultural times. A very great creator like Shakespeare will speak to us even from a considerable distance. Such creators incorporate in their experiential selves, and in their expressions, large and lasting themes: the difficulties of growing, the miseries of illness and privation, the consequences of power or poverty, the joys and transitoriness of love, the frailty and brevity of individual life, the gratefulness or ingratitude of children, the vindictiveness of the slighted, the strange largesse or cruelty of the human world, the sublimity of becoming and whole-making, and so on. That is why the works of Bach still speak to us. If someone should ask if the idiom of rock music is as good as the ecstatic configurations of Bach, our first response should be that the two in their functioning, potent surfaces cannot be compared. They give pleasure and completion to very different constellations. In this sense, they are not comparable. But when we try to understand what components go into their incomparable configurations, we can see how substantial Bach is, and how thin the other. Rock may be much closer to us, our very own, yet it is already passing quickly like other popular forms that express the very limited components of their integrated structure.

Some speak of value judgment and try thereby to invalidate a statement. It is inferred that an opinion of this sort is anybody's guess or feeling, and thus not worth considering. But one should not be taken aback by red herrings of this sort. Value judgment is the only kind one can have in the affairs of the human world. The question is what that judgment is based on, what it considers, what it leaves out, what kind of self system produced it, out of what experience it flows. To expect value judgment as unequivocal as a mathematical statement on bridge construction is not available, not possible, and perhaps not even desirable. To try to invalidate judgment that is short of such certainties makes no sense. At the same time, let us not forget that value judgment must and does contain hard facts and validated relationships, e.g., that night follows day, that summer heat does not last forever, that men die, that inflicting injury hurts others, that starvation leads to weakness, that no child is born without a father, that steel is harder than the human skull, that syphilis is not due to bad intentions, that scurvy is not due to bad family stock, that evil is not due to gremlins. Value judgment is not a personal whimsy. Judgment statements convey much of the background, the depth or poverty of the experience from which they spring.

As in so many other instances, our brain performs almost instantaneously a multitude of operations. We come to judgment almost instantly and without the slightest awareness of the partial functions that go into such an event and require a great many antecedent and carefully developed abilities. Judgment involves first the giving of resonance with some kindred patterns, the activation of some complementary characteristics and through them the apperception of the form pattern that is being considered. Then comes the discovery of components that we find to have gone into that final configuration we are judging. All these are then brought to the touchstone of our experiential view and understanding of what constitutes the present state and the present concerns of our culture. It all arises almost as a feeling tone and is put into expressive concepts and words subsequently. It is really the complex and marvelous extention of that ancient function of recognizing and appraising what helps or what hinders us.

CONCLUSIONS

Let us now summarize, roughly, some comparative characteristics of experience forms in light of the principles discoverable in the integrative transformations of the human world.

All Else Being Equal:

When components are rich and well integrated in the considered experience form, then the experience form is higher than when they are not.

When its components include the present predicament and concerns of the culture, then the experience form is higher than when they do not.

When its components include the basic desires and basic impelling drives of man, then the experience form is higher than when they do not.

When its components include trustworthy relations and regularities found and validated (and that collective enterprise is science), then the experience form is higher than when they do not.

When its components show understanding of the causes and consequences that are stunting and life inhibiting, then the experience form is higher than when they do not.

When its components include the nurturing, supporting, and vital roles of discipline, compassion, and order, then the experience form is higher than when they do not.

And lastly, when the components include the reality of spontaneity and absence of coercion, then the experience form is higher than when they do not.

COMMENTS

But must we be prigs? For that is what a constant and careful judging leads to. The labelling of deeds and experience forms, once it preoccupies us, will falsely elevate the importance of the judge and depress the great many virtues (abilities) of anyone. The basic democratic and humane conviction that the "calling of any man is deep" will be obfuscated. So what is to be done? Discrimination is necessary; it is done anyway whether acknowledged or not. To deny it leads to the very substantial unconscious hypocrisies of our time. We must try to arrange and cultivate our interpersonal life so that distinguishing and judging will not be paramount, that the fellowship and respect of others can flourish in our personal world. This requires a life space that is not too demanding. We must discriminate, but we must not blame. For to judge truly, we must see the genesis of a thing, the inevitabilities, the hardships, the sad lack of opportunities of others. To blame or ridicule someone who has a lame leg for limping — that sort of judging — is not in consonance with what we found the main theme of the competence of human transformations. Yet it is done so often. But to discern the limp, see how it arose and what it eventuates in, that can be life supporting. So discernment, yes; blame, no. To point out heresy, yes; to persecute heresy, no. To restrict evil behavior, yes; to force someone into a mold we prescribe, no.

It is evident that experience forms and instances of behavior flowing from them can be seen as a work of art and can be seen as good or bad deeds. There is a huge diversity of them everywhere. It would be a mistake to lump together those actions that offend our taste with those that deprive others of life, even if the dynamics of both are related. The equating of such far apart events is not only confusing but leads easily to condemnation of smaller negligible evils, with a disregard of the major ones. It is really sufficient to have the embodied conviction that those who live their lives realizing the present concerns of their fellow men and represent life-affirming, individuality saving, person fostering, life-inspiriting love, are superior to those who destroy life, who destroy a person or inhibit the personal growth of another.

When we examine a particular experiential system, a particular individual self, we can see that it is made up of many dissimilar and often contradictory structures. There is no easy royal road to summarize anyone. To have a theory and rational scale for judgment does not guarantee that even the simplest person can fit such categories. Life and life's creatures in their fullness defy definition. To think in terms of superior and inferior is a very schematic and constricting notion. In real life, apart from laboratories and chairs of theory, if we wish to be fair or even truthful, we must see a person in his or her full quality and richness, incredibly individual and capable of highly unexpected transformations. Categories of high and low are helpful in establishing a view on desirable development and even in establishing a modicum of personal certainty in the scheme of things, but they offer a kind of rating that is usable only in analytic criticism. Such categories are not useable to recognize or really understand the person. That requires a constant give-and-take, a basic communion. And even with that, it is an uncertain undertaking.

As suggested before, we are all made up of experience forms that have various origins, that can be composed only with difficulty, that can become competent only with a good deal of care and cultivation. Incongruities abound, not only in individuals, but in any human condition and are readily available for ridicule. For what is the nature of poking fun? It is the creation of an experiential landscape and then suddenly the showing of incongruities and contradictions, and the ineffectual or inferior nature of it. The expectations in the created tensions collapse and become free flowing; the free energy dissipates in relief and laughter. The nature of the expectation, the quality of the incongruity pointed up in a joke can be witty, vulgar, boring, or irritating; it can be an invitation to scoff and make light of most cherished convictions and relationships. It can show contempt easily, for incongruity is a lack of wholeness, a lack of inner strength and consistency, that flows from it. But once we feel part of the experiential world that is so treated, we smile with it, instead of guffawing at it. It then is humor. And that will keep at a distance the danger of becoming a blue-nosed moralizer. Emerson had breadth enough; he could afford to say that men can be divided between benefactors and malefactors. For it is hard to escape the conclusion that some men further the richness, the solidity, the cooperation, the power and reliability of a cultural landscape. Others reduce the reality of these riches. Extant morality is the normative system of present human relations and their immediate extention. There is no such human condition that does not involve good or bad consequences. In this sense there is no amorality; there is only superior or inferior morality. A sense of humor (sympathy with those who, like ourselves, are being shown up as highly imperfect beings) will keep us from becoming Moliere's Tartuffe.

In conclusion, let us note that experience forms and their encoded equivalents in the semiotic world tell us what is good, what is serviceable, what is useful for our health and happiness, and in what way we can become whole and at home in our world. So what is of paramount importance is whether or how these structures guide us in our present world, in the midst of our present predicaments. By now we can even speak of the present concerns of the human world. For with greater communication, interdependence, exchange of every sort, the cultures of the world are confronting and dealing with similar concerns. One can speak now not only of the present concerns of our own culture, as we had to do in the past, but of the present concerns of the life of man. When there is no hope that the larger issues of this situation are met competently, then arise those separatist and splintering resolutions which can hold together increasingly smaller groups on the strength of "ethnicity," i.e., of the exclusive past. But when large and competent resolutions emerge, then the loyalties of the individual go to the next overarching forms of life. I think we are at the threshold of seeing this happen.

REFERENCES

CHRISTIAN, A. B.
1973 Principles that govern the folding of protein chains. *Science* 181: 223–230.

COON, C. S.
1955 *The story of man*. New York: Knopf.

DEWEY, J.
1934 *Art and experience*. New York: Capricorn Books.
GADAMER, H. D.
1965 *Warheit and Methode*. Tubingen: Mohr.
HAYDU, G. G.
1958 *The architecture of sanity*. New York: Julian Press.
1972 Cerebral organization and the intergration of experience. *Annals of the New York Academy of Science* 193:217–232.
1977 "The nature of values and experience entities," in *Language and thought.* Edited by W. C. McCormack and S. A. Wurm. World Anthropology. The Hague: Mouton.
KATCHALSKY, A. K., V. ROWLAND, R. BLUMENTHAL
1974 *Dynamic patterns of brain cell assemblies.* Boston: Neurosciences Research Progress Bulletin.
MOORE, G. E.
1903 *Principia ethica*. Cambridge: Cambridge University Press.
RENNER, K. E.
1972 Coherent self-direction and values. *Annals of the New York Academy of Sciences* 193:175–184.
SAHLINS, M. D., E. R. SERVICE, *editors*
1960 *Evolution and culture*. Ann Arbor: University of Michigan Press.

The Shaping of Experience

H. G. BARNETT

THE URGE

Experiences are not random assortments of events that just happen and then fade away. If they are truly such, and not the undifferentiated background out of which they loom, they "make an impression," or as will be argued later, they become an integral part or attribute of the experiencing individual. They are not only directly and immediately sensed by him; they are encoded by his neural system and are retained as traces or "cortical nerve cell assemblies" (Haydu 1972:217). As such, every new experience constitutes a prototype, which is a reference datum for later sensory reports and a charter or warrant for their interpretation. The more of these that an organism possesses the more adaptable it is.

As a species, human beings have never been content to leave well enough alone. That which exists may be good enough for some people, but it is never for others. The history of world cultures is witness to man's disposition to do, think, believe, or say something different, whether out of supposed necessity or sheer folly. But that perspective yields only a hint of the extent of human ingenuity and perseverance in exploring the limits of knowledge and testing the resiliency of custom. Children are expected to be unruly; but none of us is ever completely grown up, and the more courageous among us become artists, inventors, and explorers.

There can be little doubt that we are experience prone. We seek it for its own sake. That propensity is not restricted to the desire to learn, know, or control. Often it is, especially in the modern world; but in the great majority of instances it is the urge to change a present mental or physical state to attain relief, novelty, or excitement. It is an inherent demand of an organic, dynamic system such as ours which is relatively free of instinctive constraints; but that freedom by its very nature leaves monotonous gaps. We accept the challenge of monotony because it is intolerable.

The array of available experiences varies greatly from one society to another; but nowhere do people now live, nor have they ever lived, as economic robots saving space, time, and energy without a laugh or a song. At the very least, they have had games and other diversions, imaginative myths, and endless true-false tales, some form of self-expression in art or technology which invited admiration, festivities to celebrate the reunion of kinsmen, and rituals symbolizing births, marriages, and deaths.

In stabilized communities — as distinct from nomadic bands — the incentive to create and participate in leisure time activities is augmented not only because more people generate more enthusiasm, but because a greater diversity of interests must be accommodated in an orderly manner. Every ancient society of which we have record, from China to Egypt to Greece to Rome, lauded their heroes in sacred or profane rituals, in song, dance, or oration, Many of those societies called primitive are no less remarkable for their periodic celebrations. The Kwakiutl Indians of British Columbia vacationed during their sacred season from November to March and spent most of their time watching or participating in feasts, dances, and dramatic displays by privileged groups who re-enacted their personal encounters with supernatural beings, many of which were terrifying, one a cannibal. The Aztecs of Mexico, unlike the Kwakiutl, did not pretend to be cannibals; they were. During one of their annual ceremonies war captives of both sexes were offered to the gods, flayed, eaten, and their skins worn for several days by persons afflicted by skin diseases. On other occasions virgins, snakes, and children donated by devout parents were sacrificed. Such spectacular events climaxed eighteen annual celebrations which began with solemn processions bringing offerings of fruit and flowers and continued with day and night dancing which lasted several days or weeks.

The Hopi Indians of Arizona also spent much of their time in religious activities, but they were a peaceful and benign people who gave humor a place in their devotions. They had fraternities whose function it was to make devotees laugh during periods of relaxation. The duties of these ceremonial clowns were regarded as a sacred obligation; but their humor was mundane and profane. They were privileged characters, saying and doing things that in others would be considered shameful or blasphemous. Their comedy was slapstick and, by the white man's standard, often vulgar and indecent; but for the Hopi it was both funny and devotional.

To create or to participate in a humorous situation is an enticing experience. For the Hopi and many other people it has provided relaxation from the constraint of prolonged dedicated attention. But as a distinctive form of expression, it is within itself an abrupt departure from formality. It encapsulates a breach of experience by beginning with a routine expectation, then suddenly switching to an incongruous frame of reference. To master this technique is to become an artist and so to join poets, painters, and musicians whose role it is to "break prescriptive rules about form." That is how Morse Peckham phrases it, then adds, "It takes the scientific genius, the scientific innovator, to force a rule change, which is resisted in science as much as it is in any other branch of human behavior — except art. It is clear that the artist is rewarded with praise, status, and money for breaking or violating rules, for offering discontinuous experience" (Peckham 1965:220). In doing so, he gratifies "Man's rage for chaos."

Drama — and sometimes comedy — appeals to those who are able to identify themselves with other people and thereby share their emotions and quandaries. This applies to audiences as well as actors. The convincing actor appeals because he is not playing a role; he *is* the person portrayed. And a totally involved spectator is neither himself nor the actor; he, too, experiences another existence, sometimes tearfully or gleefully. To those who have cultivated a taste for the theater, only on-stage performances can be so captivating; others can become equally engrossed by horrifying motion pictures; and readers steeped in literary classics prefer them because they are more tranquil, but entrancing nonetheless.

Imitation is a more common means of being someone else. All of us engage in it at times, whether upon our own initiative or prompted by others. Most American parents expect their children to behave as they do and are not persuaded otherwise by a child who attempts to justify his aberrant behavior by pointing out that the Jones children do it that way. There are many other approved patterns of imitation, such as a student modeling upon an admired teacher, a devoted friend acting like a well-mannered associate, and the Cub Scout who idolizes the father of his country.

Finally, we must not neglect the universal predilection to play which brings exciting experiences to adults as well as children. From the time infants are able and free to move within their environment, they explore it, and themselves, sometimes at random, sometimes with fascination. Later on they "play house" or "cops and robbers." Before long, those who have the energy and the skill engage in contests as their adult exemplars do. On these and countless other occasions from infancy to senility all of us play sly games, matching our wits with opponents for the fun of it or to secure some advantage.

It must be kept in mind that, despite the exotic or brutal character of some of them, all of the experience forms that have been mentioned — and a multitude of others — have an important function. Every new one expands the prototype inventory of the experiencing individual and thus increases his capacity to cope with other new but similar experiences.

Even those which we might regard as fantasies have this function if they are shared with others — until a persuasive eccentric experiences a new truth.

THE ORGANIZATION

There are no frames of reference in nature. All are created by humans and other creatures, from a bird's eye view to the mulitifaceted perspective of insects, from maps to "realistic" art forms. Clouds, rainbows, and an orbit of the moon do not compel us to perceive them in only one way. This is obvious when we review the history of philosophy from Plato to Descartes, and the development of science from Aristotle to Einstein.

No less striking are the various interpretations of natural phenomena by people reared in different traditions. Most Americans would agree with the statement that night follows day and day follows night; but in the Palauan world view, night is in front of day, and day is in front of night. And so it is with devices constructed to record, map, or measure natural phenomena. Ask a Palauan what time it is when the hands of a clock read twelve fifteen and he will answer "Fifteen minutes before twelve," and a half hour later he will reply, "Fifteen minutes behind one o'clock." Palauan time, like ours, is divided into past, present, and future; but unlike us, they constantly face the future and do not turn around to locate the past. The day before yesterday is always in front of them, and the day after tomorrow is always behind them.[1] Space and time were relative long before Einstein.

Humanly contrived complexes of things, events, and predicaments can be as uncharted and chaotic as natural phenomena. But human beings cannot tolerate prolonged uncertainty or ambivalence. They must establish frames of reference and order within those frames. Organisms are not only organized; perhaps because of that fact, they are also organizing beings. The order which we postulate around us is a reflection of that organic demand, and our perceptions, attitudes, and purposes are our organizing devices. We select those aspects of experience which are amenable to treatment by them and ignore or cease to struggle against those that cannot be reduced to order with their assistance. Self-projection and the self-fulfilling prophecy play no small part in this endless dialectic between man and his social and physical environment.

The means by which our experiences become a part of us is through our neural network, and since it is itself a system we should expect its translations of internal and external stimulation to be systemic. The

[1] The Palauans live on a cluster of small islands located at the southwestern tip of Micronesia in the Pacific.

"translations," however, are not autonomous. They are an integral part of that which activates them. Moreover, under normal body conditions when activated by an object or its medium of transmission, there must be a veridical correspondence between the activity *pattern* of the stimulation and the nerve cell activity pattern. There does not appear to be any other explanation of our ability to discern the boundaries and other qualities of the forms we experience.

We could continue to discuss the organization of experience in terms of forms, patterns, figures or structures, but "configuration" is a more expressive denotation in that it signifies any unified complex of experience. As such, it has four characteristics. It is a mental composition of sensory stimulation. It is complete in itself, a homogeneous unit without parts or internal structure and with uniform qualities throughout its extent. It has shape or form whether it be the perception of an iron bar, the sound of a musical tone, or the mental image of a fictitious character. It protrudes from its matrix as does a figure from its ground in ambiguous designs, or the face of a friend in a group photograph (Barnett 1953).

Configurations may be of any size; they may be elementary or composite; they may be generic or most specific and discrete. They may be vague and signify nothing more than "something" — which is to say that they may have uncertain, vacillating, fragmentary, or elusive characteristics. They may be nothing more than a setting or schema, such as is suggested by Bartlett (1950). But in all cases they are organized; their elements stand in certain definite relationships to each other so that they make up a whole.

Configurations are as varied in character as are human experiences involving organization. They can entail the ideas of things, of people, of ideas, or of the self, or of groups of things, ideas, and people. They may take an infinitude of forms. They may represent objects or groups of objects conceived as a configurated whole. They may be a rule or a principle or a theory. They may constitute a stereotype, or a dogma, or a belief that is organized on the basis of past experience. They are all normative thought patterns, even though they may have fluctuating and evolving limits. Although they may be changing and vague, they are standards of thought. They may be standardized for the individual only or for a social class or an ethnic group. Their relative character is indicated by their variation from individual to individual and from group to group. As norms they are systems of reference, datum points for the evaluation of incoming sense impressions. They mold and channel these impressions, and they order the thought processes of the individual.

The qualities of things we sense are mental compositions. A tuning fork causes air molecules to vibrate, and when their condensations and rarefactions reach our eardrums we hear *a* sound, not vibrations. The pulsations are discrete, but we do not hear them as such. We unite or synthesize them into one unbroken continuum. That configuration is a constant tone or twang. The only way that we can analyze it — and thereby dissipate the configuration — is indirectly, such as listening to the fork tapping against a membrane or by the use of an instrument called a phonodeik which translates air vibrations into a visual pattern; but that, too, becomes a configuration, the whole of which is qualitatively different from the wave segments it depicts.

Our sensations of color, brilliance, and temperature are mental compositions of imperceptible action systems external to ourselves that evoke comparable activities in our sensory apparatus. We compose them on another level. The same holds for our conceptions of hardness and permeability. The arrangement and organization, spatially and temporally, of these submicroscopic activities constitute the shape, which is the mental configuration of the thing. The patterned activities of these unsensed movements give what we call properties or qualities to some more inclusive entity. Certain organizations of atomic actions are so distinctive and important for configuration that we, by a process of abstraction, have given them names: edge, corner, surface, hole, pit, notch, bulge. More inclusive wholes also have names: spheres, squares, and tubes.

On a grosser level of analysis, combinations of configurations of lesser activity systems can similarly be organized into a larger configuration: the weight of a thing is a sensation that results from the perceptible activity of an observable whole; a chord is the sensation of a demonstrable action by a distinguishable thing. As with colors and sizes, there are degrees of frequencies of these motions of gross wholes for which we have words, such as "pounds" and "rhythms." These words name movement patterns of supra-atomic wholes which can be sensed by touch and hearing.

There are other activity systems that give properties to still more inclusive configurations. These movements are also patterned in conventional ways, such as up, down, north, and circular. For them, however, our nomenclature is limited. These action systems are so slow and ponderous that we can see them and define them geographically. We say, for example, that an object moves from A to B. The motions of the included configurations ("things") are so slow that they seem a part of the activity system to which we belong. Consequently, their action systems do not impart qualities to larger wholes, unless we fix our gaze and let them cross our field of vision — which is to say that their motion is permitted to have its sensory effect. It is obvious that a thing in motion is entirely different from what it is at rest. We endeavor to discover "what a thing really is" by attempting to nullify the effects of its motion. In effect, we analyze the larger whole of which it is a part by devices for stopping its motion. The fact that we can do this makes it appear that gross objects, with relatively slow motion, are not parts of larger wholes. In nullifying their action we make them appear to be isolated and are aware of their movement as such. We do not think of them and their activities as properties of larger configurations. And yet there are borderline cases, sensed movements of things which impart qualities to larger wholes. The flight of an arrow appears as a streak; spinning propeller blades appear as gauzy circles; spinning colors have whole properties that are quite distinct from those of the individual configurations; a waving field of grain has the quality of an undulating surface that is something quite different in its whole properties from that of the individual grain stalks. The pulsing, tapping, or oscillation of a thing can be so rapid that it presents a *surface* to the touch; and a form or configuration may be experienced as having a quality of hardness which is quite distinct from any quality exhibited by the thing at rest. Similarly for the feel of a file drawn through the fingers or the sound of a ratchet wheel whirling against the edge of a piece of cardboard.

A mentally organized group of what are otherwise considered to be distinct things in the same way can constitute a shape or a configuration on another level. Thus, a ceremony or a football play can be said to have a shape or a configuration no less than the things that go to make it up. It is not the quality of the components that imparts wholeness to the ceremony or to the football play, precisely as it is not the inert atoms that give redness or extension to an object; it is the sequence, the order, and arrangement of the activities of the components. Just as we could never know the qualities of objects except for the subsensory activities of atoms, so we could never have a ceremony or a football play without activity on a much grosser level.

Configurations, therefore, are made up of the activity systems of their component parts, the qualities of the configuration being distinctive to it and of another order than the qualities of its components. Whether, as in the case of a ceremony, we regard these large configurations as shapes or not is irrelevant. We may do so if we choose, for logically there can be shapes or configurations beyond the customary range of applicability of these terms.

Every configuration can be analyzed. It can be reduced to distinct configurations or subwholes within itself. Thus, the concept of fatherhood has its component parts, just as the idea of a table does. Conversely, every idea can be incorporated with other ideas as a subwhole in a larger configuration. The idea of a part is relative to the idea of the whole. A part is such only with reference to other parts. When perceived or conceived itself, it is a whole, a unit that is at that instant indivisible, with properties that pervade all of it. If it is analyzed, these whole properties disappear to be replaced by the properties of the individual components. Thus, when we think of family, we have in mind a whole entity with properties that are peculiar to it and distinct from the properties of husband, wife, and child. When the focus of attention is directed to any one of these components, it is evident that it has characteristics peculiar to it that tell us nothing of the characteristics of a family. The components are parts in so far as we think of them in relation to each other; but with respect to themselves they are wholes which can in turn be analyzed into still lesser wholes, each with its own distinctive characteristics. Whether a conceptual unit is considered to be a part or a whole is dependent entirely upon our level of analysis.

The purpose of the preceding discussion is to emphasize that our experience can be organized in an infinite variety of ways. One further consideration requires mention to elucidate this potential. It pertains to the relations between the components of an analyzed configuration. Relationships are extremely varied in their nature, as varied in fact as the human mind can conceive them to be, for they are mental constructs. There is no external source for their apprehension, no stimulus for high, low, first, more, less, or any other conceptual connection between the sensory elements of our experience. That which we call the relation between things is a mental attribution to the external situation. It is an interpretation. And yet it inheres in the nature of the stimulus and is immediately knowable. Spearman lists this property of the mind as one of his three cognitive principles. He calls it "the eduction of relations," by which he means that the presentation of two or more stimuli at once and spontaneously tends to evoke a knowing of the relation between them (Spearman 1923). It is an unmediated and an inseparable aspect of the apperception of things.

Despite their importance, we have a very underdeveloped terminology for the identification of relations. This is due partly to their indeterminate number and the shadings which they are conceived to have. It is also due to our bafflement when we attempt to isolate clearly those relations we feel to be present. Some of our everyday experiences involve very complex sets of relationships which we sense, but cannot express: the relationships between a parent and child, between a moving automobile and the pavement over which it moves, between a politician and constituents, between the legs and body of a running animal. An appreciation of their significance is necessary in order to comprehend the extent to which we are able to mold, shape, and control our experiences.

THE ENCOUNTER

During World War II, a newly recruited squad of American servicemen came upon a deserted village while patrolling a section of the eastern shoreline of the Solomon Islands. They decided to investigate the evacuation because there appeared to be no evidence of violent assault. While relaxing toward the end of the day, two of the men wagered another that they could launch a canoe which was stranded on the beach of the lagoon fronting the village. It was a challenge because the vessel consisted of a twelve foot tree trunk which had been hollowed except at its blunt ends; but the two men asserted that they had seen pictures of American Indians paddling such "dug-out" canoes and that they, too, could do it. They did, but since they could find no paddles they used bamboo staffs to pole the vessel. It was not until they returned to their base camp and had recounted their exploit to the amused audience that they learned it was laughable; they had been afloat in one of the giant feasting dishes common to the area.

What, in fact, was the object which they had poled around the lagoon? Was it that which it was designed to be, or what it was perceived to be, or was it both? Does it make a difference whether the object in question is not an artifact but a natural formation? What if the two men had encountered a floating log, decided to straddle and pole it around in the same spirit of playful excitement? Would it nonetheless have been essentially a log?

Philosophers in the Western tradition have debated such questions for centuries. Some have maintained that a thing is what it is by its very nature, which means its unalterable character. "A tree is a tree is a tree" was true in Plato's world; but that world was by definition perfect and immutable. Many other philosophers and scientists from Aristotle on have accepted the universe for what it appears to be; and from Galileo on for what it mathematically is conceived to be. With the development of the scientific revolution, empiricists such as Hume and Mill insisted that knowledge is limited to that which can be observed, shorn of any interpretation of it by the observer. Mach and other positivists refined this precept to its limit in declaring that the ultimate source of all knowledge is sensation. They granted that sense impressions may be analyzed, classified, and associated after they are free of cognition or other structuring. The philosopher Nietzsche labeled this "the dogma of immaculate perception."

I must confess that by Nietzsche's standard, all of my experiences are sinful, even incestuous; and I am obliged to say the same about everyone I have known. All of our perceptions are impregnated by our conceptions. It is impossible for us to accomplish an observation without cognition — and more. The act is infused with purpose, attitude, mood, and other affective dispositions. Together they create an experience. And so it is with every event we notice. The event, whether it be a flash of light or a proposal of marriage, is only one component in a complex of neural activities that fuse it with our personal interpretion of it. We must consciously or subliminally, positively or negatively, contribute to the total occasion and thus make it a part of ourselves and ourselves a part of it.

The personalization of stimulation begins early in life. This must be so due to a combination of circumstances. No two individuals, not even twins, are born with identical physical or mental potentials; no two are subjected to identical stimulation; all must react to stimulation in some fashion; and every person must realize early in life that he *is* an individual, an integrated, continuing entity, one person now as in the past and extending into the future.

In the beginning of his existence an individual is confronted with what to him must be an unorganized mass of stimuli, internal and external, out of which he must make sense. He must perforce nucleate himself out of the welter of sensations and formulate a conception of himself. He has the further need to formulate in some fashion, his ideas about other things, people, situations, and ideas. In this extremely complex and confusing process, he inevitably develops a personality with its own colorings and with its singular functional compulsions to filter experience. He develops ideas about himself, biases, moods, and attitudes that make him an entity. We need not ask whether he wants or needs, in a voluntaristic sense, to do all this — whether he has some yearning to thread his way through the "blooming, buzzing confusion" (James 1890) to some satisfying end, or whether he is free to choose his course. The objective fact is that he acts as if he were strongly motivated to bring order into his personal chaos. And it is undeniably true that he must do so if he is to survive by his own efforts. We may call these strivings subliminal self-wants or needs.

In speaking of these self-wants as subliminal it is not meant that they are inaccessible to the conscious mind. They are knowable to the individual by probing, and often they press over the threshold of awareness by themselves. Apprehension of them ranges all the way from a complete unawareness of their persuasive demands and effects to an acute distress and the terror of insanity with their frustration. Reactions to the recognition of these demands can range from not knowing that one is being oneself to an aggressive assertion of it; from uneasiness over memory lapses to self-persecution over bad judgments; from bland ignorance of the self to pride in it or disgust with it.

The importance of self-wants relates to the fact that they place the individual's imprint upon all perceptions. His understandings of other things are inevitably colored by his understanding of himself. In order to preserve his identity, if for no other reason, he is faced with the necessity of organizing the field of his experience at every moment. He does this in terms of his sociopsychological self, an entity that is the product of a unique life history in a unique social microcosm. He is continuously, and largely unconsciously, casting his environment in the mold of his past experiences through a dynamic interaction between its components and his self-conception; and since no two things (including himself) are ever identical from one moment to the next, he is constantly grouping together sensory and conceptual data that are different. Perceptual organization is not a photographic process. It is fundamentally an innovative act; it is an interactive, adjustive relationship between the perceiver and the thing perceived. The two together make up a dynamic creative whole.

These private moldings of the phenomenal world and the demands for them are ever-present occurrences. They are obvious upon introspection and appear also upon casual observation of others. In addition, there exists a voluminous literature on controlled experiments that attest to their operation (see bibliography in Bruner and Goodman 1947).

Subliminal wants may be discussed under several headings, each treating some particular aspect of the demand for integrity of the self. All operate under an unconscious desire to assert, stabilize, and maintain the ego; but each illuminates some need facet with consequences distinct enough to warrant their separate considerations. There may be a greater or a lesser number of them than are noticed below. That is not a critical question, since our interest is only in pointing out that such needs do exist and illustrating their personalizing effects.

One such need is for self-orientation. If an individual is to function as a psychological whole, he must work out some adjustment with his experiences. He must feel or know that he stands in some relationship to his natural and social environment. A great variety of self-placement scales are manifested in explicit or implicit beliefs, attitudes, evaluations, and judgments. The placements occur in all dimensions, including the temporal, spatial, and social. The relationships of the self to other phenomena may be positive, being expressed in preferences, interests, identifications, and participations; or they may be negative and so be manifested as neglect, avoidance, dislike, unconcern, or withdrawal. Whatever expression it may take, the individual's appreciation of himself gives his ego reality, constancy, and protection.

Not only must a person orient himself; if he is to survive in a universe of unremitting sense impressions, he must order them; he must assign relationships and thereby structure them. Again, the placement series, their dimensions, and the relationships appropriate to them are multiplex and personal. The confusion and uncertainty that characterize an individual's estimate of certain areas of his experience may prove too baffling for resolution, and it frequently happens that he either ignores the demands that his estimate places upon him or finally abandons his attempts to make any sense out of them. But if circumstances require that he deal with the situation, it cannot remain chaotic and undifferentiated. It has to assume some recognizable characteristics. In particular, perception demands some structuring of the perceived field, a fact that is demonstrated by the psychological effects of ambiguous figures. An observer presented with such a visual stimulus organizes it in one way or another. There may be reorganizations, and one interpretation may fluctuate between itself and another; but both cannot coexist. One cannot see an outline as the boundary of a continent and the borders of an ocean at the same time. Most importantly, some interpretation must be made if the observer is sufficiently motivated to attend to the stimulus.

The point here is that these compulsive perceptual organizations may be highly individualized. Moreover, their peculiarities commonly become stabilized to such an extent that a given individual can actually perceive no other organization than the one he has himself achieved. A person who has once visualized an ambiguous figure in one way must make a deliberate effort if he is to see it in any other way, and often he is unable to do so. This is not only true of such trivial things as experimental drawings; it holds for ambiguous life situations as well. In consequence, there are innumerable private interpretations of the same social, political, and historical data.

The subliminal striving for meaning is another important central need of the ego system. This want is closely related to orientation needs, as they have been discussed, and in some instances it would be difficult to assign a given reaction unequivocally to the one or the other category. They are intimately linked because both are drawn into an individual's unconscious struggle to understand his universe in terms of what he already knows. He configurates it and reads meanings into it in the light of the only possible frames of reference available to him; namely, those provided by his past experiences. New experiences must be integrated with the old. They must be drawn into the matrix of the known before they can have any significance. Otherwise, they remain utterly alien, detached, and incomprehensible. In short, they must have meaning if the individual is to deal with them; and if they appear to be lacking in meaning, he consciously or unconsciously assigns some significance to them.

This urge to rationalize the irrational is a well-known human reaction on the conscious level. Human beings everywhere evince a tendency to respond in this way when the logic of their routine behaviors is questioned. Still more insistent and subtle are their subliminal urges to do the same under the pressure of their own hidden doubts and dilemmas. This insidious demand for meaning is significant in the context of the present discussion because it to some extent inevitably distorts the stimulus data. The necessity of fitting a sensation into the framework of the known entails some degree of individual coloring, filtering, and evaluation. The observer molds his new experiences to suit his needs and in so doing makes of them something that to others they are not. This distortion is minimal and may have no discernible consequences when the evaluations of different individuals of the same social group are compared because they vary only within an accepted range of interpretations. Disparities become more frequent and more evident the less constrained a person is by the conventional definition given the thing, idea, or behavior in question. The freer he is of automatic acceptance of traditional interpretations, the more novel, incomprehensible, amusing, extravagant, or repellent the meanings that he reads in are likely to be. The freest of all is the person who is confronted with an isolated alien form, one which does not carry with it any clue to its meaning in the culture of its origin. Hence, we have the curious (to us) adaptations of European trade goods and cast-off articles made by natives in remote areas, and the perverted (to them) interpretations of feasting dishes by American servicemen who pole them around a lagoon as if they were canoes.

The gist of the preceding analysis is the affirmation that an experience cannot be dichotomized into what a person is and what happens to him, between person and environment, fact and belief, subjective and objective stimulus and response. At the moment of an experience, those abstracted polarizations are joined in one indissoluble whole. Consequently, waking life is an ongoing series of interactions between a person and whatever activates his sensorium.

This contention is consonant with, but has been developed independent of, a psychological approach which has been called "transactionalism." The term was proposed by the philosopher John Dewey, and the approach resulted from an amalgamation of his pragmatism with the experimental procedures of psychologists Ames, Cantril, and Kilpatrick beginning in the 1920's.

As Cantril states it: "Each transaction of living involves numerous capacities and aspects of man's nature which operate together. Each occasion of life can occur only through an environment; is imbued with some purpose; requires action of some kind, and the registration of the consequences of action. Every action is based upon some awareness of perception, which in turn, is determined by the assumptions brought to the occasion. All of these processes are interdependent" (Cantril 1950: 59).

Kilpatrick adds that "Evidence for the necessity of such an approach is to be found in the demonstrations and experiments reported in this volume, which suggest strongly that the search for absolute objectivity is a vain one. Apparently, the correspondence between percept and object is never absolute" (Kilpatrick 1961:3).

This invites a final statement about the personalization of experience; namely, that objectivity is only another possible component in the array of purposes, values, moods, biases, and attitudes which may contribute to an observation. This appraisal is not intended to denigrate objectivity, but to identify it for what it is. Objectivity does not yield unadulterated "facts"; if there are such, we can never know them. All that we can do, if we agree that it is desirable, is to train ourselves to perceive things and events as nearly alike as possible. There is no question that this approach has merit, but it nonetheless is as sinful as love and can be no less seductive to the ardent positivist.

THE REORGANIZATION

The organization of experience is indispensable for the survival of sentient creatures, but for mankind it is not enough. As has been noted earlier, human beings are almost constantly active during their waking hours. No small part of that activity is directed toward changing personal environments, and especially its humanly created artifacts. It appears that they must be shaped and reshaped, for one man's success is another's provocation, if only to gratify the human compulsion to manipulate things.

In this section we shall be concerned only with those reorganizations that attract attention because they exceed the range of routine expectations. Some strike their beholders as fatuous, too different to invite acceptance, but should be mentioned nonetheless. Others offer relief, advantage, or diversion. They establish a new frontier for organization by those who welcome and incorporate them into their lifestyles. These new experience forms include inventions, discoveries, novelties, fictions, and adaptations. The number of such salient reorganizations differ markedly in the time and place of their occurrence, but it is obvious that they pervade the course of human history. Only a small number of those for which recorded information exists will be cited in the overview that follows.

It is the thesis of this presentation that there is more unity in the seeming diversity of experience reorganization than has yet been realized, and that it can be characterized more significantly in its own terms than by resort to an analogy with a mechanical or a biological system. Obviously this unity must reside in the identity of the *process* by which the reorganization of elements takes place and not in its multiform manifestations. The analysis of a large and unselected sample of cases has led to the conclusion that this process can be described in the following way: it is that mental action by which a detached element of one pre-existing configuration (antecedent) replaces a partially equivalent detached element of another pre-existing configuration (antecedent) to form a third configuration with some properties in common with its antecedents but with others that are peculiar to itself.

An unfamiliar perspective is required in order to recognize the universality of this process. In the first place, it must be granted that a new combination is necessarily a recombination. That is to say, before it can occur there must be an analysis of two existing complexes so that a part of one can be united with a part of another. This cannot fail to be so, because nothing exists in complete isolation from everything else. The bonds between familiar things may be physically tenuous or entirely absent; but this is the least important consideration, since physical conjunctions are only one aspect of the interconnectedness which men conceive the universe of their experience to have. A thing, an event, an act, or a thought must be related to another thing, event, act, or thought; otherwise it can have no meaning. That to which it is related may be nothing more conspicuous than its background or setting, no more tangible than a thought; but there is some context from which it is disassociated preparatory to being reassociated.

It must further be granted that the analytic and associative acts which are essential to an invention are mental functions. They are imaginary or symbolic slicings of experience through the medium of ideas about it. It would not be necessary to emphasize this point were it not that in speaking of these processes as they relate to objects it is easy to evoke the image of a mechanical operation. They do not belong to that order of phenomena. Recognizing a rat is an analytic process, for it requires that the animal be distinguished from its environment; but it does not entail any physical manipulation of those referents. Thinking of a rat in place of a lion on an heraldic shield as a symbol of courage and nobility is an associative process, and it likewise has nothing to do with the physical relation existing between the two animals or a shield. This undistinguished invention illustrates the fact that analysis can occur instantaneously; and while it is logically prior to a recombination of the analyzed parts the two can be practically simultaneous. It also illustrates the important point that analysis can take place in the very act of perception without the awareness of the perceiver. As Bartlett says, "Although perceiving is rarely analytical or piecemeal in its method, yet it *is* a kind of analysis, since always there are some features of the perceptual situation which take the lead over others. These dominant details are a kind of nucleus about which the rest cluster" (Bartlett 1950:32).

The inventive detachment and recombination of elements is therefore a psychological process and its rules are those of mental and not physical action systems, regardless of how they must conform to the latter if and when they are implemented in material form. The difference between these two realms of action is a matter of everyday observation, and every inventor faces it when he attempts to build a workable model of his mental recombination. Some ideational combinations are physically impossible, others must be modified radically before they can be materialized. Contrariwise, there are innumerable displacements and combinations of elements in the natural world that have no mental counterparts; which is to say that they are not inventions. There are certainly instances when they have been conceptualized for the first time and have had intellectual or practical importance attached to them. Then we have what is commonly called a discovery. But again the newness lies in a mental reorganization of conceptual elements such that one replaces another in a habitually established set. This step is essential whether it precedes or follows the play of physical forces (Barnett 1953). Nature does not invent and therefore its mechanics should not be confused with the process of creative thinking.

It is necessary to abandon a mechanical view of invention for yet another reason; namely, that it tends to picture the recombination process as a severing and joining of things at their "natural" boundaries. This is a most misleading conception, for it is precisely in their disregard of conventional units that inventors differ most from their less imaginative contemporaries. They see complexes where others see wholes, or vice versa. Some of them, of course, actually do break things up and reassemble the pieces. Others do not; they do no more than imagine how it might be done. More importantly, they envisage partitionings that overlap and intersect object boundaries. They jigsaw their cognitive field in patterns which ignore conventionally established outlines within it. The results are primarily and fundamentally new ideas and not new things. Consequently, they take shape in accordance with psychological principles which have little or no relation to the material properties of objects or to the way in which they behave under the laws of physics and chemistry. Percepts and concepts of experience can, and frequently do, ignore completely the definition of things established by "common sense" or by the conventionalized procedures of "objective" science — and thus often lead to reinterpretation of the nature of things.

It is easy in retrospect to wonder at the stubbornness or blindness of men that has prevented them from seeing the world structured in a way other than that dictated by custom, and to note how a failure to take a thing out of a familiar context and put it in another has forestalled change. So firmly do things become bonded together by tradition that it requires uncommon insight or rashness to perceive them as complexes composed of detachable parts that remain together only because they have "always" been that way. Sails were the exclusive property of ships — until a forgotten minor genius hitched one to ice skates, a parsimonious board of directors of the South Caroline Railway Company grafted one to a railcar in 1829, and a company in California rigged one to a squat tricycle to initiate the sport of "sand sailing." Kites were for boys — and for eccentric gentlemen like Benjamin Franklin, who used one to capture electricity; and George Pocock, who obtained a patent for attaching one to a carriage in 1826; and an anonymous Solomon Islander, who tied a lure to the tail of one and succeeded in catching fish with it. For centuries horses were harnessed to lordly war chariots, not ridden and not put to the bovine task of drawing carts and plows. They were associated with aristocrats, and therefore with luxury and pomp. They continued to be linked with adventure and sport until relatively recent times. And so it was with bronze in ancient history. It was a costly metal and therefore the exclusive possession of the rich and powerful. It was so firmly entrenched with ritual, heroism, and authority that iron, when it became known in

Europe and the Near East, was rejected as a lowly instrument of peasants, ruffians, and barbarians (Kroeber 1948).

When two previously undifferentiated units are analyzed preparatory to their recombination, the parts of each are apprehended as being held together by some specific relationship. These relationships give distinctness to the parts, and they are crucial factors in the recombination process, the reason being that when one part replaces another it assumes the relationship that the replaced part had to its matrix, context, or associate — that is, the remainder of the analyzed unit — regardless of the relationship the replacing part had to its matrix, context, or associate. The significance of this characteristic of the recombination process must be emphasized because frequently it is precisely the assumption of a new relationship by one part with respect to another that constitutes the most striking feature of an invention. It makes a difference, for example, whether a passenger car rolls on top of rails or is suspended from them; whether air is compressed by or compresses a piston in a cylinder; whether a waterfall causes a wheel to turn or a turning wheel makes a waterlift; whether parents treat their children as equals or as dependents.

It is evident that when an inventive recombination occurs the elements of the new combination directly or indirectly become substitutes for the elements of the original combinations which they have displaced by their union. If AB is one pre-existent analyzed combination and XY the other, one invention made possible by their recombination would be XB. In that case X displaces and therefore substitutes for A and at the same time B displaces and substitutes for Y. But it makes a difference whether X or B is the *active* substituting element because of the relationship factor. If X is the active displacer then it joins B in the same relationship as did the displaced A. If, on the other hand, B actively displaces Y, it assumes Y's relationship to X, and that can result in a totally different XB. There are therefore two possibilities for the formation of the XB combination. The same holds true for the combination of A and Y. In addition to these four possibilities there are just two others. One occurs when both X and Y replace and substitute for A and B with the assumption of the relationship between A and B; the other when both A and B replace and substitute for X and Y with the assumption of the relationship between X and Y.

These six patterns exhaust the possibilities of the XYAB paradigm and it is believed that all inventions, as well as many creations that are not generally considered to be such, conform to one or another of them. The maximum of six is established by the relationship factor acting in conjunction with one other control. This is the inventor's conception of equivalence or similarity, for substitution takes place in accordance with his apprehension of a common factor in a pair of parts of existing combinations. In other words, recombinations are not made in random fashion; they result from the substitution of one thing for something like it. X is substituted for A because, as conceived by the inventor, it has something in common with A. Hence, it is natural that there should be a substitution. Similarly for Y and B.[2] Their common properties may be physical, or functional, or abstract. They may be generally accepted as such; or, as sometimes happens, the perception of them may be attributed to the insight of a genius; or, as more often happens, they may be regarded as fantasies of amusing, absurd, frivolous, or even dangerous people. Frequently, equations are made below the threshold of what is considered to be novel simply because the identity of X's and A's or of Y's and B's is taken for granted, as when one wheel is substituted for another wheel, their admitted differences being ignored by common agreement. This is precisely the juncture at which a division is often made between "significant" and "insignificant" inventions, and the reason why, as Kaempffert says, "many inventions seem almost disappointingly obvious when they are first disclosed" (Kaempffert 1930:17). It is also why there is a lack of agreement on what constitutes an invention. The process itself is a routine, commonplace event. All of us resort to it daily. If we do not have a hammer available when we need one, we use anything like a hammer that we have at hand; it may be our fist, a stick, a stone, or a shoe heel. Our impression of newness in what we call an invention is a function of the degree of similarity that we accord the substitute with reference to that for which it substitutes. And it is obvious that the measure of similarity is individually and culturally variable. Sometimes an inventor's conception of a resemblance and therefore of the possibility of a substitution is so "far-fetched" that he is regarded as a crackpot; in other instances his equation is so obvious that neither he nor anyone else thinks of him as an inventor.

In those instances on which we have relatively complete information, all the characteristics of the inventive process that have been described are clearly present. A good example comes from the fusions between automotive and manpowered vehicles during the 1890's. The earliest safety bicycles were too cumbersome to have wide appeal, so their manufacturers set about reducing their weight and the friction of their moving parts in all ways possible. They introduced ball bearings, tubular frames, wire spokes, pneumatic tires, and thin wheel rims. But, by the end of the century the bicycle began to lose its appeal so its designers turned their attention to its competitors, the steam, electric, and gasoline driven carriages. Into these vehicles they built their familiar tubular frames, chains, sprockets, and ball bearings, sometimes with little or no modification. Thus, the first Locomobiles (formerly "Stanley Steamers") offered to the public in Washington, D. C. in 1899 were equipped with bicycle wheels (*Life* 1952). If we apply the symbolism adopted above to this

[2] Equating X with B or A with Y would only rotate the model.

example, X would be a horse drawn carriage body, with its heavy wooden wheels Y; A would be a bicycle frame, and B its wheels. XB would then be the invention with respect to this particular feature of the Locomobile; Y the abandoned wagon wheels (which Daimler and Selden retained in their first gasoline wagons); and A the bypassed bicycle frame that was eliminated from this strain of powered vehicles. The wheel substitution and other evidences of the crossing of bicycles and buggies appeared in such automobiles as the Benz of 1885, the Franklin of 1903, the Haynes of 1895, and Henry Ford's first, as pictures of these vehicles clearly show (Kaempffert 1924). Meanwhile, marriages were being contrived between gasoline powered vehicles Y and bicycles B that bore issue in some form of "motorcycle" XB through the substitution of the motor X of the automobile for the pedal and foot power A of the bicycle. The details of this new breed of vehicles show unmistakable signs of their mixed parentage. There was, for example, the Hitchcock Quadricycle which carried a car seat for two mounted on the frames of two bicycles and was driven by a two-cylinder "naphtha" engine (Cochrane 1896:140).

Speaking in more general terms, horses X, taken from other tasks Y, have replaced men A in turning millstones, presses, waterlifts, and treadmills B. On treadmills the X's have replaced other sources of power A to drive mine pumps, railway cars, boats, and — if a Frenchman in 1825 had had his way — an airship B. Waterwheels X have relieved both men and beasts A in the operation of millstones, bellows, and machinery B; and in turn have given way to steam and electric power (new X's). At the end of the last century, compressed air X, acting on a piston B, substituted for steam A in a variety of devices ranging from railway engines to rock drills and it became a strong competitor with electricity. Then at about that time a temperamental upstart, the gasoline engine X, moved in to power the universe of rolling, floating, and flying objects B in place of steam, electricity, and air as well as the labor of men and beasts A.

There is a further involvement that must be taken into account to appreciate the generality of the proposed pattern of analysis and recombination. This is the fact that parts and functions of living organisms, including man, are often key elements in the XYAB paradigm. This has been implicit in much of what has been said about the substitution of machines for men and animals, but there are other aspects of the process that deserve particular attention. A very considerable part of our technology has come into being through human efforts to protect, strengthen, or maintain the physical or mental potentialities with which he is equipped as a member of the human species. Paleolithic humans had a very limited stock of artificial devices. In the main, they used their own bodies to contend with their environment, but they did employ sticks and stones as substitutes for their fists and teeth in battering and crushing their way to survival; and they used fire and caves as substitutes for an inadequate body covering. There have been thousands of inventive steps in between those elementary substitutions and the complicated machines that we use today to master our enemies and our environment; but there has been an unbroken continuity, one step building on others that preceded it. Many modern inventions have as their purpose the restoration of body parts or functions that have for some reason failed us. In this category would fall our prosthetic devices: artificial limbs, eyes, teeth, and hair, as well as our eyeglasses and hearing aids. Then there are the prophylactics and curatives which guard us from death and disability by substituting for deficiencies in our organic system — stimulants, vitamins, enzymes, and tranquilizers.

Some organic substitutes are made to look like the part they are to replace, but this is by no means necessary, and often a better substitute can be contrived if formal resemblances are ignored. After many centuries men who had visions of flying finally abandoned attempts to imitate the flapping of bird's wings and made their airfoils rigid. John Fitch's model of a ship driven by a steam engine imitating the rowing action of galley slaves was somewhat more realistic, but not as successful as the paddle wheels adopted by other inventors of his day (Kaempffert 1924:79). All that is required of a substitute is that it have something in common with that which it replaces; and this need be no more than the capacity to produce the same ultimate result. Adding machines and typewriters bear little resemblance to our brains or limbs, but do accomplish some of the same purposes.

Many body substitutes strike the public as ridiculous or impious tinkerings with nature. Among the more innocent invaders of the human domain in the last forty years which have been issued U. S. patents are an automatically released ammonia spray to keep the drowsing motorist awake; kits of snore stoppers for heavy sleepers; an automatic wiper for eyeglasses and for vanity case mirrors; ice cubes weighted to keep their chill away from the lips of highball drinkers; an electrically heated toilet seat; and a headband with a prong in front to hold the cigar of a preoccupied card player (*Time* 1940). All of these manifestations of progress have drawn their elements from other contexts and are designed to substitute for some attribute or function of the human organism through an equation of their effects.

As might be expected at this point, no significant distinction can be drawn between inventions which involve people and things and those which involve people only. Departures from custom with respect to interpersonal relations, communications, eating habits, and other traditional patterns of behavior originate with the equation and substitution of an analyzed part of one habitual unit of behavior with the analyzed part of another, just as with things. It is not usual to speak of novel — and sometimes shocking — behaviors as inventions; but inasmuch as the process by which they are conceived is the same as for new combinations of things, it is convenient and appropriate to have a term which will comprehend all instances. Innovation is a suitably neutral term and it will be employed throughout the remainder of this paper when no distinction is intended between the conception of new things, acts, beliefs, or ideas.

Coded communications of all sorts clearly follow the pattern of equation and substitution. In these instances, as when a number "stands for" a letter of the alphabet, the equation is arbitrary; that is, it is made such by definition. Similarly with gesture languages, shorthand transcriptions, and the many forms of gibberish such as pig Latin. Slang and teenage expressions are innovations which are more directly comparable with the "who" for "whom" substitution in that the replacement of a customary expression by a new one is an individual departure which may or may not become popular.

Strange as it may seem, imitation is innovation, for when it happens one person identifies himself X with another person A and instead of doing what he is accustomed to do Y he does what his paragon B does. The reason this seems to be anything but innovative is that in many instances no one questions the equation of the one person with the other. But there are numerous occasions on which the equation and its consequences are considered to be scandalous, presumptuous, or ridiculous. In this category would fall children behaving like adults and vice versa; natives incongruously garbed in alien finery; social climbers and passers of class boundaries — not to mention outright imposters.

Novel behaviors may originate not only through person-for-person substitution but through the equation and substitution of situations traditionally considered to be different. Not long ago the funeral of a famous band leader was attended by a display of jubilation that most Americans feel should be reserved for a joyous occasion. Oddly enough from their standpoint, that was the sentiment of the mourners too. It *was* a joyous occasion. A more consequential instance of situational equation occurred in the 1920's when a University of Illinois football player decided to become a wrestler. He resorted to the flying tackle in the ring, as he was accustomed to do on the football field, and so initiated the lunging strategies of hippodrome wrestlers today (Meyers 1931).

The categorical syllogism is a form of reasoning based on two premises which contain three and only three terms between them. It is a means of inference by which a relationship between two of the terms is deduced through the elimination of the third term which is common to both premises: all spiders are poisonous. This insect is a spider. Therefore this insect is poisonous. Translated into the symbols of the innovative formula this can be stated as A is B. X, whatever else (Y) it may be, is A. Therefore X is B. This is the form that many arguments take, and it is the basis for many beliefs and theories and the experiments that are sometimes undertaken to prove them. The members of the Indian Shaker Cult assert that medicine heals sickness; their rituals are medicine; therefore their rituals heal the sick. In the middle ages it was believed that a lodestone could attract an estranged wife to her husband as it did bits of metal. It is not easy to fathom the basis for the equation of iron filings and alienated wives, but the same may be said of many novel ideas. Possibly some of Benjamin Franklin's contemporaries felt the same about kites and church spires. At any rate he found enough similarity in them to test his premise that lightning was an electrical spark leaping to earth through an elevated conductor. Most anthropologists believe that the ancestors of the American Indians must have arrived in the New World by way of Bering Sea. They reason that a primitive people without sea-going vessels and a knowledge of navigation would not have been able to cross the Atlantic or Pacific Oceans. The first immigrants to America were such a people; therefore they could not have reached America by traveling across the open sea.

Inventors with a serious purpose have often labored to produce something that their contemporaries regard as a ludicrous contraption. It is not surprising, then, that other innovators have adopted the same formula to create something they think is funny and hope that the public agrees with them. Hence, we have inventors of chindriven fans for gum-chewing secretaries and crank-operated boxing gloves to be used for pounding the bottom of catsup bottles — again the familiar pattern of substituting an X for an A. Slapstick comedians, punsters, and cartoonists regularly resort to this formula, even though they may not be aware of it as such. Their appeal lies in the incongruity of their XB associations, which means the oddity of equating X with A. Whether this be considered an infantile form of humor or not, examples of it are not lacking in our literary classics. Witness Mrs. Malaprop's prescription for the education of a young lady:

> Observe me, Sir Anthony, I would by no means wish a daughter of mine to be a progeny of learning. I don't think so much learning becomes a young woman. For instance, I would never let her meddle with Greek, or Hebrew, or Algebra, or Simony, or Fluxions, or Paradoxes, or other such inflammatory branches of learning — neither would it be necessary for her to handle any of your mathematical, astronomical, diabolical instruments. But, Sir Anthony, I would send her, at nine years old, to a boarding school in order to learn a little ingenuity and artifice. Then, Sir, she should have a supercilious knowledge in accounts — and as she grew up, I would have her instructed in geometry, that she might know something of the contagious countries . . . (Stevens 1923:435).

So far an attempt has been made to select illustrations of innovations requiring only one substitution. In many instances, however, innovations are made by equating and substituting some Y for B as well as X for A. The particular Y that is substituted may not have been previously associated with X. In any event, by replacing B it comes to be joined to X in the same relationship that originally linked A and B. This double substitution produces what may be called an analogy. Theoretically this pattern of innovation is at least as common as is the one involving a single substitution and in practice it may be more so. It certainly predominates among mechanical inventions, the reason being that in order to combine X with B it is usually necessary to alter B so that it more nearly approximates the physical properties of Y before the new combination is workable.

Analogies are innovations not only because Y has never before been related to X as it is in consequence of the double substitution, but also because neither X and A nor Y and B are ever identical, only alike in some respect. The overall result is that the form of AB is preserved but the content or substance may be markedly different. Innovations of this kind are made throughout the gamut of things, theories, beliefs, and behaviors. In the history of science, Lenoir's gas engine was analogous, even in its details, with a steam engine. It was just different enough (as Y is to B) to run on gas X in place of steam A. Westinghouse's railway car brake was a miniature piston and cylinder Y like that of a steam engine B, but activated by compressed air X instead of steam A (Kaempffert 1924:357). Cayley's glider was modeled on a bird's body with inorganic wings X and a tail Y instead of the bones and muscles of their counterparts (A and B) (Kaempffert 1924:178). The wings and tail of this device were analogous because they represent a retention of form along with a replacement of substance. This is an extensively exploited formula for innovation. It accounts for the counterfeits, fakes, ersatzes, and imitations that, whether contrived for deception or not, look like something they are not because they retain the interpart relations but not the materials of their prototype AB. Also conforming to it are a varied assortment of translations and transpositions, as when a musical score is "arranged" by the introduction of novel embellishments on its X's and Y's while retaining the relationships (melody, rhythm, tempo) between the original A's and B's. Similarly for the substitution of content (sense data) in communication systems, as when a message is transmitted in Morse Code, not by the tapping of a telegraph key, but by light flashes or tape perforations.

The origin of the commission form of government provides an example of an analogy in the field of institutional innovations. In 1900 Galveston, Texas, was struck by a hurricane and tidal wave which so severely disrupted facilities that the bicameral city government was unable to discharge its responsibilities. During the crisis a committee of businessmen undertook to find a solution to the administration problem. They proposed that a commission of five persons (X) be appointed to govern the city (Y) as nearly as possible as a board of directors (A) manages the business of a corporation (B). It is reported that they "did not at the time realize that they were inventing a new form of political structure" (Chapin 1928:338). This is a significant observation, for it means that these men, like thousands of other innovators before and after them, were unaware of their role simply because for them X was A and Y was B. It is also significant that these innovators were businessmen, for that fact throws light on their choice of a prototype AB. In other analogies the source and nature of AB is more apparent: In 1949 a Chicago attorney founded Divorcees Anonymous; in 1923 Ohrbach's clothing store in New York instituted a "cash and carry" plan; and in 1951 the Oklahoma City Police Department established a charge account system for a select list of traffic law violaters.

Parodies, burlesques, and ballets are analogies. So are the numerous items in the proliferating business of pet care whereby owners — or alert entrepreneurs — identify dogs, cats, and other creatures X with their masters A to pave the way for pet hospitals, hearing aids, tranquilizers, crutches, eyeglasses, tooth brushes, motels, powder rooms, cocktail bars, and horoscopes — all Y's that are not quite the same as their B counterparts for human use. From the area of deliberate humor comes such analogical creations as a cartoon depicting a cave man spanking his child for "defacing" a wall of their home with a drawing of a woolly mammoth that is one of a kind now highly prized as examples of prehistoric art.

If the differences rather than the similarities between X and A and between Y and B are emphasized in double substitution the result is what may be called a parallel instead of an analogy. In these innovations it is still the similarities between X and A and between Y and B that bring them into conjunction, but at the same time their differences set them apart. Their novelty lies in the fact that their XY components are different from AB, and yet they are linked by a relationship that resides in the AB prototype. XY may already exist as a unit, in which case it is then paired off against AB as an alternative or as an insight that opens up possibilities for new developments paralleling those of AB: Infrared radiation X sensitizes lead sulfide and other chemicals Y as light waves A sensitize silver nitrate and other chemicals B, so suggesting the possibility of heat photography and infrared chemical analysis. Perhaps more often neither of the elements of a parallel exists and they must be constructed on the suggestion of their AB counterparts: The contests between married women X for the title of Mrs America Y have doubtless been modeled on the Miss America contests AB with the appropriate distinctions embodied in the new title. In science, particularly, XY parallels frequently emerge from a discrimination between subcategories of A and B: Whales A and fish X are different enough despite their resemblances to warrant their being placed in different classes B and Y; isotopes A and X are distinct enough to have radically different potentialities B and Y.

The balance between analogies and parallels is delicate, and what is an analogy to one person may be a parallel to another. At times this is intentional; XY is enough like AB to satisfy those who cannot have AB yet dissimilar enough to satisfy those who have some preclusive interest in it. Exclusive clubs and their imitations are good examples; so are the numerous dichotomous pleasures of the rich and the poor, the young and the old, the fit and the unfit. A practically universal category of parallels is the one founded on differential sex interests and prerogatives, among them entertainments, occupations, dress, and speech. Toy designers and ingenious parents exploit the potentials of parallel innovation by manufacturing diversions for the young that are enough like forms to satisfy youngsters but different enough to keep them out of mother's kitchen and father's tool box. Humorists have also discovered that the scientifically respectable device of discriminating subcategories of A and B can have a risible effect: Junior, when cautioned that he would have to improve his table manners if he were to be permitted to spend the weekend with his friend, scornfully replied, "You don't think I eat like this with *people* do you?"

If it is granted that the patterns of simple and double substitution prompted by conceptual similarities adequately characterize and comprehend the range of phenomena here called innovation, we are led to consider several implications of this formulation of the innovative process. They can be discussed with reference to their multiplicative, valued, and adaptive aspects.

One of the most obvious characteristics of social, cultural, and technological change is its quantitative variation. The change is of two kinds: variability in the number of innovations at different times, and variability in the number of people who accept an innovation over a given period. The reasons for these differentials are beyond the scope of the present inquiry; but given the fact that an innovation has been made we can directly relate the process by which it came into existence to some of the ordinary features and consequences of its acceptance. More precisely, whatever the conditions and the motivations for them may be, the process of innovation and the extensions of its use are the same. Hence trends, cycles, and pattern duplications are quantitative manifestations of the innovative process.

In a trend, either the number of people (X) accepting an innovation (B) increases or decreases; or the number of B's with which X is associated diminishes or increases; or the number of X's with which B is associated increases or decreases. Often these alternatives are simply different ways of saying the same thing. In any event, they all involve numerical variations in the equation and substitution of persons and things. If it can be said, for example, that there is at present a trend toward motorization, this means that gasoline or other engines B are being fitted to many X's, ranging from launches to lawnmowers (with perhaps baby carriages and wheelbarrows yet to come), all of which have something in common with, and can therefore be substituted for, older engine vehicles A. Sturtevant's invention of a suction fan B to remove dust from shoe factories A started a trend toward dust removal from homes and other places. And not only do such fans B now suck dust A but anything like dust, such as bulk grain X from boxcars. While it may seem inappropriate to speak of a trend in the latter instance, the impropriety derives from considerations that are irrelevant from the standpoint of the process by which the extension of vacuum power has been accomplished. Similarly with fads, those short-lived waves of mass identifications of the self (X) with another person (A) which result in sudden rises and declines in the adoption rates of A's distinctive traits B. Fads are trends from the standpoint of process even though they may be different enough in other respects to deserve separate terms. Oddly enough, obsolescence is another quantitative manifestation of the innovative process. It is simply that aspect of it that occasions the displacement of old items (A) in favor of new ones (X). Thus, after 1880, as one person after another (X) equated himself with a user (A) of electric lights (B) — or as they equated electric lights (B) with gas lights (Y) — there was a progressive obsolescence of gas illumination. Another numerical change occurs when an equation of X with A results in an extension of their identity through X's assumption of not one but many of the B's previously associated with A. From the equation of women (X) with men (A) there have emerged over the past sixty years in the United States, female lawyers, railroaders, bartenders, bank tellers, and business executives — all formerly male prerogatives (B) — and the end is not yet in sight.

As in innovation itself, trends and pattern multiplications may be viewed as either simple or compound substitutions, depending upon our estimate of what constitutes similarity and difference between X and A and between Y and B. If we think that a Standard Oil credit card is no different from a Diners Club credit card, or that Divorcees Anonymous is no different from Alcoholics Anonymous, and that Smokers Anonymous is "just another copy" of the same idea, then we declare them to be instances of simple substitution. But if we insist on their differences, as their innovators and adherents are likely to do, they are parallels.

The innovative process is an adaptive mechanism. It is the means by which human beings adjust to changes of situation, whether the changes are initiated by events within themselves or external to them. It yields artifacts and artifices; that is, man-made things and behaviors. But there are good reasons for maintaining that it is not restricted to these manifestations, and that it is not an exclusively human function.

One of the primary conditions for the survival of any organism is its ability to generalize upon its experiences; that is, to treat two or more things as the same instead of each one as unique. This is another way of saying that classification is essential to survival; and this means, in human terms, that a present experience (X) must be recognized, that is, equated to a previous one (A) before it can evoke the response — any response — B which assigns it to the same category as A. Often we must infer that this equation-substitution process takes place on the assumption that equivalent stimuli evoke similar responses in order to provide a rationale for the repetition of responses. If we accept the inference we are led to the further conclusion that animals as well as humans classify their experiences; and, presumptively at least, by the same process. Furthermore, there is no point in the scale of organic life where we can draw a line which delimits this reaction to diversity in experience. Whether in all instances it is a cognitive process is arguable, but so is the nature of cognition itself. In any event, the process seems to manifest itself in an unbroken continuum, ranging from the reflexive reactions of the most elementary forms of life, through the limited but undeniably inventive adaptations of apes, to the prodigious array of artificial devices created by man.

All of this puts a new complexion on what are ordinarily called inventions. In this broader perspective they are those manifestations of the innovative process for which praise is reserved. They are, moreover, subjectively selected aspects, for it is notorious that one man's panacea may be an absurdity to another. Hence, the difficulty of framing generally satisfactory definitions of invention. Perhaps an even more disturbing consequence of this perspective is that it humanizes inventive genius by robbing it of some of its glamor. For the truth is that every person is an inventor many times over. The problem is not to invent something, but to invent something that someone wants.

REFERENCES

BARNETT, H. G.
1953 *Innovation: the basis of cultural change.* New York: McGraw Hill.

BARTLETT, F. C.
1950 *Remembering: a study in experimental and social psychology.* New York: Cambridge University Press.

BRUNER, JEROME S., CECILE C. GOODMAN
1947 Value and need as organization factors in perception. *Journal of Abnormal and Social Psychology* 42: 33–44.

CANTRIL, HADLEY
1950 *The why of man's experience.* New York: Macmillan.

CHAPIN, F. STUART
1928 *Cultural change.* New York: The Century Co.

COCHRANE, CHARLES HENRY
1896 *The wonders of modern mechanism.* New York: J. B. Lippincott.

HAYDU, GEORGE G.
1972 Cerebral organization and the integration of experience. *Annals of the New York Academy of Sciences* 193:217.

JAMES, WILLIAM
1890 *Principles of psychology*. New York: Henry Holt.
KAEMPFFERT, WALDEMAR
1924 *A popular history of American invention*. New York: Charles Scribner's Sons.
1930 *Invention and society*. Chicago: American Library Association.
KILPATRICK, FRANKLIN P.
1961 *Explorations in transactional psychology*. New York: New York University Press.
KROEBER, A. L.
1948 *Anthropology*. New York: Harcourt, Brace and Co.
Life
1952 Article in *Life*, Sept. 8, 1952.
MEYERS, JOHN C.
1931 *Wrestling from antiquity to date*. St. Louis: Van Hoffman Press.
PECKHAM, MORSE
1965 *Man's rage for chaos*. Philadelphia: Chilton Books.
SPEARMAN, C.
1923 *The nature of "intelligence" and the principles of cognition*. London: Macmillan.
STEVENS, DAVID H.
1923 *Types of English drama, 1660–1780*. Ginn and Co.
Time
1940 Article in *Time*, July 29, 1940.

Individual and Environment: An Exploration of Two Social Networks

JEREMY BOISSEVAIN

This article grew out of my interest in the interrelation between people and their environment, between individuals, on the one hand, and social structure and culture, on the other. It seemed to me from my experience both in the university and in various research situations that people influenced their sociocultural environment, and, in turn, were influenced by it. Yet my generation of social anthropologists cut their theoretical teeth on formulations of Radcliffe-Brown (1952) and Evans-Pritchard (1940). The theoretical paradigms of these giants still dominated British social anthropology in the late 1950's, when I received my first training in anthropology. They focused on ideal systems, on the way people were supposed to behave. Their works, thus, were rule books about the generalized average man. Real people facing real problems, often having to choose between several conflicting norms were conspicuous by their absence. Moral systems impinged upon individuals, influencing and molding their character and behavior. The influence of people on such systems, on their cultural and social environment, was ignored. Nor did they display any real interest in the way such systems of rules evolved in the past and are changing in the present or in the reasons for such changes. But people are influenced by other people, as well as by moral norms. Such norms do not evolve in a vacuum. They are formulated and used by real people to cope with specific problems.

This article seeks to examine the social constraints that influence personality and behavior. It focuses, thus, on the personal networks of social relations of which people form a part. A personal network, sometimes also called a primary or first order zone, consists of all the persons to which a given person (ego) can trace a social relationship, has personally

This article is a revised version of Boissevain (1973).

met, and the interconnections between these persons (Barnes 1969: 59). I have compared the personal networks of two school teachers, one from the town, and the other from the country. My hypothesis was that the social microenvironment of the network of social relations of which a person forms a part has a strong influence on his behavior and personality and that the structure of and relations within this microenvironment, in turn, were strongly influenced by the social, geographical, and cultural macroenvironment of which it forms a part. In a sense, thus, this study also continues Bott's pioneering effort to relate variations in network structure to environment (1957: Chap. 4).

Originally I began this study on a pilot basis with two informants.[1] I planned to branch out and test findings more systematically on a wider sample. But collecting this data proved to be very difficult and very time consuming, as did its analysis. Hence, I have data on only two first-order zones. Rather than wait for years until I have the time and funds to gather data on a broader basis, I have chosen to present some of it so that its relevance may be judged and some of the problems associated with it pondered before further attempts to collect similar material are made.[2]

[1] Research on which this study was based was carried out in Malta in the summers of 1967, 1968, and 1969. It was made possible through the generous assistance of the Wenner-Gren Foundation for Anthropological Research (1967), The Netherlands Association for the Advancement of Pure Scientific Research, ZWO (1968) and The University of Michigan Mediterranean Network project directed by Professor Eric Wolf in 1969. Francis Mizzi and Anthony Cuschieri sacrificed most of the summer 1968 to the mind-deadening task of organizing and collecting the network data. Rudo Niemeijer devoted much time to processing the data, including the preparation and testing of the computer program, and Jolanthe van Opzeeland-De Tempe helped work out some of the tables. I am grateful to them and to the University of Amsterdam which made their services and those of the staff and computer of the University's mathematical center available. Thanks are also due to the Royal University of Malta for the encouragement and facilities it provided during my brief summer visits to Malta. Finally I am grateful to the various discussants of earlier drafts of this paper some of whose criticisms I have tried to meet.

[2] At this stage it would seem advisable to say a few words about how the research was carried out. I selected Malta because I had already considerable knowledge of the island and the social customs of the people (Boissevain 1965, 1969). As informants I chose two school teachers. Pietru is a village teacher, married to a city girl, with whom I had worked before. Cecil, the second informant, is a city teacher, married to a girl from a rural town. I met him the year previous to carrying out the research. Each made out lists of all the people over fourteen he knew. As each knew many hundreds of persons, this task involved considerable work. After all acquaintances had been listed, each informant filled out an information sheet on every person known. These provided data on the social background, number of shared role relations, frequency of contact, last contact, the actual contents of the relationship, and the other acquaintances held in common. Once the data sheets for each of the informant's acquaintances had been completed, I asked Cecil and Pietru to sort the forms according to the affective or emotional importance of their acquaintances to themselves. I then discussed the resulting categorization with them. I also gathered detailed biographical information from each informant and collected case histories on a variety of situations in which segments of the network had been mobilized. This data, which is continually expanding, was typed on some 200 six-by-eight inch cards. Finally, back in Holland, the data was coded and transferred to punched cards and simple cross tabulations were carried out for each pair of categories of data. A rather more complicated calculation of the density was computed for a number of relevant categories, such as zones of intimacy and activity fields.

The following study is really little more than a first attempt to map and compare two first-order zones and relate the features of these to the observed differences in the environment and aspects of the behavior and personality of the two informants. I try to answer the obvious questions about the personal networks of Cecil and Pietru, my two Maltese informants: What is their cultural and social background? What are the objective structural and interactional characteristics of their respective networks? What are their subjective characteristics? Is there a relation between objective and subjective characteristics? What is the influence of the environment on the structure and content of the networks? Finally, what is the relation between network structure and personality?

ENVIRONMENT: THE SOCIOCULTURAL BACKGROUND

Malta

The Maltese archipelago covers 122 square miles. Malta, the largest and southernmost island, is 17 miles long and 9 miles wide and covers an area of 95 square miles. There are just over 314,000 people on the islands, making the Maltese archipelago, with a population density of just under 2,600 per square mile, one of the most thickly populated countries in the world. Slightly over half of the population lives in the urban area which centers on Valletta and the Grand Harbour; the rest live in clearly separated villages and towns which range in size from 900 to 17,000 inhabitants. No village, however, is more than an hour's bus ride from Valletta, and Gozo is only 30 minutes by ferry from its larger sister island.

The Maltese see the family as the most important institution in their lives. The ideal household is composed of an elementary family living together in a separate house. Maltese have large families and reckon kin relationships equally through males and females. Each person is at the center of a wide network of consanguineal and affinal relatives. Though the kinship structure is bilateral, there is a noticeable bias in favor of the mother's/wife's relatives. Maltese families are mother-centered and the contacts with relatives beyond the nuclear family are by and large articulated through women. Kinship obligations extend out to first cousins for men but as far as second cousins for women.

Malta may be said to have a service economy. Until recently the Maltese lived off the income they derived from providing services as civil servants, clerks, soldiers, skilled fitters, semi- and unskilled laborers to the British colonial and military establishments which governed the islands for years. Although Malta has been independent since 1964, she still largely lives off the services she provides, now to the ever-increasing number of tourists visiting the islands in search of sun. The tourist boom and the expanding light manufacturing industry have created what amounts to a situation of full employment in the islands, and there is actually a shortage of labor in the construction and catering sections of the economy. Less than 10 percent of the islands' inhabitants are full-time agriculturists. Consequently most Maltese travel from their place of residence to their workplace, which is normally located somewhere in the expanding conurbation centered in Valletta.

As may be expected, there are considerable differences between town and country and, especially, between the residents of the smart, large seaside suburb of Sliema and the inhabitants of small relatively isolated villages such as Farrug. The differences are not only a matter of dialect, but also of behavior and interests. The Sliema residents regard the religious feasts, processions, and fireworks, so characteristic of village life, with abhorrence. The people in villages have a good deal of informal contact in streets and shops, but rarely visit with nonkinsmen in their homes. Sliema inhabitants, on the other hand, arrange formal visits and frequent large receptions to which nonkinsmen are invited. Villagers deliver personally the invitations to attend their more modest baptismal, betrothal and wedding ceremonies; Sliema people post printed invitations even to kinsmen who live close by. The inhabitants of Sliema consciously as well as unconsciously imitate the behavior and many of the attitudes of the English, who for so long ruled Malta's fortunes. Even the Maltese spoken in Sliema affects an English accent which contrasts sharply with the broad vowels and gutturals of the countryman. The latter, in turn, patterns his behavior on the rural elite, the doctors, lawyers, and wealthy merchants, residing in the villages and rural towns.

Most Maltese are fervent, practicing Roman Catholics, and there is traditionally a strong attachment to the Church. This has had an important influence on the moral code. It is especially noticeable in the field of kinship where the large families and strong bonds between members of the elementary family are supported by exhortations and sanctions of the Church. Nevertheless certain ideological divisions exist. These may be characterized as pro- and anti-establishment, for the Church and against the Church. In recent years this cleavage became centered on political parties: respectively the pro-Church Nationalist Party which formed the government from Independence until 1971, and the relatively anti-clerical socialist Malta Labour Party. Much the same cleavage has also been incorporated for a long time at the village level in rival moieties which compete over the festivities in honor of their respective patron saints. The establishment moiety is supported by the Church and includes most of the leading figures of the village and celebrates the official patron of the village. The anti-establishment moiety, composed chiefly of the less influential members of the parish, celebrates a local rival saint, often in contravention to the Church's regulations and interests. These ideological cleavages, at the national level or at the village level, at times actively engage most Maltese villagers. Such ideological commitments influence in no small way the relations between persons. Hence its interest.

Pietru

Pietru lives in Hal-Farrug, a village of just over 1,200. Three out of four of the inhabitants work outside the village, chiefly as skilled and unskilled industrial laborers. In Farrug there are no striking differences in the standard of living, as there are in larger villages and towns; there is nonetheless a difference in the circle of acquaintances of a teacher and an unskilled laborer. The village is divided into two ritual moieties which celebrate the feast of Saint Martin, the parish's patron saint, and St. Rocco, his local rival. These moieties in turn focus on rival band clubs. There is also a football club in the village and a number of Catholic lay organizations. The Labour Party is very strong in Hal-Farrug and is centered on the football club. Cleavage for and against Saint Martin and for and against the Labour Party cuts through the village at all levels. It flares into intense rivalry at the time of the annual feasts of the two saints and at election time every five years.

Pietru was born in 1933. Thus at the time of the first study he was about thirty-five years old. He had been married for three years and had one son. Pietru's father, a fairly well-off peasant cultivator, died when Pietru was four years old. His mother, though from the neighboring village of Qrendi, remained in Hal-Farrug and supported her three daughters and son by continuing to farm the fields she had been left. It was a difficult life and she soon sent one of her daughters to live with her unmarried sisters in Qrendi, where she lives to this day. At the beginning of the war she became ill and, unable to get much help from her husband's brother and sister, both married and by village standards very well off, she and the children moved in with her own family in Qrendi. They returned to Hal-Farrug and her bombed out house after the war.

Through a priest, a friend of the family, Pietru's mother was able to get him placed in a private school run by English monks. She paid what she could toward the cost of schooling from the meager farming proceeds.

Pietru looks upon this period as one of hard work, for every day after school he had to work long hours in the fields helping his mother and sisters. In 1953, then eighteen, he tried the highly competitive entrance examination for the Teachers' Training College. Apparently he was unsuccessful. I say "apparently" because the Rector of his school telephoned the Department of Education in his presence and heard that he had passed the examination and was to be admitted. But when the list of successful candidates was published, Pietru's name was not included.

Both he and the Rector assumed that he had been scratched from the list in favor of some better connected candidate from Sliema. The Rector then helped him get a part-time teaching job at an institution run by the same religious order. Two years later he was admitted to the training college. He successfully completed the two year course of studies in 1957.

He was fortunate and immediately found a job in the Government Elementary School in Farrug. He continued to help his mother in the fields after school. He soon became engaged to a fellow teacher, a Valletta girl, who began to occupy his spare time away from teaching and farming. After a difficult year and a half, during which his fiancée persuaded him to abandon his attempts to win a Government scholarship to study agricultural science abroad, but was unsuccessful in getting him to adopt the customs and dress of the city, he broke the engagement.

Freed from his unhappy courtship, Pietru plunged into the social life of his village. He became an extremely active fund raiser for his band club, standard bearer for the confraternity of the Blessed Sacrament, and the leader of the Catholic Action section for unmarried men. He also became the confidant of the new parish priest, supplying him with information about parish affairs and helping him with his numerous fund raising activities. During the political upheaval of 1961 and 1962, he became one of the Church's most vocal advocates in Farrug. At this time he also joined a national anti-Labour Party, Church-sponsored study group. His militancy and forceful manner alienated many of his former friends and acquaintances in this predominantly Labour village. I first met Pietru in 1960–1961 when we lived in the village. He was then in his most active period as a village organizer.

In 1962, Pietru, now 28, became engaged again, this time to a Sliema girl, who had taught for many years in the village school. Having been rather badly bruised during the three year intensive participation in village politics, he withdrew almost entirely from the village arena. He devoted his free time for the next three years to getting to know his fiancée and to building a house.

Pietru has been married since 1965. He is still as active as ever, but he has little time to devote to village politics. Though he helps the new parish priest in his fund-raising activities and takes part in the village processions as the standard bearer of the confraternity of the Blessed Sacrament, he no longer holds office in the band club and shows little interest in its activities. He still continues to help his mother in the fields and grows his own cash crops of potatoes, onions and tomatoes, which he works in the early morning before school begins and in the late afternoon. He also does most of the bookkeeping for the ever-expanding household bazaar run in his mother's house by an unmarried sister. During the last two years he has also been studying intensively to pass his "A" level examinations in Economics, Social Geography, and Maltese so that he may be qualified to teach in secondary schools and, possibly, to begin studying for a university degree. Four afternoons a week during the school year he and his wife run private classes in their home to prepare children from the village for the competitive government secondary school entrance examination. And last but not least, he devotes a considerable amount of time to his growing family and helps his wife, who has none of her own relations in the village, with many of the household tasks. He also does much of the shopping for the family. In Bott's terms, Pietru and his wife have a complementary — bordering on joint — conjugal role relation.

Owing to his retirement from the front line of the village politics, his continuing studies, and especially his excellent results in getting his school and fee-paying charges through the secondary school examinations, he has regained some of the goodwill of the village which he lost during the heat of the 1962 political battle. His standing in the villages is complemented enormously by the quiet dignity of his Sliema born wife who, as remarked, taught in the village for many years before her marriage.

Cecil

Cecil, in contrast to Pietru, is Sliema born, though he now lives in his wife's town of Birkirkara. Sliema is a rapidly expanding residential suburb of some 21,000 persons. It is divided into three parishes and merges imperceptibly with other large suburban parishes to form the northern half of the conurbation centering on Valetta. For the past forty years it has been regarded as the best residential area of the island and many from the upwardly mobile government, military, and business classes have moved there. It is predominantly a white collar residential area. Many English and other foreign residents live here. Nevertheless there is a sizeable population of industrial laborers. It is, thus, an area characterized by sharp class divisions.

Birkirkara, where Cecil now lives, is a town of some 17,000. Though it gives the impression of being part of the urban sprawl, it is in fact a separate town and is one of the island's oldest parishes, with intense parochial patriotism. It is a stratified community composed of old established noble and professional families, shopkeepers, artisans and businessmen, industrial laborers both skilled and unskilled, and a fairly sizeable population of farmers. It has two active band clubs and active political party clubs. It is the home base of the former Nationalist Prime Minister and his brother, the former Minister of Education. It has also a strongly organized labor faction which manages to return at least one member of parliament every election.

Cecil was born in 1940. He is the sixth child of a grocer and has eight siblings. He recalls that he was very strictly brought up and was not allowed to play with the children on the street. He thus had little contact with his neighbors. Some of this parental vigilance was temporarily relaxed when his family moved in summer to a seaside house outside the town. From the ages of six to twelve he was sent to a strict private elementary school. His father was very ambitious for his sons and obliged them to spend many hours closeted in their rooms studying. This regime was successful. One brother is a Franciscan priest, another a doctor and the third, who is also the only one besides Cecil to have married, is a university lecturer in England. His two older sisters have married; his two younger sisters are still unmarried.

At twelve he was sent to an exclusive, Jesuit-run secondary day-school. At seventeen he became a Jesuit novice and lived in the school as a boarder, and later transferred to the order's seminary. At twenty-one he was sent, at his own request, to the order's mission station in India. There he learned and taught Hindi as well as other subjects at primary and secondary level. He also obtained a degree in philosophy at the Jesuit university in Poona. From there he went to one of the order's secondary schools where he ended up as vice rector. In all Cecil spent five years in India before deciding to withdraw from the Jesuit order. At twenty-six he returned to Malta, where, after a period of searching, he found a teaching position in the government secondary school system.

When he returned to Malta, he lived first with his family in Sliema and that summer moved to the family's new seaside house at St. Paul's Bay. There he met the neighbors and discovered that one of them was an ex-Jesuit of about his own age. He soon became engaged to the latter's sister. The next summer, in August 1967, he married and moved to his wife's town of Birkirkara. He told me many times that he finds living in Birkirkara strange in comparison to life in Sliema. By his standards Birkirkara has much of a village about it. In the beginning he found it annoying that people from the neighborhood into which he and his wife had moved seemed to know a great deal about him. They also seemed to spend so much time talking.

In many respects Cecil is unsatisfied still. The considerable responsibility he held while a Jesuit, his education, and the general background of his family leave him with unfulfilled aspirations. For the past few years he has been trying to get a scholarship to study social psychology in England. He seems gradually to be establishing roots in his place of residence, a process helped by the birth of two children. He sees considerably more of his wife's relatives than his own.

Comparative Summary

A number of factors relevant to the study of the structure of the networks of Pietru and Cecil emerge from this very brief description of their social environment. To begin with, there is the difference between town and country. Pietru's network is anchored in the rural area whereas that of Cecil in the urban. These differences are of considerable significance.

Cecil comes from an environment in which class differences are pronounced, whereas Pietru comes from what may be called a one-class milieu in which there are differences in styles of life which do not seem to be permanent.

Pietru has climbed many rungs in the economic prestige ladder during his life; he has moved from farmer's son to teacher, married a Sliema teacher and brought her to live in the village. Cecil's upward movement has been much less pronounced.

There is also a striking difference in the range of activities in which the two informants engage. Pietru is not only a teacher who gives private lessons, he also farms, helps his sister run her shop, is a part-time student and takes part in the ritual life of his parish. Cecil's activities are largely restricted to his kinship and formal teaching roles. This difference in activity is of course partly a function of the difference between life in a small village and life in a large center, that is, of their residential and social niches in Maltese society.

There is also a difference in the geographical mobility of the two. Cecil has traveled widely, both within and outside Malta: between St. Paul's Bay in summer and Sliema in winter, and between Malta and several parts of India. Moreover, as already noted, he has now left his place of birth and moved to his wife's town.

These various factors have combined to make a difference in the personal prestige of Pietru and Cecil in relation to their immediate environment. Pietru is a relatively big fish in a very small pond, whereas Cecil is a smaller fish in a much larger pond. Pietru has high social visibility, whereas Cecil, partly because there are more teachers in his present environment, and also because he plays fewer social roles, has lower social visibility.

Pietru's immediate social environment is riven with ideological cleavages: the fiesta factions and political parties. These cleavages are not noticeable in Cecil's surroundings. Various factors have combined to influence Pietru to take an unequivocal stand and play a part in these conflicts.

Finally, the differences in personality and character are also significant. Pietru has a hearty outgoing character. He is used to personalizing social relations — in the sense of exchanging small talk and information, in short in getting to know people — with bus conductors, shopkeepers and other strangers met in the course of his expeditions beyond the limits of his village and district. Cecil, though friendly, is more reserved in his dealings with people than Pietru. He is not as garrulous and does not appear to seek to personalize relations with strangers.

STRUCTURAL, INTERACTIONAL, AND SUBJECTIVE ASPECTS OF THE TWO NETWORKS[3]

1. *Structural Characteristics*

SIZE. The size of the two networks is easily compared. Pietru's network includes 1,751 persons and Cecil's 638. These contacts are what we may call first-order contacts: people with whom Cecil and Pietru have or have had some contact. Both have included a few people of whom they know a great deal, but have not actually met. These are persons who have married close kinsmen abroad. They know that they will meet them as soon as they visit Malta. In the meantime they know a great deal about them through letters which are passed to them by family members. Although each has done his best to recall, list and discuss all the persons he knows, quite obviously some have been left out, for the human memory is fallible. Persons under fourteen have been excluded. In Pietru's case this involves at least 450 village children whom he knows personally. Except for the few unknown close affines mentioned, no second order contacts were included.

COMPOSITION. Who they know is more interesting. The networks can be compared as to sex, age, education, occupation, and residence in order to give an indication of the relative spread of contacts throughout a number of important social categories. It is also important to ask in what capacity, that is in which roles, Pietru and Cecil know the members of their networks. In other words, we must compare the relative importance in each network of various activity fields from which the members of the network have been recruited.

Sex. Pietru's network displays a more even spread than Cecil's: 42 percent of Pietru's network are women, while only 23 percent of Cecil's are. This proportion roughly holds true not only for the total network, but also for the many zones and categories into which it can be divided.

[3] In setting out the structural and interactional characteristics of the two primary zones I have been greatly aided by the discussions of Mitchell and Kapferer in Mitchell (1969). A more detailed discussion of the density, multiplexity, and degree of the personal networks of Pietru and Cecil, an indication of how they are used for problem solving, and a discussion of the forces that keep them constantly in flux are given elsewhere (Boissevain 1974: Chap. 5).

Age. Pietru's network also shows a slightly more even age distribution than does that of Cecil. Pietru knows slightly more young persons — fourteen to nineteen year olds — than Cecil. He also knows slightly more older people — seventy years and older — than Cecil. But except for the very old and very young, the distribution of the two networks according to age is similar.

Table 1. Age

Age	Pietru	Cecil	Malta[a]
14–19 yr.	9.4	4.5	15.7
20–29	20.7	24.3	21.3
30–39	28.3	31.5	16.9
40–49	15.7	16.6	15.3
50–59	11.0	13.0	12.5
60–67	7.4	8.3	11.9
70 +	2.7	0.6	7.4
unknown	4.8	1.1	–
Total %	100	99.9	100
Number	1751	638	220,447

[a] *Source: Annual Abstract of Statistics: 1968* (Malta: Central Office of Statistics, 1969), p. 14, Table 4. The figures given for the first age category are 15–19 years rather than the 14–19 years I used.

Education. The educational composition of the two networks differs considerably. More than half of Pietru's network has had either no schooling or primary schooling only, as compared to only 14 percent for Cecil. By contrast 65 percent of Cecil's network has had secondary or higher education (seminary, teacher training, university), compared to only 32 percent of Pietru's network.

Table 2. Education

Education	Pietru	Cecil	Malta[a]
Unknown	10.5	21.0	0.6
None	21.8	2.4	16.6
Primary	36.3	11.3	62.1
Secondary	16.4	32.2	18.6
Higher	15.1	33.1	2.1
Total %	99.1	100	100
Number	1744	637	215,316

[a] Based on unpublished material from the 1967 census kindly made available by the Central Office of Statistics. The figures for Malta are for the population over 5 years not attending school, which is compulsory from 5 to 14 years. Those for Pietru and Cecil show only persons over 14 years, some of whom may still be at school.

Occupation. Considering that Cecil's network is on the whole better educated than Pietru's, it is no surprise that religious and white collar workers form just over half Cecil's network, but make up only one quarter of Pietru's network. It is interesting to note, too, that though Cecil knows many more priests than Pietru (145 as compared to 90) the latter knows more in Malta than the former (83 compared to 57). Both, however, know about the same proportion of skilled workers. Unskilled workers, on the other hand, are more important in Pietru's network than in that of Cecil. Not surprisingly, while four out of every hundred persons Pietru knows are farmers, Cecil knows virtually none.

Table 3. Occupation

Occupation	Pietru	Cecil	Malta[a]
Religious	5.1	22.7	31.6 (Religious and Professional and white collar combined)
Professional and white collar	19.2	29.6	
Skilled service (police, broker, shop, etc.)	12.7	12.2	41.9 (Skilled and Unskilled service combined)
Unskilled service	28.2	21.5	
Skilled industrial	3.3	2.2	18.7 (Skilled and Unskilled industrial combined)
Unskilled industrial	9.6	0.5	
Agricultural	4.4	0.3	7.7
Other	3.3	1.4	–
Don't know	14.0	9.6	–
Total %	99.8	100	99.9
Number	1751	638	94,367

[a] Based on unpublished material from the 1967 census for the gainfully occupied population, kindly made available by the Central Office of Statistics.

Table 4. Place of residence

Place of residence	Pietru	Cecil	Malta[a]
Farrug (Pietru) or	48.3	–	0.4
Birkirkara (Cecil)	–	20.2	5.5
Villages around Farrug	13.2	0	5.5
Sliema	5.6	21.5	9.0
Urban Malta	13.3	15.0	35.5
Rest of Malta	9.8	10.2	44.1
Abroad	7.3	24.5	–
Don't known	2.6	0.6	–
Total %	100.1	100	100
Number	1751	638	314,216

[a] The figures show the total population and are based on unpublished material from the 1967 census kindly made available to me by the Central Office of Statistics.

Residence. Almost half of Pietru's network live around him in Hal-Farrug. In contrast, only one fifth of Cecil's network live in his place of residence, Birkirkara. Moreover, Pietru's network is clearly rural based, for 73 percent of it is located is rural towns and villages, while only 10 percent of Cecil's is. As can be seen from Table 4, most of Cecil's network is located in various urban centers in Malta, while a full 25 percent is situated abroad. Although this last table shows that the geographical spread of Pietru's network in Malta is greater than that of Cecil, it is even more pronounced than is indicated: Pietru knows people in forty of Malta's forty-four parishes, while Cecil has acquaintances in only twenty-seven. Moreover, Pietru knows a number of persons in Gozo, while Cecil knows none.

ACTIVITY FIELDS. It is now important to ask in what capacity Pietru and Cecil know the members of their networks. Fellow villagers make up the largest segment of Pietru's network. Cecil appears to know most persons in connection with the work that they do or he does. That is, they are part of each other's occupational role sets. Kinsmen, though forming the second largest activity field for both Pietru and Cecil, occupy a relatively larger place for Pietru, for 24 percent of the members of his larger network are kinsmen. In contrast, only 17 percent of Cecil's network are kinsmen.

Table 5. Activity field

Shared activity	Pietru	Cecil
Occupation	19.9	41.4
School/College	0.5	11.6
Kinship	23.9	16.6
Association and similar interests	6.1	4.1
Village	40.6	0.4
Neighbor	4.8	11.0
Miscellaneous	1.5	14.5
None	2.6	0
Total %	99.9	99.6
Number	2020[a]	821[a]

[a] These totals exceed the number of persons in each network since some relations derive from more than one activity field.

We must not lose sight of the fact that conclusions so far have merely been numerical statements about the composition of the networks. They indicate only that a relation exists between the central ego and a member of his network which may be expressed in terms of a particular activity field. This tells us nothing about the quality of that relation; that is, what the relative importance of that relation is. Phrased differently, how well do Pietru and Cecil know the members of their networks? Is there a relation between the size of a given activity field and the importance of the persons in it to Pietru and Cecil? I attempted to answer these questions by examining the content of their relations in terms of what is exchanged.

DENSITY. By density I mean the extent to which members of a person's primary zone are in touch with each other independently of him. Density is thus a measure of the *potential* communication between members of his network, not of the *actual* flow of information (or exchange content). The density of each network was calculated on the basis of the formula $200\,Na/n(n\text{-}1)$ where *Na* refers to the actual number of links (social relations) in the network excluding those with Ego, and *n* to the total number of persons in the network including Ego (cf. Mitchell 1969:18).

The density of Pietru's network is 23.7 percent and that of Cecil's 5.2 percent. Briefly then, more members of Pietru's primary zone are linked to each other independently of him than is the case with the members of Cecil's primary zone. At least this is the way each perceives the density of his personal network, for it is important to remember that only Pietru and Cecil, and not the members of their primary zones, were asked to indicate which persons knew each other. Thus, this density index is their subjective perception of how communication/exchange could potentially take place.

2. *Interactional Characteristics*

The type of exchange relations which Pietru and Cecil maintain with members of their network is set out in Table 6. From this it would appear that Pietru seems to maintain more personal relations with more members of his network than Cecil, though he ignores a greater portion of his network than Cecil. He also seems more personal; that is, he exchanges conversation, visits and gifts more than Cecil. He seems less noncommittal: while 57 percent of Cecil's network are persons with whom he exchanges only greeting or civilities, these categories include only 28 percent of Pietru's network. In contrast, Pietru exchanges conversation, information, and visits with 40 percent of his network, while Cecil maintains these types of exchange relations with only 20 percent of his much smaller network.

In Table 7 exchange relations are compared to certain activity fields. It is clear that the activity fields which numerically seemed so important to Pietru (village) and to Cecil (occupational role sets) are considerably less important in terms of exchange relations. While the village is proportionally important for Pietru in terms of his total network, he exchanges

Table 6. Exchange relations

Type exchange	Number Pietru	Number Cecil	Percent Pietru	Percent Cecil
Ignore	537	138	30.7	21.7
Greet only	288	191	16.4	29.9
Civilities only	204	174	11.7	27.3
Conversation/Information	560	69	32.0	10.3
Visits and gifts	140	59	8.0	9.3
Total	1751	638	100.1	99.4

conversation and visits with only one out of every four fellow villagers. Thus, with three out of four persons in his village, he maintains only nominal relations. Pietru also appears to maintain closer relations than Cecil with the persons with whom he shares occupational role relations. Although this category is numerically the most important in terms of Cecil's total network, he maintains only nominal relations with or ignores (does not greet) 80 percent of the persons within it. Pietru, in contrast, maintains nominal relations with or ignores only one third of the persons with whom he shares occupational role relations and exchanges conversation and visits with two thirds of them.

Table 7. Exchange relations and activity field

Type exchange	Occupation Pietru	Occupation Cecil	Kinship Pietru	Kinship Cecil	Village Pietru	Village Cecil
Ignore	3.2	20.3	21.3	8.8	41.2	–
Greet only	6.5	29.1	15.7	24.3	23.0	–
Civilities only	21.4	30.1	8.1	36.8	10.8	–
Conversation/information	63.8	14.4	26.9	6.6	21.9	–
Visits and gifts	4.7	5.3	24.8	21.3	2.9	–
Don't know	2.0	0.6	2.9	2.2	0.2	–
Total %	99.8	99.8	99.7	100	100	–
Number	401	340	483	136	846	3

It is also obvious in Table 7 that Pietru maintains much more intimate exchange relations with his kinsmen than Cecil. In fact, he exchanges conversation and visits with 51 percent of his 483 kinsmen, while Cecil does so with only 28 percent of his 136 kinsmen.

Finally, from Table 8 it is evident that most of the people whom Pietru ignores (i.e., does not greet when they meet) are fellow villagers, while most of those Cecil ignores are people whom he knows because they are members of each other's occupational role sets. While Pietru and Cecil visit more often with kinsmen than with persons from other activity fields, fully three-fourths of the persons Pietru visits are kinsmen, while just under half of Cecil's visits are with kinsmen.

Table 8. Activity field and exchange relations

Shared activity	Ignore		Greeting		Conversation		Visits	
	Pietru	Cecil	Pietru	Cecil	Pietru	Cecil	Pietru	Cecil
Occupation	2.2	41.0	8.0	44.0	36.2	45.8	12.0	22.5
School/College	0.4	9.0	0	5.8	0.1	20.6	0.7	10.0
Kinship	17.9	7.2	23.3	14.7	18.4	8.4	77.3	35.0
Association/similar inter.	4.2	4.2	1.8	2.7	9.8	7.5	2.0	6.3
Village	67.0	0	58.6	0	25.3	0	1.3	0
Neighbor	2.5	16.9	5.5	12.0	5.4	14.0	6.0	6.3
Miscellaneous	0.2	21.7	1.5	20.9	2.5	3.7	0.7	20.0
None	5.6	0	1.2	0	1.4	0	0	0
Total %	100	100	99.9	100.1	99.9	100	100	100.1
Number	552	166	326	225	707	707	150	80

3. *Intimacy Zones*

Each informant was asked to separate from his total network those persons who in some way or another meant something special to him. This decision was made on a purely subjective/emotional basis. Although I shall not discuss these categories in any detail here, it is interesting to note that Cecil and Pietru, quite independently, each selected 132 persons from his total personal network. Moreover, each classified ten persons in a first, the most intimate, category. These include spouse, parents, and most siblings. Beyond this intimate cell there is a second zone of approximately twenty persons with whom they maintain especially close relations. While the first zone for both consists mostly of relatives, the second zone, though composed chiefly of relatives for Pietru, is made up largely of nonrelatives for Cecil.

Who are the 132 people in each of the zones of intimacy? They reflect closely the relative importance of the various activity fields and interestingly enough, exchange relations set out in tables 5 and 6. Kinsmen form the largest category for Pietru (45 percent compared to 21 percent for Cecil), whereas persons with whom he shares an occupational role are more important for Cecil. Pietru is also more personal with the members of his intimate network than Cecil: he exchanges visits with 52 percent of them, while Cecil visits with only 29 percent.

DISCUSSION

The salient differences between the two networks which emerge from this very cursory comparison include the following. Pietru's network is larger than that of Cecil and appears to have a more even distribution of the sexes, age groups, educational, and occupational categories as well as a more even geographical spread in Malta. Kinsmen appear to occupy a more important place in Pietru's network than in that of Cecil, and the density of the former's network is higher than that of the latter. Finally, Pietru has a tendency to personalize a greater proportion of his social relations in the intimate as well as the nonintimate zones of his network than Cecil.

The striking difference in size of the two networks may be explained partly by the fact that Pietru is seven years older than Cecil and thus has had the time to meet and maintain contacts with a large number of persons. But it does not account for the fact that Pietru knows almost three times as many persons as Cecil. A second explanation might be that Pietru plays more occupational roles than Cecil. He is after all schoolteacher, farmer, tutor, informal assistant of the parish priest, member of a confraternity and band club as well as father, son, husband, and kinsman. It is apparent, however, that many of these roles derive from his niche in a small village. Furthermore, because he is socially prominent, people seek contact with him, and do not let him forget them. Moreover, by simply being a native of the village, his network is larger than Cecil's by some 846 persons — his fellow villagers. Cecil, as a city dweller, has no comparable ascribed segment in his network. Without this ascribed segment their networks are more nearly comparable: 805 to 638.

Another important factor in explaining the difference in size is Pietru's tendency to personalize his relations to a greater extent than Cecil. The difference in size between the two networks is perhaps also a function of a difference in memory. The explanation for this is not biological but social: Pietru personalizes more relations, therefore has more guidelines to aid his memory. The details he learns through conversation help to fix persons in his memory. Moreover, by personalizing relations, he creates multiplex ties. Through these he more easily meets new persons. Expressed in another way: people seem to be more important to Pietru than to Cecil; therefore he knows more.

Finally, it is important to note that Cecil's training as a Jesuit entailed a high measure of isolation, a systematic effort to eliminate or curtail contacts with non-Jesuits — especially family and former friends — and considerable geographical mobility. These all served to limit the number of persons with whom Cecil could maintain contact and remember.

The more even spread of Pietru's network in terms of sex, age, education, occupation and residence can be explained by the fact that he is a villager. This even spread must be seen as an ascribed attribute of his network. There is in the village an even distribution of sexes as well as of ages. Pietru is well educated, most of the villagers are not, though an ever larger proportion of village teenagers are going to secondary school. Besides this, villages like Farrug are reservoirs of skilled and unskilled labor. Pietru grew up with these persons and still lives at close quarters with them. Cecil has no such bloc of persons who are not white collar workers in his network. Similarly Pietru knows more people in other villages than Cecil, because he is a villager himself: village teachers form blocs in opposition to town teachers in the training college and associations of which Pietru is or was a member (and also in the secondary school where Cecil teaches). Furthermore, his own large kin network is anchored in the villages and through the many kinship ceremonies he meets people from other villages. Moreover, the custom of celebrating the patron saints of villages attracts people from other "rival" villages who come to enjoy the spectacle and compare notes. Fiestas are occasions on which old acquaintances are "serviced" and, through them, new contacts are made. This is perhaps a cultural explanation, but an important one nonetheless, for Pietru attends ten to fifteen fiestas a year. Cecil only attends one, that of his wife's parish, and that only since he has been married. Sophisticated inhabitants of Sliema, especially of his generation, regard it as provincial to show great interest in these fiestas.

How can one explain the greater number of kinsmen in Pietru's network? To begin with, there could be a demographic reason: In his genealogy he listed the descendants of eight married uncles and aunts of his parents. They produced no less than 182 living second cousins. Cecil in contrast could name only two married siblings of his grandparents, and they have produced only three living second cousins. This sound demographic reason for the greater number of kinsmen in Pietru's network would perhaps be sufficient if it were not for two additional factors. First, I strongly suspect that Cecil's grandparents had more married siblings than he in fact knew and therefore listed, for he noted three persons of whose exact relationship he was unsure but whom he guessed were probably second cousins. Secondly, Pietru consistently listed more relatives in every genealogical category than Cecil. For example, Pietru knows eighty-eight of his wife's relatives, whereas Cecil knows only thirty-one relatives of his wife. Even though Pietru has been married eighteen months longer and could have met more affines, it does tend to support my contention that kinsmen are more important to Pietru than they are to Cecil and therefore he knows more. This is also supported by the higher rate of exchange relations that Pietru maintains with his kinsmen (see Table 8) and the more important place they occupy in the intimate portion of Pietru's network (45 percent as opposed to 21 percent for Cecil).

The question we must answer therefore is why are kinsmen more important — in the sense of a higher social investment — to Pietru than they are to Cecil? Can this be explained by Pietru's more gregarious nature? It could be argued that because people in general seem to be more important to Pietru than to Cecil, and that kinsmen are a special category of people, that they therefore receive proportionately greater attention from Pietru.

Have we reached the point where the essential differences between these two networks have been reduced to a difference in personality between Cecil and Pietru? I think not. In fact, I see personality here as the dependent rather than independent variable. As the differences in sex, age, education, occupation, and residence were explained in terms of Pietru's village niche and Cecil's town niche, so the difference between village and town environment also explains why Pietru's network scores higher on density than Cecil's and why kinsmen are more important to him than to Cecil.

Pietru's network scores higher on density than that of Cecil primarily because it includes a compartment of no less than 846 persons that has a density of 100 percent, namely his fellow villagers, all of whom know each other. If these villagers, and those persons for whom no data was available, are removed from the two networks, it is interesting to note that their respective density scores become almost the same: 5.4 percent for Pietru's, and 5.3 percent for that of Cecil. But in spite of the similar density scores, each member of Pietru's corrected primary zone (excluding villagers) is linked with a mean of 42.8 other persons in the primary zone, while the corresponding figure for Cecil is only 31.7 persons.

The greater degree of interconnection between the (nonvillage) members of Pietru's network may be attributed principally to the presence in his first-order zone of large, densely interconnected segments or compartments. These include blocs of 170 primary school teachers, 54 policemen, 60 members of the clergy, and nearly 500 relatives.

In short, the greater density of Pietru's network is due principally to the place that fellow villagers occupy in it, as well as to the important place assumed by kinsmen. For if almost half of his primary zone is composed of fellow villagers, almost one-quarter consists of relatives, which in turn form segments (his mother's, his father's and his wife's relatives) which have high internal density.

Why are relatives more important to a villager than to a townsman? To answer by saying that kinship is more important in village subculture is to beg the question. To plead culture, I think, is an admission of defeat. It is merely an excuse for stopping analysis. To some extent culture is the normative rationalization of patterns of behavior. Though there is feedback, of course, the rationalization is of less interest than the underlying sociological explanation of those behavioral patterns. Kinship is important in villages because they are moral communities in which membership (in the case of those born there) is ascribed. As a kin network is built up of relations which normatively are morally sanctioned (by church as well as public opinion) — *this is also true in the city* — the kin network is a moral, ascribed segment of the total network. This moral network segment in turn is partly embedded in a larger moral community. This in combination with the many-stranded relations which exist between inhabitants of a small village indicates some of the reasons why the enforcement of kinship norms is much more efficient in a village than in a town. Although the kinship segment of Cecil's network is also a moral network, it is not embedded in a larger moral community of which he is a member. Thus, if he does not honor his kinship obligations, and as I could demonstrate he is indeed very cavalier about these, his offended kinsmen cannot appeal for justice to nonkinsmen who can exert influence on him in several other roles because they are members of the same moral community. In contrast, Pietru lives at relatively close quarters with many of his kinsmen. If he offends them they can appeal to persons who form part of the moral village community of which he and they are members.

Moreover, following Kapferer's line of argument, (1973) a failure to discharge obligations to kinsmen could well jeopardize relations with other villagers, who might well assume that he would also betray the relations he shares with them.

These various reasons combine to ensure that Pietru meets his kin obligations more punctually than does Cecil and that his own kinsmen do likewise. Thus, the fact that Pietru lives at close quarters in a village with a large number of his relatives ensures that he knows more relatives and sees them more often and that they meet their obligations to each other more faithfully than is the case with Cecil.

Why then does Pietru personalize his social relations to a greater extent than Cecil? Is it because their basic personalities differ? I think that their personalities do differ. The reason for this is less genetically than socially determined. I suggest that the gregarious personality of Pietru is found more often in villages than in towns and, conversely, that the more reserved personality of Cecil is more common in large towns with mobile populations than in nucleated villages. Why do villages personalize their relations? This is chiefly because most villages are highly interconnected communities. In a village any person with whom one has dealings knows all the other persons in the village (beyond 3,000–4,000 this is no longer true: Farrug has but 1,200). They are therefore used to associating with people who have similar backgrounds and know the same people. They, thus, can exchange information about shared experiences, problems, and acquaintances. They do not do so all the time or with all villagers, but it does mean that there are no anonymous persons in the village: everyone knows everyone else.

City people are used to persons with whom they have only single-stranded service relations. In a city there is a category which can be called the anonymous service fringe. These are the people with whom one has relatively frequent dealings, such as bus conductors, shopkeepers, neighbors, etc., but about whom little is known. It is significant, I think, that

Cecil did not know the names of fifty nonrelatives with whom he had regular dealings while Pietru listed only one case. Pietru begins talking easily to strangers, such as bus conductors and shopkeepers. Cecil does not. What I am suggesting is that because Pietru is not used to single-stranded, anonymous relations, he cannot handle them. Consequently he takes steps to ensure that relations do not remain single-stranded: he begins conversation and searches for common ground. Cecil, on the other hand, used to traveling and the social relations in a relatively impersonal urban center, is accustomed to such impersonal relations. They form a familiar category. He can handle them. He, therefore, does not seek to personalize single-stranded relations.

The difference in gregariousness between Cecil and Pietru is similar to the difference between the latter and his city-bred wife. Pietru's wife does not have the personalizing nature of her husband. On her shopping expeditions through the village she does not talk a great deal, although she is not taciturn. Her reserved relations with the shopkeepers and neighbors are regarded as rather unusual by the people of Farrug, while they see nothing unusual about Pietru's talkativeness. In contrast, Pietru's long conversations with strangers are looked on as exceptional behavior in Sliema, while his wife's reserve is not. I have already pointed to this same contrast in talkativeness between Cecil and his village-bred wife. Cecil is somewhat reserved, and his wife, according to his standards, is very talkative. I know the cases discussed are few. They suggest, however, that there may be evidence to associate an outgoing, talkative personality with a person who is brought up and lives in a highly connected moral community of no more than 3,000 to 4,000 and a reserved personality with a person who has been brought up in an (urban) area with a relatively low density.

CONCLUSION

Given the limited data I collected, the conclusions drawn concerning the differences in the network profile of townsman and countryman, the explanation of these, and their effect on personality cannot be taken as proven. They are hypotheses that must be tested in greater detail. But the data do suggest the nature of the relation between individual behavior and personality, on the one hand, and macroenvironment, on the other. This relation is mediated through the personal network, which is partly constructed and partly ascribed. It is strongly influenced by personal factors, including personality, and by elements from the macroenvironment, such as place of residence. This personal network of persons who to some degree depend upon ego, and upon whom he also depends, provides the basic social field which encapsulates him. But it is a dynamic field: It impinges upon him, but he also influences it. It reflects his personality, for he constructs it in part; in turn his behavior and personality are influenced by it.

This paper has explored some of the interrelationships between macro-environment and the individual. Incidentally it also has shown how culture — an intense devotion to kinship in Maltese villages — emerges and is maintained. Considering the critical part played by such personal networks in influencing behavior and personality, it would seem essential to examine these in detail in order to be able to understand why people are the way they are.

The insights into human behavior and personality that can be obtained from network analysis can have more than just theoretical interest. For example, this type of analysis could furnish leads for those interested in psychiatric or development work. Would it not be reasonable to examine how far different types of mental illness vary with differences in network structure? If it is found that they do, as I suspect, therapy could conceivably be obtained by manipulating network structure through the environment. Similarly, the developer seeking to introduce innovations might be well advised not to attempt this first in towns of less than 4,000. In these the networks of the inhabitants (unless there is a relatively high degree of mobility) could be such that social control over the innovator/entrepreneur would be so considerable that he would not be able to make profitable decisions which contravene traditional norms and/or adversely affect his relations with others.

REFERENCES

BARNES, J. A.
1969 "Networks and political process," in *Social networks in urban situations*. Edited by J. Clyde Mitchell, 51–76 Manchester, England: Manchester University Press.

BOISSEVAIN, JEREMY
1965 *Saints and fireworks: religion and politics in rural Malta*. London: Athlone.
1969 *Hal-Farrug: a village in Malta*. New York: Holt, Rinehart and Winston.
1973 "An exploration of two first order zones," in *Network analysis: studies in human interaction*. Edited by Jeremy Boissevain and J. Clyde Mitchell. The Hague: Mouton and Co. for the Afrika Studiecentrum.
1974 *Friends of friends: networks, manipulators and coalitions*. Oxford: Basil Blackwell.

BOTT, ELIZABETH
1957 *Family and social network*. London: Tavistock.

EVANS-PRITCHARD, E. E.
1940 *The Nuer*. Oxford: Clarendon.

KAPFERER, BRUCE
1973 "Social network and conjugal role in urban Zambia: toward a reformu-

lation of the Bott hypothesis," in *Network analysis: studies in human interaction*. Edited by J. Boissevain and J. Clyde Mitchell, 83–110. The Hague: Mouton.

MITCHELL, J. CLYDE
1969 "The concept and the use of social networks," in *Social networks in urban situations*. Edited by J. Clyde Mitchell. Manchester: University Press, for the Institute for Social Research, University of Zambia.

RADCLIFFE-BROWN, A. R.
1952 *Structure and function in primitive society*. London: Cohen and West.

Aspects of Environmental Explanation in Anthropology and Criticism

DENIS DUTTON

Few anthropologists writing today would be inclined to appeal to natural environment as something from which cultural forms generally can be deduced. Even so stout a defender of the ecological perspective as Roy Rappaport allows that ". . . we cannot predict from the geographical particulars alone of any region what will be the character of the culture prevailing there" (1971: 246). In light of the fact that the vast majority of researchers would accept this remark as a virtual truism, it is surprising how often ethnographers invoke a tribe's natural habitat when pressed to account for either the character of a culture or a particular social practice. Such invocations frequently explain far less than they appear to. Moreover, they are associated with problems that have long been recognized in other disciplines. The notion that a particular action or patterned activity is to be explained by relating it to some larger context is as old as historical and critical studies themselves. Historians and art critics have always realized that the activities of peoples —tribal or "civilized" — are influenced by both their natural and cultural setting. Perhaps it is one of the marks of the discipline's character as a "new science" that its researchers have repeatedly fallen into methodological traps that many historians and critics through long practice have learned to avoid. In order to understand the nature of these difficulties, I shall examine some of the possibilities for misuse of appeals to social environment in aesthetic criticism and then show how what is learned there might be applied to appeals to natural environment as practiced by anthropologists.

The critic who sets out to write about a work of art may use one or a combination of a number of critical approaches. He may compare the work with others in the same genre. He may pursue historical studies, relating the work to others in the artist's *opera*, tracing the influences that affected the artist, or placing the work in its general cultural context. Or he may concentrate his attention strictly on the sensuous surface of the work, analyzing its formal characteristics. While none of these approaches is ever exhaustive in relation to aesthetic understanding, each has its uses and misuses. As an example of what I take to be misuse, let us look at the essay on understanding the music of Bach by Gilbert Highet (1954). According to Highet, in order to attain an understanding of Bach's music, we must "think ourselves back into" the baroque age and "try to define its ideals . . . then, only then satisfactorily, [can we] appreciate the art it produced." Highet goes on:

> I speak from experience. For nearly forty years I have been playing the piano and listening to music. But it is only in the last twenty that I have come to understand the work of Johann Sebastian Bach. All through my teens and twenties I thought he was a dry old stick who had written some peculiarly difficult puzzles for the piano and organ and some tediously monotonous religious utterances for the choir. Now I think he is the greatest composer who ever lived. This change was not simply a matter of growing up and getting more sense. No, it sprang from a new understanding of the age in which Bach lived (Highet 1954:43).

There follows a discussion of the ideals of the baroque age, which Highet tells us are tradition, symmetry, and control.

Now I would not want to say that Highet's analysis is without value. It involves a legitimate kind of approach familiar to all who have enjoyed (or endured) courses in music appreciation. However, there is a latent element in Highet's discussion that can tend to mislead the reader. It is one thing to say that a study of Bach's times can serve as an introduction to the study of his music or that such inquiry can be an aid to coming to appreciate his music. But it is quite another thing to suggest that an understanding of Bach cannot be had without such study or to claim that such studies can give us an insight into Bach's musical art. Highet holds that only when we have understood the ideals of the baroque period will we be able to comprehend the greatness of Bach. But this does not follow.

Highet treats Bach's music as a social and historical phenomenon. In part, it most certainly is. It is hardly an accident that Bach composed music using the forms and styles of his era and was to that extent a "product of his times." So also were such men as Vivaldi, Schütz, Telemann, and even Frederick the Great (the latter, incidentally, a sometime composer and performer of vapid flute sonatas and other pieces which show at least that he was a competent amateur, if nothing more). The difficulty here is that considered as a social or historical phenomenon, Bach's music is utterly indistinguishable from, for instance, Frederick the Great's, while *musically* there is no comparison between the two. Moreover, a grasp of the music of Bach as a social or historical phenomenon cannot guarantee a musical understanding of it. If, on the other hand, one possesses the sensibilities requisite for a musical understanding of Bach's art, then historical studies may be relevant and helpful along the way, but they are not absolutely necessary. Many fine critics — D. F. Tovey would be a good example — are able to say much of importance about Bach's music with scant reference to history or biography. Highet's essay, on the other hand, is largely historical though it is ostensibly about Bach. Yet everything that Highet says about Bach *qua* baroque composer applies equally as well to Telemann or Frederick the Great. The net effect is that we are being told nothing about the peculiarities of Bach's art. Rather, we are presented with an essay on baroque music purporting to give us insight into the greatness of Bach.

It is sometimes supposed, particularly by social scientists, that the more general an explanation, the greater its value. For example, in his defense of the importance of ecological studies in anthropology, Rappaport makes the following remarks:

> Whereas cultural anthropology has generally taken as its starting point that which is uniquely human, an ecological perspective leads us to base our interpretations of human existence on that which is not uniquely human . . . Moreover, to emphasize first man's status as an animal makes available to anthropological explanation the generalizations of ecology and other biological disciplines which, applying as they do to all life, are of broader scope than any generalizations which anthropology may provide itself on the basis of its own observations, limited as they are to single species. Other things being equal, explanations of greater generality are to be preferred to those of narrower range because they allow us to introduce more order into our comprehension of the universe (Rappaport 1971: 244).

One can only wonder what the qualification "other things being equal" is supposed to mean. But Rappaport's final remark would seem to overlook the fact that there are forms of explanation that suffer not from being too narrow but from being *too broad*. Of course, an explanation that accounted for the movements of Jupiter on alternate Tuesdays would presumably be inferior to one accounting for its movements at any time. Its generality would be broader still, and its explanatory power thus greater, if it explained the movements of other planets as well. Ultimately, the best possible explanation of the movements of Jupiter would be one that accounts not only for the movements of that particular planet, but of celestial bodies in general. Part of the greatness and power of Newton's law of gravitation lies in just such generality.

But surely we ought not to be deluded by this consideration that it therefore follows that the more general our analysis of a subject matter, the greater its explanatory power. It all depends on what we want to explain. Newton's law, for example, applies ultimately to all bodies in space, since objects of any mass exhibit, and are subject to, gravitational effects. Yet it is precisely this generality that limits its worth to one who is interested in fundamental differences between objects to which it applies.

Planets, spacecraft, stars, and human beings are all equally subject to gravitational effects, but it is exactly this fact that makes laws covering gravitational effects so useless in telling us about some of the interesting differences between, say, a comet and Neil Armstrong. Thus, *contra* Rappaport, there may be little reason to prefer an account of a human phenomenon simply because it can be as easily applied to the behavior of a baboon. Such generality may deplete important content that may be relevant to some of the significant distinctions between human beings and baboons.

Highet's essay on Bach is a good example of just this sort of error. He attempts to explain Bach's music in terms of its social milieu. The difficulty here is that if a set of social conditions can be said to account for Bach's music, it must likewise apply to every other baroque composition. Thus a discussion that starts out looking as though it is going to tell us something interesting about the great musical art of J. S. Bach disappoints us in the end by telling us just as much about every other baroque composition. Highet's considerations explain too little because they cover too much. After having read his generalizations we are as far as ever from understanding the gulf that separates Bach's art from that of Frederick the Great.

Of course, it may be entirely worthwhile somehow to relate or connect particular pieces of art with the artistic or social milieu in which they were produced. Critics have always done this and will continue to do so. But these connections have little explanatory force; they simply enrich our knowledge of the art of a given period. And critics in general do not invoke socioenvironmental considerations as a principle of aesthetic explanation. To the contrary, such considerations are never supposed to allow us to deduce or predict, beyond the most superficial aspect of formal characteristics, how a particular piece of music, painting, or literature will be put together. Still, it is always possible to relate any work of art to its social milieu. This fact provides a convenient diversion for critics unwilling or unable to direct their attention to the specific art object at hand. But the greater the generality of the considerations brought into relation with the work of art (and socioenvironmental considerations — applying as they do to all the works of a period — are general indeed), the greater the danger that the critical account, will slide into vacuity.

This danger has its counterpart in the subject of Rappaport's remark: anthropologists' explanations of cultural practice in terms of natural environment. Not surprisingly, appeals to natural environment are among the most plausible candidates for explanations of cultural practice. And why not? It can hardly be imagined that it is an utterly arbitrary matter that some tribes live in houses built of ice blocks, whereas others live in huts of grass. Yet again, the problem is one of how the practice itself is to be connected with the general conditions that are supposed to account for it.

Ice is unavailable in some parts of the world and grass unsuitable for shelter in others. Knowledge of such facts may count as a necessary condition for understanding why houses are built as they are in some cultures. But such facts are hardly sufficient to explain the enormous diversity in form and style of shelter over the face of the earth. The same goes for many purported explanations of cultural phenomena in terms of physical environment. Despite the fact that the demands of virtually any physical environment are never as varied as the cultural possibilities for, in the fashionable term, "coping" with that environment, the tack often taken by anthropologists resembles that exhibited by Highet's essay on Bach. The anthropologist starts with the claim that knowledge of the physical conditions of a culture is a necessary condition for understanding it. Soon, however, what was initially touted as a necessary condition for understanding the culture is being implicitly invoked as an explanatory principle — in the sense of constituting a sufficient condition — accounting for the existence of any number of cultural traits.

If there ever was a school of "environmental determinism" in anthropology, it has been allowed a quiet death. No major figure in anthropology today subscribes to the notion that environment alone is a particularly useful tool in accounting for culture. Yet despite the abandonment of physical environment as a basic theoretical concept in anthropology, it is still persistently invoked as an *ad hoc* device to account for otherwise inexplicable cultural factors. This persistence is nowhere more evident than in the anthropological literature on the native Southwest.

Why, for instance, are the Navaho so inclined toward mutual accusations of sorcery? Kluckhohn and Leighton provide many reasons, among them this: "Much of the tension among The People may actually be traced to the uncertainties of making a living in a difficult environment with the technological means at their disposal. Since the caprices of the environment are not controllable by the society, the worry related to this is attributed to witches who, living as individuals, can be dealt with" (1962:244). Or take another example: In his monograph, *Old Oraibi*, Titiev provides a general description of life on one of the Hopi mesas. The last page of his section on religious practices purports to give "the meaning of Hopi religion." Here is his conclusion:

> When it is reduced to its barest essentials Hopi religion loses its distinctive flavor and turns out to be no more than a local manifestation of universally sustained religious beliefs. Everywhere primitive societies seek to guarantee their stability and permanence; and lacking the material means to counteract the effects of a hostile desert environment or of the inescapable ravages of death, they turn to the supernatural world for assurance that they will not be destroyed. It is, therefore, primarily for the attainment of this goal that the entire apparatus of Hopi religion has been devised (Titiev 1944:178).

So all of Hopi sacred drama — those moving texts and elaborate ceremonies, those magnificent dances — can be understood merely as an apparatus to cope with the threat of the hostile desert environment.

Here is yet another example: In her discussion of Hopi socialization, Goldfrank tells us that "large scale cooperation" seen among members of the Pueblo tribes is "no spontaneous expression of good-will or sociability," but results from a "long process of conditioning" required by trying to engage in irrigation agriculture in a desert environment: "To achieve the cooperation necessary for a functioning irrigated agriculture, the Zunis and Hopis strive from infancy for 'a yielding disposition.' From early childhood, quarrelling, even in play, is discouraged. . ." (1945: 527). Thus the demands of the hostile desert environment are supposed to cause the culture of the Pueblos to encourage cooperation and discourage, wherever possible, friction and disagreement. Contrast this with still another example: Eggan's account, entitled "The general problem of Hopi adjustment." She begins, like so many others, by informing us that "it seems reasonable to assume that man's reactions are shaped in part by pressures which derive from his total environment," and that an understanding of the Hopis' physical environment, is a "prerequisite to an examination of their emotional reactions." As it turns out, those emotional reactions are characterized by what Eggan calls "mass maladjustment," which she explains as a state of pessimism in which (curiously, in light of Goldfrank's thesis) "*friction* predominates in personal relations" (1943:357).

And so it goes. Why are the Navaho so concerned with witchcraft? asks an anthropologist, who learnedly informs us that it is because of the strain of living in a hostile desert environment. Why this vast and rich spectacle of Hopi sacred life? asks another, who wisely tells us that it is all just a device intended to counteract the hostile desert environment. Why are the Pueblo tribesmen so marvelously cooperative with one another? asks yet another, who answers that it is all brought about by the demands of trying to carry out irrigation agriculture in a hostile desert environment. But why, conversely, is so much friction and maladjustment found on the Hopi pueblos? asks still another researcher, who knowingly proclaims that it is an emotional reaction to pressures from the hostile desert environment. It is not long before one begins to get the impression that taken as an explanatory concept, the notion of "hostile desert environment," like the concept of vital *élan* in its bygone day, is being made to do too much work.

Of the above examples, Eggan's is particularly instructive because her paper erroneously seems to presuppose that there is a physical environment that can be usefully identified independent of the culture it supposedly affects. Just on the face of it, we might imagine this poses no particular problem. Yet even here, there lurks a dangerous ambiguity.

Read Eggan's own description of the habitat, an understanding of which she tells us is necessary to explain the Hopis' "mass maladjustment":

> Surveying these external conditions briefly, we find in northeastern Arizona a ragged plateau of sand and rock from which a series of arid mesas, averaging six thousand feet in altitude, descend to increasingly arid plains some thousand feet lower. It is a land of violent moods, of eternal thirst and sudden devastating rains, of searing heat and slow, aching cold. The flora and fauna have long since become discouraged and are sparse. Neither was it a promising environment for human habitation, yet for perhaps fifty generations the two thousand or more constituents of the Hopi tribe have lived, entirely surrounded by their traditional enemies [the Navaho], on a "reservation" where several neighboring mesas disintegrate into the plains below.
>
> Here the seasons marched by with scant regard for man's trampled ego. The Hopi could only sit and wait while sandstorms cut to ruin the young plants that meant "life for the people." Through long hot summer afternoons they daily scanned the sky where thunderheads piled high in promise and were dispersed by "bad winds." When at last the rain fell, it often came in destructive torrents which gutted the fields and drained away into the arroyos (Eggan 1943:359).

Well, no wonder the Hopi are maladjusted! Who would not be, trying to eke out a living in such a "hostile desert environment." Eggan describes for us "searing heat," "aching cold," "eternal thirst," "devastating rains," and so on; why, even the flowers and animals have become "discouraged," while the march of the seasons "tramples" men's egos. This, then, is the "understanding" of the natural environment of the Hopi that Eggan claims is a "prerequisite" for further understanding of their culture.

But we can go far astray if we try to impute to a primitive people *our* notion of their natural habitat. For it is not their physical environment *as we experience and understand it* which influences them, but that environment *as they experience and understand it*. Rappaport suggests that we make this differentiation using the terms "cognized model" of the environment for the tribe's view and "operational model" of the environment (for our, presumably more scientific, view). It is an important, in fact crucial, distinction to make; if we bear it in mind we can readily see the confusion implicit in Eggan's account. In point of fact, no Hopi would ever describe the desert in the terms she has. One detects that Eggan does not like the desert at all. Surely she does not know the desert well or she would not say such as that it "discourages" flora and fauna. As any naturalist — or any Hopi Indian — fully realizes, the desert teems with life. One must only know where and how to look for it.

Eggan's remarks are particularly enlightening, precisely because their dismal description of the desert is so obviously at odds with what we might expect from a Hopi Indian. Many observers have repeatedly stressed how the Hopi sees himself as cooperating with the land and the weather to produce the food that means "life for the people." This is connected with

Eggan's absurd contention that the desert elements "trample man's ego." Only one who feels estranged from nature can so fear the possibility of being "trampled" by it. But such alienation is unknown to a Hopi who, by accepting nature and accepting his dependence on it and by affirming through his intense rituals and dances his belief in nature's dependence on him, does not see the desert as something to be feared or hated.

Compare Eggan's remarks with the following made by Walter Hough:

> It is surprising to find such a general knowledge of the plants of their country as is met with among the Hopi. No doubt this wonder arises among those who live the artificial life of the cities. The Hopi is a true child of the desert and near the desert's heart. His surroundings do not furnish clear streams, grassy meadows, and mossy trees; there is much that is stern and barren at first glance, and there is a meagerness, except in vast outlooks and brilliant coloring. Here nature is stripped and all her outlines are revealed; the rock, plains, and mountains stand out boldly in the clear air. Still, in all this barrenness there is abundance of animal and vegetal life which has adapted itself to the semi-desert, and if one becomes for the time a Hopi, he may find in odd nooks and corners many things delightful to both the eyes and the understanding (Hough 1915:57).

Hough, who tells us on the first page of his book that he is moved by "affection and respect" for the Hopi, is willing to try to see things from their point of view. Note his interest in the Hopi vocabulary for the articles of their natural surroundings. He recognizes that they know the desert well, and he realizes that they make their life according to its demands. He does not claim that the natural surroundings of the tribe determine important aspects of their lives: "Out of this environment the Hopi has shaped his religious beliefs. . .And in a like manner has he drawn from this niggard stretch his house, his pottery, baskets, clothing and all the art that show how man can rise above his environment" (Hough 1915:15).

Some may wish to argue that Hough's remarks cannot be taken very seriously since he wrote in 1915, before anthropology became truly "scientific." Yet I would claim that this *is* accuracy in ethnological description: Hough wants to tell us what the Indians have *done* out in the desert, by which I mean that he wishes us to know what they have achieved there. Some too will claim that what is really needed here is a scientific or objective description of the physical environment of the Indians against which we may put the Hopi description of the environment. To this it should be said that if the description of a natural environment by a people who have made a living out of it for thousands of years is not "objective," it is difficult to know what would be.[1] They are the ones

[1] In this respect, Rappaport's distinction between the "cognized" and "operational" models of the environment, valuable though it surely is, must be qualified: The possibility is far from remote that, despite animistic or superstitious elements, the cognized model of the primitive may surpass in various ways the ecologist's model in both complexity and accuracy.

whose vocabulary outstrips the botanist's in classifying local flora. They are the ones who have developed an agriculture on land that would be the despair of any European farmer. But it should be said that however interesting or accurate (by whatever criteria) the scientist's description of that desert environment might be, it still is not the Hopi's description. And it is exactly the Hopi's description that is in question, if our wish is to understand how his habitat affects him. The tendency in these matters is to take one's own view of things too much for granted, in the way that Eggan apparently takes for granted that the desert is a disagreeable place in which to live.

Analogously, we could never expect that it would be our view of historical environment of Bach's time that would in any sense "determine" his work. To whatever degree his music is at all affected by a social or musical environment external to it, it will be Bach's view of that environment that will have the influence. Thus, we cannot even usefully speak of an influencing environment independent of Bach's understanding of it. Nevertheless, many anthropologists in their investigations continue to overlook these considerations by asking the question: How does the physical environment of this tribe affect its culture? They ask this question without realizing that it is not the environment by our description that affects culture, but the environment by the tribe's description.

Human activities in general, whether the kind studied by art critics or by social scientists, always stand in some sort of relationship to a context — social and natural — in which they arise. Nothing I said should be construed as denying that it is an important task to take note of that relationship. But there are limits on what can be learned from such study and these limits are frequently narrower than we may realize. I think there is without doubt much about the origins and forms of human social life in general, and tribal life in particular, that will remain forever mysterious to us. Surely we ought to be on our guard against those who, desperate to say something significant about the myriad forms of social activity, invoke considerations of natural environment, implicitly, treating them as sufficient conditions to an explanation of the existence of those forms. The endless variety of intelligent human activity is not the product of environments working on minds; it is the creation of minds working with environments.

REFERENCES

EGGAN, DOROTHY

1943 The general problem of Hopi adjustment. *American Anthropologist* 45: 357–373.

GOLDFRANK, ESTER

1945 Socialization, personality, and the structure of Pueblo society (with particular reference to Hopi and Zuni). *American Anthropologist* 47: 516–539.

HIGHET, GILBERT
1954 *Talents and geniuses*. New York: Meridian.

HOUGH, WALTER
1915 *The Hopi Indians*. Grand Rapids: Torch.

KLUCKHOHN, CLYDE, DOROTHEA LEIGHTON
1962 *The Navaho*, revised edition. New York: Doubleday.

RAPPAPORT, ROY A.
1971 "Nature, culture, and ecological anthropology," in *Man, culture, and society*. Edited by Harry L. Shapiro. Oxford: Oxford University Press.

TITIEV, MISCHA
1944 *Old Oraibi*. Papers of the Peabody Museum of American Archaeology and Ethnology 22. Cambridge: Harvard University.

Identification and the Curve of Optimal Cohesion

GORDON E. MOSS

The societies of the world form an interlinked social network through which changes reverberate. The notion that the pre-existence of some social patterns puts the individual at the mercy of his culture, having no influence on it and tyrannized by it, fails to recognize that culture is a dynamic process continually being reaffirmed, negotiated, and discarded. Each generation makes its unique contribution as it tests its culture, visits others, and negotiates social relations. And, as Skinner (1972:122) points out, "A person is not only exposed to the contingencies that constitute a culture, he helps to maintain them." But not all of the relevant information and contingencies of a given milieu are controlled by any given group. Other cultures and the nonsocial environment also provide information and constraints. It is possible for individuals to have insights or see social patterns that are not a part of their culture and to introduce them and thereby change and enrich their own culture.

Culture is whatever information is being exchanged and whatever the members are doing, their shared patterns of the moment. No one participant is able to comprehend, monitor, or control the whole of such a dynamic process. Dubos noted this:

> The emergence of a new culture is rarely if ever the result of a conscious choice with a definite goal in mind. What happens rather is that the social customs of mating and raising children, providing shelter and means of subsistence, developing natural resources, protecting the land against enemies, enjoying life, or worshiping God interact and become organized into unique patterns (Dubos 1968:143).

Man is what the social system helps him become and as he behaves he in turn shapes the society. As Etzioni suggests:

> The social entity is not an oppressive reality hovering above the individual, constraining his acts. It is far more penetrating than this, for it is a part of what he views as his irreducible self, encompassing his streaks of disaffection and rebellion as well as his periods of compliance. The individual more readily can participate in transforming a social entity, making it closer to his image than he can engage in a fully individualistic act (Etzioni 1968:3).

It is interesting that the three quotes above are from a psychologist, a biologist, and a sociologist, and all seem to have grasped an aspect of the social dimension of a larger theme. This theme has been admirably expressed by the amateur biologist and great novelist John Steinbeck (1971:218):

> . . . all life is relational . . . One merges into another, groups melt into ecological groups. . . And the units nestle into the whole and are inseparable from it. . . one can come back to the microscope and the tide pool and the aquarium. But the little animals are found to be changed, no longer set apart and alone. And it is a strange thing that most of the feeling we call religious, most of the mystical outcrying which is one of the most prized and used and desired reactions of our species, is really the understanding and the attempt to say than man is related to the whole thing, related inextricably to all reality, known and unknowable. This is a simple thing to say, but the profound feeling of it made a Jesus, a St. Augustine, a St. Francis, a Roger Bacon, a Charles Darwin, and an Einstein. Each of them in his own tempo and with his own voice discovered and reaffirmed with astonishment the knowledge that all things are one thing and that one thing is all things.

It is this living oneness of the earth and all of its people; of nature, body, and society which I attempt to embrace in my theoretical model of biosocial resonation (Moss 1973). The focus of this paper is on the application of this model to social relations producing and maintaining recognizable social groups, giving order to human behavior and to individual self-concepts. It is important to remember that the processes discussed here articulate with the rest of the environment, though these connections cannot be thoroughly traced in this paper.

HUMAN SOCIAL BEHAVIOR AS INFORMATION COMMUNICATION NETWORKS

Information Communication as the Substance of Social Interaction

The oneness of people and their societies can best be grasped when we focus on information processing. Social behavior is people interacting with one another, and this is primarily information communication. "As Dewey emphasized, society exists in and through communication; common perspectives — common cultures — emerge through participation in common communication channels" (Shibutani 1972:164). The advent of electronic communications and rapid transportation has freed human communication channels from the limits of face-to-face contact and the immediate geographical region. As Shibutani points out (1972:165): "Culture areas are coterminous with communication channels; since communication networks are no longer coterminous with territorial boundaries, culture areas overlap and have lost their territorial bases."

Every social network has its distinctive vocabulary. These vocabularies embrace the unique information and behavioral patterns of that network. The richness of a person's appreciation of his world depends heavily on the richness of the language of his networks. Control over access to the meaning of these vocabularies permits the existing group members to control who shall become members of the network and provides a means of recognition between participants who have not met before. Within the network, increasing access to the network's special knowledge is a reward for conformity and acceptance of responsibility (Moore and Tumin 1949). Each network has a set of rules or patterns organizing some aspect of life, and each is inclined to assert that its own pattern is the best.

Since social communication networks are the continual interaction of people, the culture is not static but is the ongoing behavior of people — whatever they are doing and saying today. The norms of social groups are rarely firmly established and rigid. In most cases there is considerable flexibility. Improvisation by role players is usually required. The norms which compose a particular role take some of their content from incumbents and the emergent qualities of the ongoing situation (McCall and Simmons 1966). As people interact they may change each other's behavior. As participants change, the social system is changed, its information modified by just that much, modest though it may be.

Networks vary in their normative organization and are experienced by the participants as different kinds of groups. The network may be highly formalized and very large, such as a company: it may be only mildly formalized and small enough for many of the members to know each other, such as a ski club; or it may be a small, intimate group, such as a family or friendship clique. The size of the network and the frequency of membership change influence considerably the form of communication patterns.

Social Communication Network Rhythms and Clusters

When people stop communicating and interacting, the social system ceases to be. It comes into existence only to the degree that people are communicating in that network at that time. It is inaccurate to portray social systems as if they were like the brick buildings through which they often flow — permanent, stable, unchanging. They are transient, usually undergoing rhythmic fluctuations in communication activity. If the members are home in bed or are participating in other networks the network is temporarily not in existence. If one wanted to enroll in a university, one would be thought foolish to attempt it at 3 a.m. The relevant communication networks of the university do not exist at that hour. Most networks' information and norms include specification of regular times for coming together and engaging in the network activity, and these are the times the network comes into being. Without the clock, the flow of order in modern societies could not continue.

In modern societies, social communication networks also cluster. There are two types of clustering: interdependence and multiple membership. The first type involves the intertwining of communication channels between several networks all of whose information is required to complete a specific information content and to provide skills necessary for enactment of the information. This is the specialization and division of labor of a given body of information. Networks in a given cluster may either overlap or intersect. Networks overlap to the degree they share the same people as members. Networks intersect to the degree the networks provide regularized channels of contact without common membership. The medical information cluster is composed of hospitals, professional associations, supply and drug companies, government bureaus, research centers, and so forth. Some of these networks, such as the hospital, are important arenas in which clustering networks, overlap and intersect. Nurses working in the hospital are an example of overlapping networks, and the drug salesman's visits to the hospital pharmacist is an example of networks intersecting.

The second type of clustering is the overlapping of networks due to the multiple memberships of individuals. Virtually everyone belongs to more than one network. Simultaneous membership in certain networks is part of the normative expectations of Western societies. For example, one is expected to have a job, raise a family or at least participate in the one into which he was born, attend school at certain ages, and to pay taxes. Participation in these networks is necessary in order to provide the necessities of life. Networks that compete with each other, such as religions, have virtually no overlap. Networks that reinforce each other may have considerable overlap. Nurses, for example, tend to be drawn disproportionately from fundamentalist Christian religions (Mauksch 1972).

Incompatibility between the information of two or more overlapping networks is often experienced by the individuals having joint memberships as a personal problem. Sociologists have studied such incompatibilities through research on conditions such as status inconsistencies and role conflict. Gibbs and Martin (1964), for example, studied the overlap between marriage, or the family network, and various occupational networks in relation to suicide rates. They suggested that the fewer married people found in a given occupation, the more incompatible those two networks must be, and the less integrated the roles between the two. They hypothesized that the less integrated a person's marriage and occupational roles, the higher the suicide rates. They further suggested that the role conflicts would lead to a reduction in conformity to socially sanctioned expectations and thus a reduction in the cohesiveness of the network relations. Their research supported these hypotheses.

The form and stability of any given network depends considerably upon the cluster of networks of its members. The compatibility of these networks is crucial for both the continuing order of a given network and the self integration of the participants. A picture of the clustering of networks helps us to distinguish and locate distinctive information content and to predict groups in which given individuals would encounter behavior patterns and ideas that confirm or challenge their own. The consequences of social change are better understood when the clustering of the networks which are changing is known, since networks that cluster together also resonate more sensitively to changes in each other's information.

SOCIAL COMMUNICATION NETWORK COHESION IS IDENTIFICATION OF MEMBERS

Some Theories of Cohesion and Solidarity

Most sociological discussions of cohesion and solidarity begin (and unfortunately often end) with Durkheim's organic and mechanical solidarity, Toennies' *gemeinschaft* and *gesellschaft*, and Cooley's primary group concepts.

Durkheim's mechanical solidarity applies to groups in which cohesion is the consequence of consensus and members are very similar in their behavior and beliefs, even to the point where there is little to distinguish them from each other. The division of labor is simple, and the skills required to survive are relatively complete in each person. The burden of maintenance of cohesion is on continuing consensus. Durkheim noted that the laws in mechanical solidarity societies emphasized penal or repressive laws, often based on religious dogmas, in which retaliation for breach of taboo is sometimes enacted. He emphasizes that such solidarity grows inversely with the distinctiveness of the individual. Mechanical solidarity "... can be strong only if the ideas and tendencies common to all the members of the society are greater in number and intensity than those which pertain personally to each member (Durkheim 1964:129)."

Organic solidarity is the result of the interdependence of members due to a well-developed division of labor. The "collective conscience" or consensus accounts for less of the person's behavior, with a greater expression of unique individuality. Each person expresses himself through specific activity within the division of labor. He cannot enact all of the society's patterns himself, only those of his position. The division of labor largely replaces common sentiments as the basis of solidarity. Laws are restitutive rather than repressive. In his later years, Durkheim saw that the interdependence required by the division of labor was not enough to maintain solidarity, and that there would still have to be a system of common values. This reaffirms our belief that if clusters of networks are to be considered as cohesive bodies, they must have a shared body of information and form distinctive networks, such as a nation or hospital, even though they are made up of many other networks.

In the following quote we can see Durkeim is struggling with the same phenomena we are trying to grasp with different types of clustering:

Society is not seen in the same aspect in the two cases. In the first, what we call society is a more or less organized totality of beliefs and sentiments common to all the members of a group: this is the collective type. On the other hand, the society in which we are solidary in the second instance is a system of different, special functions which definite relations unite. These two societies really make up only one. They are two aspects of one and the same reality, but none the less they must be distinguished (Durkheim 1964:129).

Cooley finds within modern society, with its division of labor, associations not unlike Durkheim's mechanical solidarity. Cooley refers to such associations as primary groups.

By primary groups I mean those characterized by intimate face-to-face associations and cooperation. They are primary in several senses, but chiefly in that they are fundamental in forming the social nature and ideals of the individual. The result of intimate association, psychologically, is a certain fusion of individualities in a common whole, so that one's very self, for many purposes at least, is the common life and purposes of the group. Perhaps the simplest way of describing this wholeness is by saying that it is a "we;" it involves the sort of sympathy and mutual identification for which "we" is the natural expression. One lives in the feeling of the whole and finds the chief aims of his will in that feeling (Cooley 1961:315–316).

Primary groups give the individual his first and most effective experience of social unity. Cooley sees primary groups as more stable and unchanging than other, usually more complex relations. Devotion for the group and affection for other members, despite some internal competition, is placed above self-interest. Cooley rather idealistically saw this sensitivity and selflessness expanding to include the community and even the nation.

For Cooley, society is a whole. He emphasized that in order to understand it we must see it as a "vast tissue of reciprocal activity" which if we cut up "it dies in the process (Coser 1971:307)." The individual properties we think of as human nature were included in this whole: ". . . human nature is not something existing separately in the individual, but a *group-nature* or *primary phase of society*, a relatively simple and general condition of the social mind . . . Man does not have it at birth; he cannot acquire it except through fellowship, and it decays in isolation" (Cooley 1961:318). Toennies' *gemeinschaft* is a sense of community exemplified by kinship, neighborhood, and friendship. Like Cooley's primary group and Durkheim's mechanical solidarity, *gemeinschaft* emphasizes the common sentiment. In *gesellschaft*, relations are guided by convention rather than interpersonal familiarity. It is characterized by secondary relations such as in the market place (Toennies 1961).

We will have much more to say about these theories as we develop our model, but it should be noted here that all three fail adequately to take into account the consequences of network clustering, especially overlap, for the maintenance of cohesion.

The most common theme in psychological discussions of social cohesion is group attractiveness. In a review of social cohesion Schachter observed: "Whatever definition of the concept is favored, virtually all definitions have the clear implication that cohesiveness varies with the attractiveness of the group for its members" (1968:542). Attractiveness may vary with expected or observed success of the group in achieving some desired end, either within or outside of the group, or the activities of the group may themselves be rewarding (Schachter 1968; Secord and Backman 1964). Secord and Backman (1964:269) also emphasize the importance of comparisons between "outcomes available in alternative relations outside the group." Groups that are more successful in obtaining various rewards for the same or less cost presumably will be the more attractive. Cohesiveness is not necessarily the same as effectiveness in task performance, the literature of labor-industry relations provide ample evidence of cohesive work groups providing norms for low performance of tasks.

Of considerable interest from our perspective of biosocial resonation is the evidence of physiological covariation in groups being uncovered by physiological psychologists. There is evidence that when groups are cohesive and the members feel rapport with each other, their physiological processes, especially neuroendocrine responses, may converge (Leiderman and Shapiro 1964). Caudill (1958) cited rowing crew studies showing similarities in physiological responses to the racing experience of more solidary crews. Mason found the 17-OHCS excretion levels of B-52 bomber crews working in the back of the plane together were similar, even though these levels were lower and quite different on nonflight days (Mason and Brady 1964). Similar results were found in studies of young adults in small groups who volunteered to live in a hospital ward. Leadership and conflict in group relations have also shown physiological processes to be responsive to social interaction as it proceeds (Kaplan, Burch, and Bloom 1964; Bach and Bogdonoff 1964).

From the leads provided by studies showing galvanic skin potential, plasma free fatty acids, and the urinary excretion of 17-hydroxycorticosteroids to be related to small social group interaction, "it seems altogether likely that CNS-mediated changes in variety of endocrine and autonomic functions will be discovered in the next few years under conditions of strong affect engendered in group situations" (Leiderman and Shapiro 1964:x). These fluctuations in physiology may also provide additional means of measuring group cohesion, perhaps even permitting monitoring of cohesion in small groups.

Individual Information Repertoire and Information Incongruities

We are convinced that group cohesiveness is best understood in terms of information processing and exchange in social communication networks. The human body is exquisitely constructed to seek and process information. The sensory and motor organs of the body form perceptual systems used in exploring the environment for patterns in the stimulating energy (Gibson 1966). While one acquires considerable information through direct perception and can use this to correct network information, the world in which a person lives is largely determined by the social communication networks into which he is born and socialized. Associates in the network provide the individual with a vocabulary that calls attention to some things in the environment but not others; the individual conceptualizes the relations of the parts in a manner characteristic of the language (see for example Tung-Sun 1970). The bulk of information about the empirical nature of the world comes from reports, instructions, and assertions of other members of the network. Since values and norms are essentially manmade and arbitrary and can take an almost infinite variety of forms, one can acquire values and norms only from one's social networks.

A person's information repertoire is the result of direct contacts with the environment and social relations. This repertoire is located in the physiological processes and structures of the central nervous system. Information processing has physiological consequences through the operations of the central nervous system, including the influence of the autonomic nervous system and endocrine glands on virtually every physiological process of the body (see Moss 1973 for discussion of the physiological aspects of social information processing).

A person's information repertoire is an amalgam of information from his complement of networks and direct perceptions. Information from one network is frequently tested for validity in others in which the person participates, or in direct contact with the environment. To the degree that a communication network's information is continually confirmed to the person's satisfaction, it becomes part of his perspective of his world. In time it may become so familiar and taken for granted as to be part of the background of experience, without much attention being paid to it (Moss 1973:132).

In one's contacts with the natural environment and other people, one enters each situation with expectations, whether articulate or vague, based on past experiences and information learned in social networks. If the event proceeds as expected, the information of one's repertoire is reaffirmed and strengthened. If, on the other hand, expectations are disconfirmed in the situation, the individual has experienced an information incongruity.

The significance of subjectively encountering information incongruities has been observed by a large number of behavioral scientists. Festinger's theory of cognitive dissonance (1957), Heider's cognitive imbalance, Newcomb's asymmetry, and Osgood and Tannenbaum's incongruence all focus on such experiences (Shaw and Costanzo 1970). It is through information incongruities that we become aware of rules and norms. Indeed, only after such contrasts point to irregularities in patterns is one able to formulate rules or patterns at all (Whorf 1956; Burke 1965). Incongruities may produce insight into the inaccuracy of one's information repertoire, and even though such insights may be uncomfortable, new horizons of understanding can be opened. As Maslow (1968:16) has observed, the loss of "illusions" while painful at first may be "ultimately exhilarating and strengthening." The socialization process often requires the socializing agent to produce through sanctions an incongruity to call the recruit's attention to a norm he is not familiar with.

People become aware of different social norms when they are caught in successive situations where conflicting demands are made of them. Since these conflicts are alternative ways of defining the same situation, the contrast calls attention to the different perspectives offered by different networks (Shibutani 1972). People whose clusters are incompatible will be more aware of the variety of norms and norms as such than those whose clustering networks are completely compatible.

The information in one's repertoire derived from social networks provides both the values from which one's self-image is constructed and the substance of the network itself. When people enact the information learned in the network, they bring the network to life and are often expressing their selves in the process. If members change the way they enact their roles due to encounters with information incongruities, their selves and the network are both changed. It is when network members are exposed to information incongruities that problems of loyalties arise and social cohesion is threatened.

Types of Subjective Involvement with Social Communication Networks

A person does not relate to every network in his membership cluster and those with which he comes into contact with the same degree of commitment, and he does not contribute to the cohesion of all of his networks to the same degree. It is only when he is identified that he contributes to cohesion, and then only to the degree of his identification. There are four types of involvement a person may display in relation to a particular network: identification, autonomy, alienation, and anomie. A given person may exhibit all four with the networks in his cluster and show varying degrees of both identification and alienation, for example, with the same network.

Our typology of subjective involvement with social communication networks has two key dimensions. First is the proportion of the network's information the person has learned and incorporated into his social behavior and self-concept. Second is the degree the person subjectively perceives the network's information he has learned as accurate and working effectively. Encountering information incongruities may change his type of involvement.

Table 1. Types of subjective involvement with a communication network

Information repertoire, perceived as congruous when utilized in the communication network and its environment	Information repertoire dependent upon the communication network	
	yes	no
yes	Identification	Autonomy
no	Alienation	Anomie

Source: Moss 1973:141.

This typology reflects both the member's evaluation of the congruity of the network's information and the degree of rapport or subjective feelings of involvement the person experiences with that network. In terms of the network, the typology also reflects the degree the person fits into the network: his level of conformity and affection for the network (Moss 1973). Identified types are the most conforming and have the greatest rapport with the network; the alienated and autonomous can conform but have less commitment to and affection for the network; and the anomic are not able to conform, do not fit in, and have little if any affection for the given network.

The *identified* person follows the network's norms, uses its information to manipulate the environment, and is satisfied with the results. He feels rapport with those with whom he communicates in the network. The social practices, goals, and rewards of the network are meaningful to him and are capable of delighting him. He will incorporate his network's values into his self-concepts and is likely to identify himself in terms of his role in the group or the group itself. For example, if asked to respond to the question "Who am I?" he might reply, "I am a Catholic," or "a foreman at Ford," or "a doctor" (Kuhn and McPartland 1953). For the identified, loss of their role or of their membership in the group would mean the loss of an important dimension of their very self. The identified members are the source of the cohesion of the network. They are the ethnocentric believers, doers, and defenders.

Only the identified experience the euphoria of the "we" feeling of belonging and contributing to some human endeavor larger than themselves, which in fact they do. As Steinbeck suggested, this feeling is one of the most prized and sought after of human experiences. Cooley's primary group. Durkheim's mechanical solidarity, and Toennies' *gemeinschaft* all capture aspects of this dimension of identification. However, in our perspective every group has members who are identified, not just the primary, mechanical, or *gemeinschaft* types of groups. The network could not continue as a cohesive body unless there are identified members. The notion that the communication networks in societies, having a high division of labor and high degree of bureaucratization, cannot experience identification is incorrect. The differences lie mainly, as in Durkheim's mechanical and Toennies' *gesellschaft* type of societies, in the consequences of multiple memberships in networks that only partially overlap in their common memberships. The greater variety of personal information repertoires resulting reduces the proportion of people in the society that one can feel the "we" rapport with, but it does not eliminate such feelings. Identification is with selected networks of the cluster that forms the whole society. However, within these selected networks the valued rituals and events are still capable of eliciting from the identified of that network the full measure of emotional expression and gratification. This is one of the greatest rewards for identification and maintenance of social cohesion — to belong to a group that gives meaning to life.

The identified can be, and frequently are, identified with several of the networks of their membership cluster at the same time. Obviously, the compatibility of the information of the networks in the person's membership cluster increases the likelihood that the person will be identified with all of the networks in it.

In identification there is the fusion of the member and the network through the incorporation of the network's information into member's self-image. As he is socialized, he learns to apply the network's values to himself and rate himself as he perceives himself being rated by others of the network (Cooley 1972). The properties he pays attention to are those to which the network calls his attention and sanctions. The member knows himself largely as he is known and valued by the members of his networks. If the network's values applied to him indicate he is of considerable worth and provide him with a positive self-image, the member will more actively reinforce and maintain the network's values and maintain his own positive self-image in the process. If the network's evaluation of him is more negative, he may try to conform to the norms more effectively in order to improve his value in the network.

The theme of the self emerging from social interaction and of the content of social patterns becoming the content of one's self is a common one in sociology. It is at the base of symbolic interactionists theories, such as those of Blumer (1969), Mead (1964), and Cooley (1972) (see also Goffman 1959; Manis and Meltzer 1972). Parsons believes that Freud, Durkheim, Cooley, and Mead all converged upon this insight from different directions to produce one of the "truly momentous developments of modern social science" (Parsons 1970:49). Parsons writes of Freud:

> Indeed it follows from Freud's whole main treatment of the process of socialization — and was, at least at one point, explicitly stated in his writing — that the major structure of the ego is a precipitate of the object-relations which the individual has experienced in the course of his life history. This means that internalization of the sociocultural environment provides the basis, not merely of one specialized component of the human personality, but of what, in the human sense, is its central core (Parsons 1970:49).

Parsons also notes, as we have earlier, that there is more to the person's information repertoire than what he has learned in social interaction, a person's experiences with his own body and the nonsocial environment also make contributions, including some checks on the network's information content.

Identification with networks and network clusters having varying information leads to a variety of personalities. Riesman's tradition, inner, and other directed types are examples. All three are labels for behavior characteristic of identified members of particular societies. They "respond in their character structure to the demands of their society" and they "fit the culture as though they were made for it, as in fact they are" (Riesman 1961:241–242). While there is evidence that in general conforming to the network's information and experiencing meaningful group membership is related to better mental health (Leighton et al. 1963; Cassel 1970), there is also evidence that some networks' norms can lead to behavior and personalities both physically and mentally unhealthy (Friedman and Rosenman 1959; Caffrey 1967).

The *alienated* were once identified but encountered information incongruities that led them to abandon important portions of the network's information. The alienated have concluded that the world is not the way the network told them it was, or that the norms either do not work or are not "right." There are degrees of alienation, ranging from feelings of vague dissatisfaction and apathy to ambivalence, anger, and a sense of betrayal. The alienated may try to change the network, find an alternative, or just live with their dissatisfaction. The process of alienation requires the simultaneous rejection of network information and that portion of one's self-image that was based on it. Network alienation is necessarily to some degree self-alienation. For the alienated the network has often become "disenchanted," though they may never be able to completely free their repertoire of some of the network's information. To the freshly alienated, the strongly identified may seem narrow and shallow in their intense and often single-minded performance of the network's patterns. Strong satire is written by the alienated, not the identified.

Alienation is the major source of disruption of social cohesion. The rejection of the network's norms, reduction in participation, or the abandonment of the group by large numbers of members could severely reduce social cohesion, perhaps to the point where the group would dissolve. The distribution of the alienated within the group is also important. If one subgroup only becomes alienated it can often be contained and perhaps it will split off. However, if widely distributed alienated members remain apathetically active, they can reduce cohesion by interfering with the identified members. Such interference could produce incongruities for the identified who are unable to "do things right" and may lead to their alienation from the network as it exists when weighted down with apathetic baggage.

The difficulties that accompany the experience of information incongruities bedevil the alienated, and they may find their health impaired (Moss 1973). In addition, they lose the resources, both social and material, that the network could have provided. And thus they become even more vulnerable to further information incongruities.

Autonomy is the third form of subjective involvement. The autonomous have developed a repertoire of information peculiar to them, based on what they have found to work and accurately represent the environment. Typically, autonomy is the end result of a series of identifications and alienations which have led the person to the crucial insight upon which autonomy is based. That crucial insight is the realization that all norms are arbitrary and that there are no "right" or "true" social norms. One simply selects his particular game and plays it without considering the judgment of others and their values. The autonomous may incorporate information from the network but they rearrange and qualify it to fit their interpretations. They are able to understand and effectively use the network's information. They are able to get what they want from the network yet do not accept completely the network's information nor do they identify themselves in terms of the network. They are more likely to identify themselves in terms of personal qualities such as "I am intelligent."

True autonomy is quite rare (Maslow 1954). Mild autonomy involves little more than a rather unique collection of norms from many networks, with perhaps a synthesis of some conflicting norms into a different form that embraces the best of both. In a sense, such a person is still identified with portions of his old network information but no longer can accept any network as the final statement of human behavior. Since the autonomous have their own unique patterns, identified members may find them difficult to understand and predict. Extreme autonomy, of the type approached in Herman Hesse's *Steppenwolf* (Hesse 1969), occurs when even the social norms composing the unique blend of the autonomous individual's information repertoire are rejected. Since meaning in life is derived primarily from social values and norms, arbitrary though they may be, the utter rejection of all social values is tantamount to the abandonment of meaning. As for Harry Haller in *Steppenwolf*, life becomes a constant struggle to discover a reason for not ending one's life. The intense ethnocentrism and euphoric group experiences available to the identified are not available to the completely autonomous because no social norms can excite in them such devotion. Clearly, the autonomous do not contribute significantly to the cohesion of the network but rather are sojourners. They may, however, contribute insights that may lead to improvement of the network's information or alienation of some members.

Anomie is the fourth type of involvement. Anomic persons do not understand the communication network's information, cannot interact effectively with its members, and are uninvolved. They may have difficulty finding and entering the communication channels. For example, they may be isolated from communication channels because they are members of a disfavored group. There are few people who are completely anomic; the severely mentally retarded would be an example. No one is familiar with the information of all networks. Everyone is anomic with regard to some of the networks in their society. As a rule anomic persons in any given network are identified with other networks and take their information from them but find themselves within the given network, such as immigrants who never learn the language. The anomic have little impact on the social cohesion of the network unless they become antagonistic towards it.

The Maintenance of Identification

Social cohesion and identification are the same phenomenon. Social processes for maintaining social cohesion will be those for maintaining identification with the network. Since information incongruities can produce alienation, the maintenance of both identification and group cohesion revolves about preventing the encounter of information incongruities and the resolution to the participant's satisfaction of such encounters as they occur.

The network members have two means of preventing each other from encountering information incongruities at all. One is by providing congruous information. In this case, the network provides participants with correct information, adequate skills, and resources. There can be no information incongruities to the degree no incongruity exists. The second means of preventing contact with information incongruities is through social isolation. Prominent examples are censorship, intra-group marriage, private schools, geographical isolation, and scheduling of large numbers of activities so members have little time to participate in other networks.

While both techniques can maintain identification, the quality of the interaction and information communicated varies as one or the other dominates. To the degree a network's information is congruous with its natural environment and other networks in its cluster, members find their information reaffirmed and supported in their interaction. These positive experiences strengthen identification and thus cohesion and encourage members to have broader contacts with the network cluster. This contributes to compatibility between clustering networks and maximizes the amount of correct information each has about the others in the cluster. Network members will experience less anomie in other networks of the cluster than would be the case if isolation were the dominant means of maintaining cohesion.

As a communication network finds its information incongruous with networks in its cluster or is faced with stiff competition from similar networks for the same pool of members, particularly if the competing networks have more accurate or successful information, social isolation may increasingly be used as a major means of maintaining cohesion. Networks may encourage isolation to different degrees from different groups depending on the degree of incompatibility and competition. A church may discourage contact with other churches while encouraging members to work and be active in the community. Networks relying on social isolation would have a more closed communication pattern and members would have fewer multiple memberships. They are more likely to substitute unflattering stereotypes for accurate descriptions of the clustering networks, decreasing the possibility of cooperative interaction.

The information repertoire of identified members of networks using social isolation would be more limited than those of networks using information congruity. Their self-image will be more heavily invested with fewer networks with fewer options when they do encounter disturbing incongruities. They will be more likely to attempt to explain away the incongruity than to examine it and thus are less likely to grow and adjust to the changes in the social milieu than are members of networks relying more on congruous information for maintenance of cohesion. Whether a network faced with incongruities changes in the direction of congruity or greater social isolation depends in part upon which offers the more desirable self-images to the identified members. Change is resisted if no positive alternative self-image is offered.

In a pluralistic modern society the probability of encountering information different from one's own is very high, especially since it often occurs as one conforms to the required norms of larger networks, such as attending public schools. Every network, then, must handle the incongruities members encounter when isolation and information congruity fail. If the incongruity is one in which the member lacks correct information in his repertoire but the network has it, then the resolution is simply to instruct the member in the correct information. If the incongruity is the lack of skills or resources on the part of the member but which the network has, these can be provided, sometimes in the person of experts trained in these skills. Obviously, the more correct information, skills, and resources a network or cluster has, the more it will be able to resolve the incongruity in this manner.

If, on the other hand, the network does not have the information and resources, or if the incongruity is the presentation of incompatible values and norms by another network, the members will have to resort to explanation and justification to resolve the particular member's incongruous experience. In explanation the members try to give reasons why the event was experienced by the member as incongruous even though the network's information is "really" correct. Some examples would include: the member was experiencing a temporary perceptual distortion such as being high on drugs and misperceived; the failure of the network's information was due to the member's incompetence rather than a shortcoming of the information; some outside group sabotaged the network's activities making the information look inadequate; and discounting the source ("You can't believe him he's just a . . .").

If explanations are not effective in explaining away the incongruity, and the member persists in believing there is indeed incongruity, the other members may have to use justifications. In justifications, the argument is essentially, "You're right, there is an incongruity but it is all right, even necessary, because . . ." Common justifications include: it was necessary for the security of the group; it is God's test of our faith; and this is the way we've always done it (Moss 1973).

If another member is the source of the incongruity, particularly if he has violated the norms, he may be required to explain or justify his actions to the satisfaction of the other members. He will be asked "why" he did it. There are a number of culturally acceptable replies he can use, such as illness, accident, superseding obligations, or ignorance. If he cannot appropriately account for his actions and thus resolve the incongruity of his actions for other members, he may be negatively sanctioned or even removed from the group (Scott and Lyman 1968).

Being identified does not mean that one never encounters incongruities. Every network has normatively structured patterns that even when conformed to may lead to incongruities. Some of these are required status changes. Among the rural Zulu of South Africa child bearing is the major role and source of status among females. When menopause occurs the women lose status and this is reflected in an increase in hypertension (Scotch 1963). In our society, retirement and the disengagement of the elderly are examples.

Another source of incongruities within the network is the stratification system. Stratification is a reflection in part of the degree to which individuals are able to conform to and effectively utilize the network's information. Those who are more effective in general, have more money, prestige, inside information, and access to power. Those who are less successful receive less and perhaps even negative sanctions, leading to a less positive self-image. The incongruity between aspirations and actual achievements can be a constantly grating incongruity (see for example Kornhauser 1965). The network must find a way to give positive sanctions to those who conform without losing the less able as identified members. The network also must prevent the alienation of those who conformed but now find that their status is dropping or that their aspirations will never materialize. We have labelled the means of resolving these problems "social therapy" (Moss 1973).

Social therapy serves to relieve the symptoms and tensions of the information incongruity without resolving the incongruity itself. It is utilized when the network cannot or will not resolve the incongruity. If social controls are to be effective, the negative sanctions applied must be experienced as such. Social therapy is helpful in soothing the sanctioned member without removing the sanction. There are also many unresolvable incongruities — such as the loss of loved ones, the uncertainties of human life, and illnesses — that produce tensions and anxieties that social therapy processes ease. Funerals, good luck charms, prayers, and understanding doctors can all provide social therapy, as can an understanding friend or relative. Social therapy may contribute strongly to social cohesion by helping the lower status members adjust their self-images to incorporate the societal judgment while reassuring them they are of value in and of themselves (see for example Goffman 1952).

All of these processes — congruity, isolation, explanation, justification, and social therapy — contribute to the maintenance of identification and cohesion, and if they fail to prevent an incongruity from leading to alienation, cohesion may be impaired.

THE CURVE OF OPTIMAL SOCIAL COHESION IN CLUSTERING NETWORKS

Cohesion is Not Tension Between the Individual and Society

Social cohesion for Durkheim was a dynamic tension between absorption of the individual into the group and the maintenance of individuality. The curve of solidarity or cohesion found its optimum in a balance between individuality (personal qualities that distinguish one from any social system) and group involvement; in his terms, between individual and collective consciences (Durkheim 1964;130: Coser 1971).

In his classic study, *Suicide*, Durkheim argued that both too much and too little individuation leads to suicide. Both egoistic and anomic suicides represented too much individuation. Durkheim believed that man was basically insatiable. Without social values to tell him when he should be satisfied, he had no way of judging when he had enough and would exhaust himself in a futile search for satiation. In anomic suicide individual desires are freed beyond a level tolerable to the person. "In anomic suicide, society's influence is lacking in the basically individual passions, thus leaving them without a check-rein" (Durkheim 1951:258).

Egoistic suicide could well be described in our terms as the encounter with an incongruity that leads to alienation. In his discussion of the relationship of suicide to increased education and membership in Protestant religions, Durkheim concludes that greater individuation is the result of increased reflection. "Reflection develops only if its development becomes imperative, that is, if certain ideas and instinctive sentiments which have hitherto adequately guided conduct are found to have lost their efficacy" (Durkheim 1951:158). This increased reflection, and perhaps alienation in our terms, leads to a weakening of the person's collective beliefs and thereby eliminates the person's main source of a reason for living. "Egoistic suicide results from man's no longer finding a basis for existence in life . . ." (Durkheim 1951:258).

Altruistic suicide represents too little individuation and an overdominance of the social group. In our terms it is the result of being completely identified with primarily one network (the military or a primitive tribe). In this case, one's own life has little value in comparison with the preservation of the group, and if the norms of the group encourage suicide for religious or other reasons, one will not hesitate to sacrifice oneself.

Durkheim's model, we feel, is based on an incorrect assumption about man's nature. We do not see man by his very biological nature as insatiable in his desires for socially produced goods and services. These desires must be learned in social interaction and if anything are the product of identification, not extreme individuality. While there are undoubtedly many physiological requirements that must be met and desires generated by one's unique experience with one's body in the environment, still the bulk of one's values and all of one's social behavior are learned as a participant of a social network. From our point of view it is nonsense to set the individual against society since, as we have emphasized, the individual *is* society and the society *is* the individual. Conceptualizing cohesion as a tension between individuality and society is like saying the ingredients of a chocolate cake are fundamentally at odds.

The theme of individual versus society takes a more convoluted twist when we turn to the *gemeinschaft-gesellschaft* themes. A number of sociologists have taken the position that more primary, *gemeinschaft* types of relations are better and that the *gesellschaft*, secondary, rationalized, routinized, and formalized relations (i.e., of bureaucracies) are questionable. Cooley, Simmel, and Weber are but a few to take variations of this position (Coser 1971; Nisbet 1966; Zeitlin 1968). The arguments often suggest that in the impersonalized bureaucracy or similar organization one cannot be his "true" self, and is treated like a cog in a very large piece of machinery. Some suggest that only in the cozy comfort of primary or *gemeinschaft* types of groups can one really be appreciated as an "individual," as special.

From our point of view, each person has an information repertoire acquired in social interaction which provides him with the substance of his self. When he is participating in the network from which a particular value and pattern were internalized, that aspect of himself is paid attention and rewarded by others. When there is a high degree of overlap in common membership between the networks of a cluster, a larger proportion of the people one comes into contact with in different networks share a larger number of values. Because of this, in any given interaction a larger proportion of those values and patterns one feels to be his "self" are appreciated. We commonly say things like "I can really be myself with those people" when having such contacts. If, on the other hand, there is little overlap in membership between the networks of the cluster, then participants are more likely to interact only in terms of shared values derived from their common network. The individual may experience this as impersonality and lament that only one small aspect of his self was appreciated.

We see the difference between Durkheim's mechanical and organic solidarity and Toennies' *gemeinschaft-gesellschaft* dichotomy as polar positions on a continuum of degree of network overlap in given clusters. The differences between them are a matter of degree and not kind. Gross (1970) has argued that all social relations are by nature constraining, not just the more formalized ones. In fact, the primary, *gemeinschaft*, mechanical solidary types of relations are the most tyrannical of all since they pass judgment on all of the person's behavior and not some specified segment (see also Thomas and Znaniecki 1961; Coser 1971:136). As the overlap of networks increases, the proportion of one's information repertoire relevant in each interaction increases, and conversely as the overlap decreases, the proportion of one's self-image and relevant repertoire decreases. While one can "be himself" more with others to the degree network overlap increases, he is also constrained in more aspects of his behavior by the same people. As network overlap decreases, an individual may not be able to "be himself" or have an intimate personal relation with everyone he meets, but at the same time he has a much greater range of alternative networks he can participate in and still be acceptable in any given network; he can belong to any church he wants and still keep his job.

Frequently the individual versus modern society argument implies that one can have feelings of belonging only with primary, *gemeinschaft* types of networks. We believe that feelings of oneness may occur with any network with which people are identified. Identification does not require complete self-expression in the network. One can be identified with several networks having relatively little overlap and in which one expresses only the relevant parts of one's self shared with others of the network. While much is made of being understood as a whole person, we would raise the question of whether this must be done by one person or in one small group, or whether support for all aspects of a person is all that is required, even if the support comes in parts from a variety of people in different networks.

Identification and the Curve of Cohesion

It might seem that the relationship between the vitality of the group and the proportion and intensity of identified members would be a linear one. This may be true of single network societies, but in modern societies the complex clustering of networks changes the relation between identification-cohesion and group vitality to a bell-shaped curve.

As social cohesion moves from the optimum to the low end of curve, there is a smaller proportion of members identified and the degree of identification is reduced. The members decreasingly provide rewards for conformity or punishments for violation of network norms. Without these sanctions the norms must disappear and with them the network dissolves. Members turn increasingly to other networks for their information and self-concepts.

At the optimal level of cohesion, ethnocentrism is high but there is a tolerance and respect for other networks and a willingness to modify the network to improve harmony in the cluster. There is willingness to adopt individual insights and innovations and to permit the network's information to metamorphose rather than insist on rigid adherence to established patterns. The individual's self-concepts are more likely to be a compatible blend of roles and information from the cluster of networks.

This is the best of all worlds for the individual. He is harmoniously identified with a number of networks, providing him with large amounts of information, skills, and resources for understanding and manipulating his world and giving meaning to his life. Since there would be a minimum of conflicts between social communication networks and those that occur would be worked out by adjustments between the networks, there would be a minimum of personal problems stemming from conflicting information from incompatible networks. Alienation would be low.

As social cohesion moves from the optimum toward the high end of the curve, intolerance toward other, especially competing, networks increases. A hospital is not just nurses, doctors, and administrators, but also contains Catholics, Protestants, and Jews among its workers. If the religious groups became extremely ethnocentric, it would not only lead to conflict between these networks, but would also interfere with the functions of the hospital. Extreme cohesion disrupts the interaction patterns of the networks with which highly cohesive networks overlap sending reverberations throughout the cluster and spoiling the performance of identified members.

The networks of the cluster can be expected to resist the development of a high level of ethnocentric cohesion in any component network. To permit it would mean a loss of members from the cluster. The high ethnocentrism of a component network necessarily requires rejection of the norms and information of other networks, and in some cases may lead to active proselyting. The other networks in the cluster are forced to defend their own information from the incongruous information being presented by the extremely cohesive group, if they are to retain their members and their crucial level of identification. If their defenses are successful, they not only retain their identified members but curtail the extreme ethnocentrism of the component network which is unable to draw its members completely out of the clustering networks. There may be internal checks as well. If demands for money, time, and reduction of activity in other networks that accompany increasing cohesion exceed comfortable levels given the requirements of other networks and resources, members may become alienated, or at least less identified. The network may then have to modulate its cohesiveness.

If all the networks of the cluster were to attempt to become maximally cohesive, the cluster would be literally torn asunder, with some networks destroyed and the remaining becoming self-contained, isolated, and aggressively suspicious. Only conflict and negotiation would remain as means of internetwork interaction. As networks move toward the extreme of high cohesion, we expect an increased reliance on isolation, and larger amounts of energy spent on explanations, justifications, and social therapy as members who still have multiple memberships struggle with the differing interpretations and conflicting norms of their networks. Those who have positive self-images through identification with multiple networks of the cluster are likely to withdraw from the overcohesive networks, eventually leaving only a hard core cadre of extremely commited members, many of whom have remained because they were already alienated from the other networks of the cluster. It is not surprising that extremely cohesive groups are often hostile. They reflect not only the crusading combativeness of absolute righteousness but sometimes the retaliative anger of the alienated, who may also be the rejected.

In general, we would expect to find more autonomous people in those clusters where there is conflict between networks which are somewhat overly cohesive. The basic insight of autonomy, the relativity of all norms, is best learned in such settings where one's norms are impressively called to one's attention by the conflict with different ones. A person trying to resolve the growing conflict between his networks may experience a personal conflict between two parts of his self. This is in fact what is happening. Two parts of his information repertoire are no longer compatible. The individual may seek counselling and lament their internal turmoil. Both in this case, and when there is too little cohesion, members may abandon the troublesome networks. These are the proselyter's paradise. Those participating in optimally cohesive network clusters are poor bets for conversion.

If the participants of a network are too varied in their network memberships, the weakening of ethnocentrism may be the result of the inability of the membership mixture to come to any negotiated consensus. In such groups, large amounts of energy will be expended on negotiation of roles and norms through committees, conferences, legislative bodies and the like, and less on carrying out goal-oriented activities. The compromised norms that arise may not excite the membership to the level of identification of the optimally cohesive group.

On the other hand, the continual negotiation of a mixed membership network could over time move the network toward a larger synthesis of the norms of all component parts as each is subjected to the scrutiny of all and gradually the "best" is selected. The determination of which norms win out is in part the consequence of what information is found by most to work effectively, and also on which groups have the most sanctions (i.e., money, force, etc.) and thereby the stronger bargaining position. If some of the networks in a cluster are more ethnocentric than others but not too far beyond the optimum, we would expect these networks to dominate the cluster. Their members will be more successful in negotiating the norms of their other networks along lines more compatible with the more cohesive network.

The constant assessment of group information and negotiation of roles within clusters and clusters of clusters contributes to the elimination of information that does not work and the increasing adjustment to changes in the environment and other networks. This constant modification means that the content of the networks' information is also constantly changing and with it the content of the self-images of the identified. If a person has left the network with which he has been identified and returns a number of years later, he may discover the network has changed and that he no longer fits in. He is identified with what the network was, and can no longer accurately predict others' behavior.

We hypothesize that in general as we move to either extreme of cohesion beyond the optimum, the accuracy and effectiveness of the network's information will decline. Both extremes may result from the discovery and response to the incongruities produced by inaccurate information or norms rejected by the other networks of the cluster.

The elements of the model become clearer when applied to concrete situations. We have selected Liebow's *Tally's corner* and Lofland's *Doomsday cult* as examples of the extremes of cohesion. By contrasting these two it is also easier to understand the optimum.

Cohesion below the Optimum: Tally's corner

Tally's corner (Liebow 1967) is a study of streetcorner interaction of negro men. Liebow emphasizes that the Negro "culture" is less a distinctive subculture in conflict with the rest of society than a consequence of repeated failure of Negroes especially men, to find and hold steady work providing adequate family income. The common myth that lower-class persons band together to help one another and are highly cohesive would not follow from the model we have presented. We would expect that since the Negro communication networks do not provide them with enough effective information to successfully manipulate the environment and obtain skills, information, and resources that could be exchanged for a job or money, the network members would experience considerable information incongruities and become alienated, producing a low cohesive network. This is in fact what Liebow found.

> Thus, this streetcorner world does not at all fit the traditional characterization of the lower-class neighborhood as a tightly knit community whose members share the feelings that "We are all in this together." (Liebow 1967:219).
> . . . From this perspective, the streetcorner man does not appear as a carrier of an independent cultural tradition. His behavior appears not so much as a way of realizing the distinctive goals and values of his own subculture, or of conforming to its models, but rather as his way of trying to achieve many of the goals and values of the larger society, of failing to do this, and of concealing his failure from others and from himself as best he can (Liebow 1967:222).

From our perspective, Liebow has described a distinctive communication network even though he emphasizes it is not a self-perpetuating subculture. The most valuable information the network offers pertains to the handling of the incongruities of failure to find a suitable occupation, and failure to establish and maintain a family. The network could not provide accurate information for obtaining these desired goals, although it does suggest probable failure. The network cannot isolate itself for lack of material and other resources. It must rely upon explanation, justification, and social therapy as the main substance of its information and interaction patterns as well as the main source of cohesion and participants' self-images.

Liebow found the Negro man's entry to the streetcorner network marks the transition from one who is still trying, to one who sees failure upon his hands and is looking for a means of salvaging his self-image.

> At the moment his streetcorner relationships take precedence over his wife and children he comes into his full inheritance. . . This is the step into failure from which few if any return. . . The streetcorner is, among other things, a sanctuary for those who can no longer endure the experience or prospect of failure. . . failures are rationalized into phantom successes and weaknesses magically transformed into strengths (Liebow 1967:214).

A major element of the explanations and justifications used in any network is the redefinition of the incongruous situation so that it is either explained away or becomes more acceptable. Liebow shows us some of the redefinitions the black males used to handle their unsuccessful marriages.

> On the streetcorner, the man chooses to forget he got married because he wanted to get married and assume the duties, responsibilities and status of manhood; instead, he sees himself as the "put-upon" male who got married because his girl was pregnant or because he was tricked, cajoled or otherwise persuaded into doing so. He explains the failure of his marriage by the "theory of manly flaws." Conceding that to be head of family and to support it is a principal measure of a man, he claims he was too much of a man to be a man. He says his marriage did not fail because he failed as breadwinner and head of the family but because his wife refused to put up with his manly appetite for whiskey and other women, appetites which rank high in the scale of shadow values of the streetcorner (Liebow 1967:214–215).

Their attitudes toward work was of the sour grapes variety. They emphasized they did not want to work anyway. These redefinitions are based on the values of the network, and in the case of the streetcorner world these values place those things the black male can do (i.e., sex and drinking) on the high end and those things he cannot do (i.e., marriage and steady work) on the lower end. Much interaction was apparently given to the reaffirmation of these values, which was simultaneously a reaffirmation of self-worth of the male. These public presentations of self, however, often did not match their actual behavior. For example, the streetcorner men claimed it was good to be a ruthless "exploiter of women." Liebow found that in practice these males were much more humane and considerate than their streetcorner boasts would suggest.

As Liebow laments, the streetcorner network's information and values were pitifully poor because they neither provided accurate descriptions of the real world nor anything more than a film of self-respect maintained more by emphasis than fact. The lack of material resources left only approval as a major sanction for maintaining these values and the cohesion of the group. Given these conditions, it is not surprising that the streetcorner male communication network was very unstable and low in cohesion. Liebow reported:

As if in anticipation of the frailty of personal relationships — to get as much as he can from them while they last and perhaps hopefully to prolong them — the man hurries each relationship toward a maximum intensity . . . The fluidity of personal relationships appears, at another level, as the fluidity in neighbor and kin groups and in families and households which are built up out of these personal relationships. Indeed, transcience is perhaps the most striking and pervasive characteristic of this streetcorner world (Liebow 1967:217–218).

The low accuracy and effectiveness of the information of the streetcorner network was not enough to maintain and form any kind of long-term associations or goals. This network's cohesion was so low it is barely classifiable as a communication network. As is often the case, the cohesion was low and the information and sanctions sparse in part because of conditions in other networks with which the streetcorner society clustered.

Cohesion above the Optimum: the Divine Precepts

The Divine Precepts were a Christian religious group originating in Korea. They believed themselves to be led by the returned Christ in the person of one Soon Chang. Missionaries were sent to various other countries including the United States. The study by Lofland (1966) focuses on one branch in San Francisco. The members of the Divine

Precepts had few other active social communication network memberships with the exception of some rather low status jobs, such as waitressing. They were not well received by other networks that religions ordinarily cluster with, such as families and community. Many of them including women who abandoned their children and husbands, left their families. Indignation and harassment from members of the community in which women abandoned their families for the cult contributed to the cult's move to San Francisco. Because of their active proselyting efforts and practice of exploiting other churches' goodwill during the early periods of their stay in a community, they also antagonized competing religious networks.

Lofland found the conversion process followed a seven step sequence, with crucial social and personal conditions required at each step. In our terms, they had to have experienced a serious incongruity in their existing networks, such as job failure. Their information repertoire already had to be heavily based on religious interpretations of experience, and as a consequence of the incongruity they had to define themselves as religious seekers. If conversion was to occur, they had to meet the Divine Precepts at the period when either the incongruities in their networks were leading to alienation or they were completing a major life phase. Divorce, illness, and graduation are examples. Thus, conversion was most successful among those who were passing out of or had been uprooted from important communication networks. Ties with others outside the cult had to either be nonexistant or neutralized for conversions to culminate. Finally, one had to become a full-time participant not just a member by verbal agreement.

For cult members to maintain their identification and group cohesion, some means of handling other people's continuing negative judgments of them and their message had to exist. The Divine Precepts did this by placing the evidence of information correctness in the realm of the supernatural. "Empirical" evidence was seen as a mask for the "real" explanation which was supernatural, and which only their information permitted one to correctly interpret. They believed that all events were the reflection of the cosmic battle between God and His forces and Satan and the forces of evil. Such a body of information is capable of being remarkably tenacious since as Lofland indicates: "The elegance and power of this view reside in the fact that whether plans go right or not, whether expectations are fulfilled or not, the believer cannot lose. He derives confirmation from any outcome. Everything is somebody's move in the cosmic battle" (Lofland 1966:197). No matter what happens, the members perceive their network information as congruous.

Some examples of this logic are their explanations of other religious groups' activities. "If a service was ill-attended, people were falling away from the churches in these last days; if well-attended, people were seeking for truth, but not finding it there. If it was housed in a new building, the church was resorting to external appearance to compensate for its inner death; if in an old building, people were falling away" (Lofland 1966:203). If local spiritualists claimed the spirits renounced the Divine Precepts, the cult would explain it in terms of levels of goodness of spirits, with the report coming from low level spirits working with the devil. If their prospective converts were few in number, it was Satan's work, not any flaw in the message itself. Even doubts about the cult itself were openly treated as the work of the forces of Satan against which one must constantly be on guard.

Unfortunately for the cult, this nearly impregnable body of information had a loophole through which an information incongruity sufficient to produce alienation could slip. They had made two predictions both of which could only be "proven true" by actual empirical events. One was that God intended for many converts to be made in the United States in preparation for the beginning of the millennium. The second was that the millennium would begin in 1967. The cult's conversions were very low, never reaching even two hundred during the period Lofland observed them. The members had considerable difficulty obtaining serious investigators and holding converts. Lofland revisited the group some months after his study and found only a handful of those in his study still active. This, compared with their expectations of tens of thousands of conversions, presented cult members with their most persistent and troublesome information incongruity, particularly as the millennial morn was only months away.

The handling of this incongruity is of considerable interest. There were two common explanations used by the members: (1) the "Korean pattern" which argued that since conversions in Korea were very low for a number of years and then suddenly mushroomed, the American members should not expect too much early success; (2) the notion of "restitution" which suggested that the nonconversion period was due to an imbalance in the influence of evil over good and eventually God would counterbalance Satan and exercise more than the normal amount of power in the cult's behalf. In other words, there is no real incongruity, things are still going as expected. Their major justification was that conversions were indeed low but for a good reason, to test the members and discover the ranges of their abilities. There was considerable social therapy in the earlier periods of the cult when they lived as a family in the same house. However, these efforts to handle so dangerous an incongruity were not enough.

Lofland observed cycles in which the members first mobilized their enthusiasm and self-confidence for a proselyting campaign by turning inward and focusing on group projects such as buying and renovating an old house to be used as their center, translating and preparing for publication a book of their beliefs, or the development of a new program such as bible week. They then began the campaign and experienced initial successes in interesting people, which they always inflated. But the investigators drifted away, leaving hardly any lasting converts and the group would plunge into despair as the failure became obvious. This despair was often followed by either a migration, near migration, or dispersion of members to other towns, perhaps offering more "fertile" fields. Not infrequently, the leader would reprove the members for not trying hard enough (the "You're incompetent not the message" explanation). It is unfortunate that Lofland did not continue to observe and report on the group until the fateful 1967, we can only speculate on what probably happened to the cohesion of the cult when both the millennium and the thousands of expected converts failed to materialize.

While the Divine Precepts explained their failures in terms of the Devil's interference, we would explain it in terms of the degree of incongruity of the cult's information with that of the clustering networks, making conversion of large numbers of people virtually impossible. The costs to one's self-image and personal organization of the loss of identification with all of one's networks simultaneously and the substitution of a network rejected by all of one's former associates is so outrageous that only those lacking interpersonal relations and identification with other networks ever "converted" to the Divine Precepts. Only by changing the practices and beliefs of the cult to become more compatible with the networks of prospective converts could the cult hope to succeed.

THE CONTINUITY AND CHANGE OF COHESION: IN SUMMARY

We have presented a model in which there is a unity between people and their social groups. As people interact they are simultaneously enacting the social network and their selves. As the people composing the network change the network changes, and as the conditions of network change the people change.

The substance of social interaction is information communication. The information that gives life meaning, guides behavior, and composes the self is learned in social interaction in social communication networks. The degree these networks overlap, their clustering patterns, and their compatibility with other networks all influence the content and compatibility of the elements of the individual's information repertoire, and thus the organization and stability of personality. The cohesion of the networks is the identification of participants in terms of the network's information. Preservation of identification and thus cohesion involves manipulation of information perception, especially of information incongruities.

The ongoing interaction between members of the same network, of different networks, and with the rest of the environment provides constant assessment of a person's and thus the network's information accuracy and effectiveness. This leads to constant change and adjustment which Buckley refers to as morphogenesis (Buckley 1967). The network is seen as the same by members but the information has actually been transformed. If information is invalidated, the members may become alienated and major social changes occur that lead to a new network or the radical modification of the old one.

Information incongruities have physiological consequences which may be detrimental to health. Cohesion in various groups may lead to different physiological patterns for their members as a result of different norms and physiological convergence. A fascinating and important direction for new research is the mapping of different groups' physiological patterns and their relation to social cohesion.

The cohesion of the network and the identification of the members follows a curve, with the optimum at a level capable of maintaining the network but not producing conflicts with other networks of the cluster. In a very real sense, maintenance of the network is self-maintenance, and personal problems are network problems. We are being only mildly facetious when we say "happiness" is multiple membership in a network cluster of optimal cohesion.

REFERENCES

BACK, K. W., M. D. BOGDONOFF

1964 "Plasma lipid responses to leadership, conformity, and deviation," in *Psychobiological approaches to social behavior*. Edited by P. H. Liederman and D. Shapiro, 166–171. Stanford: Stanford University Press.

BLUMER, HERBERT

1969 *Symbolic interactionism: perspective and method*. Englewood Cliffs: Prentice-Hall.

BUCKLEY, W.

1967 *Sociology and modern systems theory*. Englewood Cliffs: Prentice-Hall.

BURKE, KENNETH

1965 *Permanence and change*. New York: Bobbs-Merrill.

CAFFREY, B.

1967 "A review of empirical findings," in *Social stress and cardiovascular disease*. Edited by S. L. Syme and L. G. Reeder. *The Milbank Memorial Fund Quarterly* 45:119.

CASSELL, J. T.

1970 "Physical illness in response to stress," *Social stress*. Edited by S. LeVine and N. A. Scotch, 189–209. Chicago: Aldine.

CAUDILL, W.

1958 Effects of social and cultural systems in reaction to stress. *Social Sciences Research Council Pamphlet* 14.

COOLEY, C. H.
1961 "Primary groups," in *Theories of society*, volume one. Edited by T. Parsons, E. Shils, K. D. Naegele, and J. R. Pitts. New York: The Free Press.
1972 "Looking-glass self," in *Symbolic interaction: a reader in social psychology*, second edition. Edited by J. G. Manis and B. N. Meltzer. Boston: Allyn and Bacon.

COSER, L. A.
1971 *Masters of sociological thought*. New York: Harcourt, Brace and Jovanovich.

DUBOS, R.
1968 *So human an animal*. New York: Scribners.

DURKHEIM, E.
1951 *Suicide*. New York: The Free Press. (Translated by J. A. Spaulding and G. Simpson.)
1964 *The division of labor in society*. New York: The Free Press. (Translated by G. Simpson.)

ETZIONI, A.
1968 *The active society*. New York: The Free Press.

FESTINGER, L. A.
1957 *The theory of cognitive dissonance*. Stanford: Stanford University Press.

FRIEDMAN, M., R. H. ROSENMAN
1959 Association of specific overt behavior pattern with blood and cardiovascular findings. *Journal of the American Medical Association* 1286–1296.

GIBBS, J. P., W. T. MARTIN
1964 *Status integration and suicide*. Eugene: The University of Oregon Press.

GIBSON, J. J.
1966 *The senses considered as perceptual systems*. Boston: Houghton Mifflin.

GOFFMAN, E.
1952 On cooling the mark out. *Psychiatry* 15:451–463.
1959 *The presentation of self in everyday life*. Garden City: Doubleday/Anchor.

GROSS, E.
1970 "Work, organization, and stress," *Social stress*. Edited by S. LeVine and N. A. Scotch. Chicago: Aldine

HESSE, H.
1969 *Steppenwolf*. New York: Bantam. (Translated by B. Creighton.)

KAPLAN, H. B., N. R. BURCH, S. W. BLOOM
1964 "Physiological covariation and sociometric relationships in small peer groups," in *Psychobiological approaches to social behavior*. Edited by P. H. Leiderman and D. Shapiro, 92–109. Stanford: Stanford University Press.

KORNHAUSER, A.
1965 *Mental health of the industrial worker: a Detroit study*. New York: John Wiley.

KUHN, M. H., T. S. MCPARTLAND
1953 An empirical investigation of self-attitudes. *American Sociological Review* 19:68–76.

LEIDERMAN, P. H., D. SHAPIRO, *editors*
1964 *Psychobiological approaches to social behavior*. Stanford: Stanford University Press.

LEIGHTON, D. C., J. S. HARDING, D. B. MACKLIN, A. M. MACMILLAN, A. H. LEIGHTON
1963 *The character of danger*, volume three. New York: Basic Books.
LIEBOW, E.
1967 *Tally's corner*. Boston: Little, Brown.
LOFLAND, J.
1966 *Doomsday cult*. Englewood Cliffs, N. J.: Prentice-Hall.
MCCALL, G. J., J. L. SIMMONS
1966 *Identities and interactions*. New York: The Free Press.
MANIS, J. G., B. N. MELTZER, *editors*
1972 *Symbolic interaction: a reader in social psychology*, second edition. Boston: Allyn and Bacon.
MASLOW, A. H.
1954 *Motivation and personality*. New York: Harper and Row.
1968 *Toward a psychology of being*, second edition. Princeton, N. J.: D. Van Nostrand.
MASON, J. W., J.V. BRADY
1964 "The sensitivity of psychoendocrine systems to social and physical environment," in *Psychobiological approaches to social behavior*. Edited by P. H. Leiderman and D. Shapiro. Stanford: Stanford University Press.
MAUKSCH, H. O.
1972 "Nursing: churning for a change?" in *Handbook of medical sociology*, second edition. Edited by H. Freeman, S. LeVine, and L. G. Reeder. Englewood Cliffs, N. J.: Prentice-Hall.
MEAD, G. H.
1964 *On social psychology*. Edited by A. Strauss. Chicago: University of Chicago Press.
MOORE, W. E., M. M. TUMIN
1949 Some social functions of ignorance. *American Sociological Review* 14: 787–795.
MOSS, G. E.
1973 *Illness, immunity, and social interaction: the dynamics of biosocial resonation*. New York: Wiley-Interscience.
NISBET, R. A.
1966 *The sociological tradition*. New York: Basic Books.
PARSONS, T.
1970 "Social structure and the development of personality: Freud's contribution to the integration of psychology and social systems, in *Personality and social systems*. Edited by N. J. Smelser and W. T. Smelser. New York: John Wiley.
RIESMAN, D.
1961 *The lonely crowd*. New Haven: Yale University Press.
SCHACHTER, S.
1968 "Social cohesion," in *International encyclopedia of the social sciences*, volume two. New York: Macmillan.
SCOTCH, N.
1963 Sociocultural factors in the epidemiology of Zulu hypertension. *American Journal of Public Health* 53:1205–1213.
SCOTT, M., S. LYMAN
1968 Accounts. *American Sociological Review* 33:46–61.
SECORD, P. F., C. W. BACKMAN
1964 *Social psychology*. New York: McGraw-Hill.

SHAW, M. E., P. R. COSTANZO
1970 *Theories of social psychology*. New York: McGraw-Hill.

SHIBUTANI, T.
1972 "Reference groups as perspectives," in *Symbolic interaction: a reader in social psychology*. Edited by J. G. Manis and B. N. Meltzer. Boston: Allyn and Bacon.

SKINNER, B. F.
1972 *Beyond freedom and dignity*. New York: Bantam.

STEINBECK, J.
1971 *The log from the Sea of Cortez*. New York: Bantam.

THOMAS, W. I., F. ZNANIECKI
1961 "On disorganization and reorganization," in *Theories of society*, volume two. Edited by T. Parsons. New York: The Free Press.

TOENNIES, F.
1961 "Gemeinschaft and Gesellschaft," in *Theories of society*, volume one. Edited by T. Parsons. New York: The Free Press.

TUNG-SUN, C.
1970 "A Chinese philosopher's theory of knowledge," in *Social psychology through symbolic interaction*. Edited by G. R. Stone and H. A. Faberman, 121–139. Waltham, Mass.: Xerox.

WEBER, M.
1947 *The theory of social and economic organization*. New York: The Free Press. (Translated by A. H. Henderson and T. Parsons.)

WHORF, B. L.
1956 "Science and linguistics," in *Language, thought and reality: selected writings of Benjamin Lee Whorf*. Edited by J. B. Carroll. Cambridge, Mass.: MIT Press.

ZEITLIN, I. M.
1968 *Ideology and the development of sociological theory*. Englewood Cliffs, N. J.: Prentice-Hall.

Comments Concerning the Interpretation and Role of Myths and Fairy Tales

JULIUS E. HEUSCHER

In recent years we have observed a rapidly growing interest in folklore. Not more than twenty years ago most Western individuals, young or old, would have smiled at the idea that folklore could be profoundly relevant to modern society, or they would have been embarrassed to recognize within themselves an intense longing for this type of literary fare.

Today we often notice the opposite: any irrational outpouring, any antiscientific attitude is uncritically promoted. Thus, we find genuinely serious endeavors as well as various questionable forays into high speed Oriental mystical enlightenment, astrology, numerology, graphology and thoughtography, political anarchy, meditation, religious fundamentalism, satanism, exorcism and spiritism, psychological quackery, and drug-induced psychedelic bliss. These frequently unhealthy, immature, or commercially exploited preoccupations with the "mystical" aspects of life must *not* be taken as evidence that the entire new orientation away from — or rather beyond — an atomistically conceived, natural-scientific world is a sad or silly mistake. We should rather suspect that they are meaningful, painful, and often harmful symptoms, arising from the continued radical or exclusive application of natural-scientific methodology and philosophy to all aspects of existence — material, biological, psychological, and spiritual. These symptoms are, as are almost all symptoms, a composite of at least two competing processes. One consists in the emergence of new elements that would limit or balance those orientations that had become all too exclusive. The other is an unhealthy contamination of these new elements with these one-sided orientations. Thus, new and potentially valuable ideas are defended with puritan intolerance, new "-isms" are ruthlessly exploited commercially, and the charismatic leader's ego inflates while he preaches a new way of transcending one's petty, everyday sense of identity.

In the ebb and flow of conservativism and radicalism, the latter has asserted itself so fanatically that the emphasis on the here-and-now rejects almost any cultural achievements of the past. This radicalism, wedded unhealthily to the traditional Western faith in rapid progress, tends to discredit the genuine, tremendously important new orientations that our epoch seems to be called upon to develop and follow. Were we less inclined to disregard everything except the most recent past and were we less anxious to achieve quick enlightenment, we would be able to recognize that a positive, critical attitude aimed at placing the reductionist, atomistic sciences in a proper perspective had already begun to evolve prior to the nineteenth century. We need only mention a few names, such as Goethe, Herder, Novalis, Hegel, Fechner, Scheler, Kierkegaard, Brentano, and Dilthey. In the first half of the present century these endeavors were refined and expanded by committed thinkers from various academic disciplines: Husserl, Heidegger, Bergson, Sartre in philosophy; Buber and Tillich in theology; A. Portmann and Paul Weiss in biology; Jaspers, Binswanger, and Boss in psychiatry. Today an increasing number of serious investigators are willing to integrate their own efforts and insights with the finest achievements of those who preceded them.

Individual or cultural growth — though outwardly seeming to proceed in spurts — is always quite slow, and must be couched in the past which it affirms or opposes. This is especially true for the transition from a mechanistic-atomistic age which has lost meaningfulness to an age of meaningful values, feelings, ideas, and commitments. This cannot be achieved simply by "objective" reappraisals, by logic, or by mass movements, but only by a profound "inner" reorientation that involves considerable personal growth (May 1958).

In fact it is this "inner" psychologic growth of the human being that largely determines his view of the world, thus shaping his culture which in turn feeds his continued psychologic existence and development. While this was always so, it has probably been only recently — mainly since the Renaissance — that human beings have become keenly aware of the fact that they are not only carried by an ever-changing culture, but also able to create the very patterns — in philosophic thoughts, artistic symbols, literary expressions, and social models — that stimulate and direct the development of that culture. Yet, as humanity becomes increasingly aware of its ability to set its own ideals, there arises the danger of an overly precipitous forward rush, leading toward an abyss. Campbell (1972) thus stresses the necessity of carefully selecting some structures, so that freedom is enriched rather than degenerated in confused and ineffective license. Real poise is to know when, how, and how long to internalize, incubate, and consolidate whatever we have newly achieved before risking a new journey. To this effect, we must learn to listen to an internal, variable rhythm.

The human view of the world cannot be reduced entirely to separable, measurable physical and instinctual factors. It is characteristic of the human being that all the processes that involve a modicum of consciousness are *not* simply blind, predictable reactions to measurable environmental stimuli. As Cassirer (1944) has shown so eloquently, it is the fate of human beings never to respond to a shared, objective environment, but always to the way they see it, the way they invest it with symbolic meanings. This investment with a symbolic content is part of the circular, feedback relationship between humans and their culture. We can, like Campbell (1959), compare human responses to the symbols projected upon the world with the responses to specific "sign stimuli" that trigger in animals the corresponding "innate release mechanisms" (IRMs). There are some correspondences and some significant differences. How does the newborn chick's automatic flight at the sight of the image of a hawk compare with the human response to a symbolic image, such as the Yin and Yang, or a picture of the Madonna and Child? The animal's IRMs are genetically determined, inflexible, and common to the entire species. The human response to the symbolic image (or to any situation since any situation always takes on some symbolic connotations) has probably some genetic determinants (cf. Jung's view of the archetypes), but it is highly flexible, variable from individual to individual, and strongly influenced by the personal history of the individual as well as by the culture that sustains him. The animal reaction is immediate; the human response is complex. In fact, the human being can — with or without deliberation — improve upon the sign stimuli that normally affect him, thus shaping "supernormal sign stimuli" that become basic challenges in his and his culture's progress. It now becomes clearer that this nonimmediate response to an environment, which is given meaning by the human being living in it, opens up the possibility of an infinitely diverse, wide and rich experience of the world. The particular type of experience, then, is dependent both on the particular trend a culture follows and on the specific stage in the development of this trend.

This certainly does not imply that any single view of the world is either arbitrary or meaningless. On the contrary, whatever contents the environment reflects as a result of being symbolically structured *will* be significant; and the succession of new experiences of the environment (in other words, the development of human consciousness) obeys in each particular trend certain inexorable laws. The images of myths and fairy tales, as well as masks, symbolic designs and rites, can be viewed as normal and supernormal sign stimuli of central importance in maintaining and furthering a culture.

These preliminary considerations may now lead us to a critique of various interpretations of myths and fairy tales. Let me begin with a few slightly oversimplified, but nevertheless relevant, observations (Heuscher 1974).

When we evaluate other cultures as "sadistic," "barbaric," "neurotic," or "inferior," we focus either upon decadent remnants of formerly magnificent civilizations, or we blindly apply both our own cultural or individual values and our idiosyncratic modes of thinking to the beliefs and behaviors of foreign ethnic groups, thereby grossly misunderstanding them. This is most strikingly the case in those innumerable instances where the mythologist or folklorist interprets the rich and colorful images of ancient or "primitive" people as resulting from the transformation of ordinary phenomena of nature. He assumes that these transformations are expressions of the unusually lively fantasy of these "simple people," or the result of their need to explain naively — and usually anthropomorphically — those events in their environment for which scientific knowledge was inadequate or nonexistent. He, furthermore, assumes that these explanations are largely necessitated by fear of the unknown: A fantastic explanation is better than none at all, and it places the helpless human being in the center of creation, thus substituting self-worth for weakness and insignificance. Ultimately, thus the reasoning proceeds, the fantastic, imaginary explanations are arranged into an orderly structure and become the religion of the group that accepts them. Yet at closer view we move in this type of reasoning from one unproven assumption to another. In fact, we look at all the transmitted folklore treasures as if we ourselves had created them right here in the framework of our culture. Then we ask ourselves how we arrived at creating them; and finally we take our answers and conclude that the ancient or primitive populations arrived at their images in the same way. Thus, we superimpose our cultural framework on another culture instead of attempting to understand the basic dimensions of the culture's own framework. A behaviorist, for example, will view mythology as prescientific attempts at social control by people who have not yet achieved the absolutely objective behaviorist orientation (Skinner 1971).

The more we grasp the dimensions of the world of the people that created the early civilizations or that of contemporary primitive populations, the more we are forced to recognize that fantasy — as we commonly experience it — is an insignificant element in their imaginative view of their existence. What they see, feel, experience, and describe are realities to them; and if these realities appear grotesque, odd, illogical, childish, or silly to most of us, there may be a number of reasons for this reaction. There is, however, little doubt that this connotation of strangeness, exaggeration, and irrationality characteristic of the world of our ancestors as well as of most contemporary primitive civilizations is at least partly due to our inability to experience their world the way they experienced it. We are caught in the narrow, rationalistic framework of our so-called advanced epoch; we cannot free ourselves from it fully even for a moment; we cannot immerse ourselves into a world where measurable time and space, causality, and materiality were experienced only dimly as secondary attributes, as shadowy abstractions of a wider existence; we are unable to venture into a world where the individual's experience of subjectivity and identity was radically different from ours. The originators of myths and fairy tales had no propensity for abstract thinking. Therefore, as long as we are inclined to view symbols erroneously as some sort of abstraction, it is wrong to say that folklore uses symbols to explain various aspects of the world. Similarly, it is misleading to speak of a "play on words" in regard to the multiple meaning of some terms in myths and fairy tales. In our culture "plays on words" are only clever, abstract constructs, whereas in genuine folklore the multiple meanings of a term, or the similarity of terms, refer to *experienced*, meaningful realities.[1]

Indeed, while some folklore themes mirror external, natural phenomena, or symbolically reflect psychological processes, their content is never exhausted by naturalistic or psychoanalytic explanations. Other themes, however, are so pregnant with meaning that any naturalistic or psychoanalytic explanation could barely skim their surface. Two examples from the Finnish epos, *Kalevala*, will suffice to illustrate this.

[1] This can be illustrated with one of the tales from Brunner-Traut's (1965) excellent collection of Egyptian folklore. As the human beings become stronger and more arrogant, their creator, Re, grows weaker. But who is Re, the god who now sends his Eye in the form of the goddess Hathor to destroy the human race? He is the son of Nun, of Chaos, or of the Formless All; Re is equally the god who — like the ouroboric serpent — created himself. He is the first formed entity, and his form is concentrated in his Eye. The iris of the Eye — like the Chinese symbol, Wu-Gi — is the "perfect round" out of which "all that is" was germinated, and which can again consume everything. This Eye of Re is alternately and simultaneously sense organ and sun. As sun it is the source of all life, but as sun it also can destroy the life of the people in the desert. The creative aspect of this ouroboric "perfect round" is also apparent from the similarity, in the Egyptian language, of the words "creating" or "making," and "eye." These are *not* plays on words; they only seem to be to those for whom the terms "eye," "create," "circle," and "god" are sharply different concepts. To the original storyteller the images of the god Re, of Re's Eye, of the goddess Hathor as the revenging Eye, of the sun, of creating, must have been confluent if not identical. This was so, not because of an unbridled fantasy of the storyteller, but because of his viewpoint which led him to the intuition or immediate conviction that these images were interchangeable, just as our modern, entirely different frame of reference compels us to group and to distinguish the physical data as well as the concepts of our world in specified ways. In the Egyptian language the terms for "human being" and for "tear" were almost the same. Yet, again, it was not a play on words when the human beings were called "the tears of Re." To the ancient bard the human being *was* the self-created god's tear; separated from the Eye of god like a tear, man soon becomes god's sorrow. At first Re, living amongst mankind, wants to preserve his all-pervading power; therefore, when the human beings strive towards independence, he orders his Eye to consume them. The heat of the Sun-eye would dry all the tears and restore the former strength of the god. Yet while the tears separated from the Eye, god became divided within himself. Now his wish for self-contained perfection battles with his pity for man. This leads to the division of everything. Finally, Re dwells no longer on earth, but — having outwitted Hathor who has become his feminine counterpart — assumes a throne upon the heavenly cow, which becomes the sky. Day separates from night, and the people on earth separate into two inimical factions.

The blacksmith, Ilmarinen, must forge a magic mill out of the tip of a swan's feather, the milk of a sterile cow, a tuft of a sheep's wool, and a grain of wheat. Four times he fails, first obtaining a deadly golden bow, then a warfaring golden boat, then a cow that loses her milk, and subsequently a plow that ravishes the land. Finally he succeeds in creating the magic mill, the Sampo, that simultaneously produces corn, salt and coins, that is able to nourish itself by means of powerful roots, and that continues to exercise its beneficial influence through its splinters when it is shattered.

Or take the *Kalevala's* beautiful description of the origin of iron:

Strolled the maids with faltering footsteps
On the borders of the cloudlets,
And their full breasts were o'erflowing,
And their nipples pained them sorely.
Down on earth their milk ran over,
From their breasts' o'erflowing fullness,
Milk on land, and milk on marshes,
Milk upon the peaceful waters . . .
Where the black milk had been dropping,
There was found the softest iron,
Where the white milk had been flowing,
There the hardest steel was fashioned,
Where the red milk had been trickling,
There was undeveloped iron (Runo IX, *Kalevala* 1969).

Many of these images are so strikingly absurd from our everyday vantage point that they can only appeal to whatever sensibility we may still retain for experiencing a world of meaning, of *Being*, of creative forces, designated by Tolkien as "secondary world" (Tolkien 1966). Since this sensibility is obfuscated in most of us mainly by our rationalist, natural-scientific training, we must recapture and develop it.[2] While serious intellectual approaches may give us at least an initial awareness of the depths contained in these images, it is the method of "amplification" through other, related themes and through anthropological contributions that will take us a step further. This method of amplification that has been fruitful for Jung (1954), von Franz (1973), Meyer (1935), Neumann (1971), Lentz (1971), Kerényi (1963), Bock (1951), and others always implies a simultaneous development of our phenomenologic view. What do I mean by this?

In the most elementary terms, the phenomenologic view implies an endeavor to remove more and more from our experience all presuppositions or prejudices that — largely for very good reasons — accompany

[2] It may be superfluous to add that some of the obfuscation derives from our lack of familiarity with objects and practices mentioned in folklore: anvils, wells, mills, spindles; threshing, forging, spinning, etc. Obviously such terms must be clarified for the reader or listener beforehand.

and guide us in our daily life. Such presuppositions comprise not only cultural biases concerning the structure of our physical universe, the origin of various natural phenomena, the nature of the soul, or the characteristics of God, but also much more fundamental dimensions, such as our experience of time and space, our separation of subjective and objective data, our view of causality, our perception of materiality, and our particular convictions concerning human self-identity (Heuscher 1971). As we become able to distance ourselves from our specific presuppositions that are necessary for our everyday transactions, we recognize features in our world that were previously excluded by these presuppositions. In his phenomenologic (eidetic and transcendental) reductions that eventually reveal "the things themselves," Husserl aims for a satisfactory methodology that would allow the experience of unadulterated *essences* (Natanson 1966). Both his voluminous, often passionate writings, and the continued controversies among phenomenologists bear witness to the enormous difficulties that beset this method. And I am convinced that these difficulties are not so much of a formal, technical nature, but are due to the fact that an ever more reliable phenomenologic view is contingent upon a continuous psychologic development that eventually leads to a radical reorientation.

The increasing ability to set aside the presuppositions of our own culture allows a gradually more precise understanding of the meaning of folklore themes and narrations as well as the specific sets of presuppositions that entered their structure. For the sake of illustration we may mention the psychotherapeutic relationship that shows significant parallels to this. The therapist (like the folklorist confronting a tale) must be willing and able to accept the uniqueness of the patient's personal world (often termed "world design") that would be undermined by any one-sided attempts at generalization (Heuscher 1971). Similarly, each myth or fairy story is a unique jewel that loses some of its luster when it is viewed exclusively as an arbitrary variation of common themes. However, the human being, like the facets of the jewel, reflects in one way or another the entire world and thus cannot be thought of as entirely autonomous, as separate from the macrocosm that contains, and is mirrored in, all the jewels. Thus, on the one hand, we show disrespect for a patient if we do not let him exist, shine forth in his own uniqueness. Yet, on the other hand, the patient is also couched in the collective psyche of which we all partake, in the transcendental realm in which our true selves merge without thereby being obliterated. Thus we may then agree that it is this common realm that makes it possible for us to experience genuine sympathy, to be open to the individual world that each and every individual constitutes for himself in his interaction with his environment. Similarly it is this openness that allows us to become cognizant of the worlds and events and people — with meaningful, unique as well as common features — that are portrayed in ancient myths and in the tales of contemporary, vastly different cultures.

Let us not conclude, however, that these ancient and different cultures experience the world of their stories as identical with their everyday surroundings. Being sensitive to the narrations' imaginative language, they experience it as something that links them to the past, that kindles their awareness of the relevance of their lives, and that points to future challenges in their individual and collective growth. Modern fairy tales in Western civilization (such as written by Wilde, Carroll, Collodi, Thurber, Sendak, Leoni, Tolkien, etc.) equally express an awareness of wider worlds which the authors try to communicate to their readers. And in addition to these tales that follow closely the conventional Western folklore style, our current culture offers other literary products (westerns, science fiction, etc.) that also seek to transcend the shared limits of our everyday world. However, it is especially the existential novel or play that puts into brackets and temporarily suspends our cherished presuppositions, making us aware of yet indistinct realms that loom behind the experience of absurdity and nothingness that the bracketing first evoked. Instead of making heavy use of more or less traditional symbolism, the existential authors grope for a revitalization of language in order to penetrate the "secondary worlds" (Tolkien 1966). Especially in the West, language has suffered from the same diminution of meaningfulness that has affected art and folklore (Heuscher 1970). More than ever before the "word" has become abstract; it is abstracted from a real, meaningful, powerful, efficient content. Except for the "precise" sciences and business contexts where words retain their usefulness as accurate signs, language — no longer experienced as beautifully alive — has become a means to hide rather than to reveal truth, a means to hide rather than to reveal genuine feelings and honest commitments. With the weakening of a language that could help us communicate deeply personal meanings, there also occurred a loss of the sense of self (May 1953). The sense of self, indeed, is tied to language. Only a new, deep experience of language can again reveal the creative power of words, a power that eventually furthers human growth. Thus, we must recognize that genuine language is not so much a system of *signs*, but a system of *symbols*: and symbols always point beyond themselves to ever new meanings (Cassirer 1966).

The distinction between signs and symbols can be clarified by paralleling or comparing it with Koestenbaum's (1969) distinction between "precise facts" and "fringe facts." The first embrace mostly the facts of the natural sciences and of economy, tend to be public, can be analyzed or reduced to smaller fragments, and are more easily verifiable because of their measurability. Fringe facts, on the other hand, tend to have a connotation of privacy; and while they are not measurable *precisely*, they can be described *accurately*. Precise facts can be broken down to their atomistic elements, while fringe facts have a Gestalt and tend to suffer by analysis. Furthermore, precise facts have always fringe qualities, while fringe facts do not necessarily contain precise fact elements. Koestenbaum emphasizes and demonstrates that both must be considered as *bona fide* facts and can thus be the objects of scientific investigation.

This distinction can be amplified by paralleling it with the idea that there is a polarity of molecular interaction and system (or Gestalt) function. This concept has drawn the attention of prominent, contemporary researchers (Koestler 1969), especially in the biological sciences. Some (Frankl 1969; Weiss 1969) are willing to accord the "system" a factual reality equivalent to that of the molecular functions, though of a different order. In psychology the need for phenomenologically established systems, viewed not as abstractions but as factual realities, has become more and more pressing. And as such, psychological Gestalts (such as Berne's "parent," "adult," and "child," or the Jungian distinction of "shadow," "persona," "animus," "anima," etc. (Berne 1964; Jacobi 1949; Jung 1954) are gaining acceptance. They are becoming also more evident in the images and events of folklore.[3]

Figure 1 is a schematic illustration that roughly portrays the relationship between molecular interactions and a hierarchical systems structure similar to the relationship between precise facts and fringe facts, or between sign-words and symbolical words.

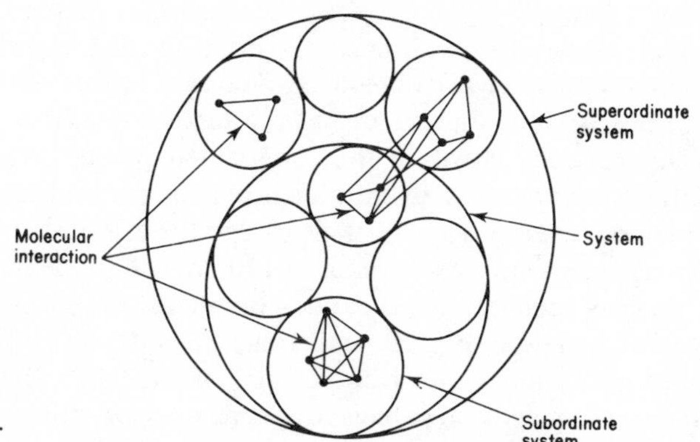

Figure 1.

[3] When we move exclusively in Heidegger's (1960) *everyday world* — the impersonal world in which *one* lives — we end up with atomistic facts; when we proceed with a radical questioning of our everyday presuppositions, we end up with encompassing meanings, with *essential qualities*. The precise, analyzed and analyzable facts are experienced as nothingness when we become lost in them; yet the holistic, encompassing view is experienced as a diffuse and diffusing phantasmagoria if it severs its relationship with both, though we may experience more keenly one or the other. Once passively couched in a phantasmagoric mythological world, Western man has swung all the way to a world where he is but the accidental result of interacting particles. A science must be promoted that can embrace both viewpoints. Then humanistic and analytic psychology, religion and natural science, factual descriptions and myths will enhance rather than exclude each other.

When Campbell stresses the need for symbols — old or new — that can awaken the life-giving, eternal life energies that have become inaccessibly locked within us as a result of our one-sided rational orientation, he refers to certain fringe facts. In our culture they must be experienced consciously in their dynamic interaction with the atomistic precise facts that modern science is so magnificently pursuing. Yet, because these fringe facts are very often expressed in words, their effectiveness is dependent on the human being's ability to use and experience a vital language.

A conscientious phenomenologic orientation, then, will make us progressively more perceptive of the meaning contained in the symbols and themes of folklore and newly responsive to the rich meaning of language. This is completely consistent with Heidegger's view that stresses language as an essential dimension of the authentic human being (Heuscher 1966).

Finally, by considering the hypothesis that links the origin of folklore to dreams, we can see more clearly how the phenomenologic method demands a radical reorientation of our viewpoint. For instance, if we look without bias at folklore and dreams, we recognize many basic similar features. In the dream we also find that time, space, materiality, causality, and even subjectivity as well as self-identity become more flexible qualities. Von Franz (1973) advances the hypothesis that folklore themes originated in dreams. This may be somewhat inaccurate even though there are dreams that would make beautiful tales and though the first building blocks of many fairy talelike literary products probably originated in the modern poet's dream. However, a more radical phenomenologic reduction that brackets — or sets aside temporarily — the separation of subjective and objective would tend to see both the tale and the dream as visions that transcend the limits established by the individual and his culture within the experiential realm.[4] While this limitation led to consciousness, to knowledge of good and evil, and to the ability to toil in our everyday world, it also led to the loss of the original, unselfconscious experience of wholeness, of Eden, and to the *sundering* from God that has been described as the "original sin."

For some time after the original sundering, the gods still visited man and bestowed upon him various boons: fire, crafts, rites, etc. Thus, humans passively received and perceived ever new opportunities to grow beyond their instinctual and cultural limits. They not only responded to the sign stimuli to which they were sensitized by various culturally reinforced IRMs, but they discovered, as if god-given, supernormal sign stimuli that led them repeatedly to transcend their own and their culture's limits. As the sundering from the gods becomes complete, however, humans grow aware that they themselves, from within themselves, can

[4] To a certain extent Jungian psychologists imply this, when they see the collective unconscious from which all these images spring as both transsubjective and transobjective.

create and establish these supernormal sign stimuli, that they themselves can affect the direction of their own and their culture's future. Yet they must caution themselves against hybris, the false pride to which so many Greek heroes fell victim.

Maybe this will be the principal characteristic of current and future myths: to portray the challenge toward deliberate self-creation. This challenge arises with the twilight of the gods and eventually, without loss of individual authenticity, must lead to a new integration of human kind in a meaningful cosmos. Through the old myths and fairy tales, we begin to fathom the infinitely rich depths of our origins. We recognize the foreshadowings of future challenges contained in these depths. Through our new myths, we may more clearly and consciously experience these challenges: to transcend the narrow limits of our natural-scientific world that made possible the development of a sharp ego consciousness; to accept the responsibility for the values, the circumstances and directions of the world with which we are evolving; to sense that this responsibility and transcendence can overcome the past sundering and lead to a new experience of wholeness in times to come.

The following illustrates these points. In a recently collected Romansh tale (*Rätoromanische Märchen* 1973) a father is willing to sacrifice his youngest daughter for the devil's gold. When the "Green Man" comes to fetch her, she protects herself by sprinkling some of the parish priest's holy water upon herself. The next time the holy water has been hidden, and now she makes the sign of the cross over ordinary water which likewise protects her. When the father removed all the water, she uses potato-peels to moisten herself, and by making the sign of the cross she protects herself for the third time.

We, thus, see how the young soul finds, more and more independently and resourcefully, a protection against the seducer.

Eventually, however, the Green Man induces the father to cut off the girl's arms with the axe. She is near death when a young prince rescues her. But just as the arms were separated from her by her father, so she and her twins are soon separated from her husband-prince; and finally she is banished by her parents-in-law. However, she now has developed such inner strength that she is able to meet the godmother who encourages her to restore her arms in the clear water of a pond. Though free to wish for whatever she desires, she continues to create various things rather than to enjoy luxurious idleness; and soon her estranged husband is restored to her. Returned to the castle, parents-in-law, prince, princess, and twin boys are reunited in a harmonious whole.

There are people who feel that too much is read into the themes of myths and fairy tales. These people are both right and wrong. The contents of folklore cannot be viewed in the same way as the contents of a box of sardines or of a scholarly scientific paper. A treatise on metallurgy, botany, or geography clearly and accurately contains all the data it wishes to present; and the limited number of inferences it allows are available to anyone with a somewhat flexible intelligence. On the contrary, the images, themes, and plots of folklore are only superficially a series of data and events that stand for themselves. They are more importantly stimuli that interact with the human psyche and change in content as the human psyche changes. They are, metaphorically speaking, like windows through which we can peer into a world that becomes ever richer as our ability to view it improves. Thus, the question is *not*, whether the original teller of the tale saw everything in this world that we believe we see. Rather, the original storytellers, sensing and seeing, vaguely or clearly, some aspects of the world beyond the immediate one, were able to shape for us the windows that can orient us towards various regions of this broader realm. The riches and depths of this new world are, I believe, inexhaustible. And if my own experience of it is vague, pale, inaccurate, or erroneous, the weakness lies neither in the images of the tale nor in the nature of this broader world that appears through these images, but in my own insufficiency. As Spiegelberg (1966) has stressed so effectively, we must not make the mistake of condemning phenomenology because of the conflicting vistas of different phenomenologists. Our perception and understanding become more accurate only when we train our thinking and our feelings to become reliable organs of perception, obfuscated and distorted neither by inner conflicts and desires nor by our common presuppositions or prejudices. Only then is this broader world seen as far richer than the variegated world of my everyday life; only then do I recognize in it ever new dimensions that eventually are seen to include my everyday existence as well. The latter, then, is like my hand before my eyes, obstructing a fuller view, like the hand which becomes part of this fuller view as I hold it at a distance (Buber 1949). Or, to quote Tolkien (1966), the secondary and the primary world become one.

Just as the images of myths and fairy tales are seen here as a series of windows towards a broader and more meaningful experience of life, so I would like you to view this modest paper not as a comprehensive exploration of the topic proposed, but as a very limited stimulus that would challenge the reader to proceed on this journey. It is the journey of the human being who must combine a maximal, precise awareness of his everyday world with the willingness to perceive accurately the meaning behind it. Without this willingness, the best meant endeavors to cope with the present problems of our culture are likely to go astray.

REFERENCES

BERNE, E.
1964 *Games people play*. New York: Grove Press.
BOCK, E.
1951 *Apokalypse*. Stuttgart: Verlag Urachhaus.
BRUNNER-TRAUT, E.
1965 *Altägyptishce Märchen*. Düsseldorf-Köln: Eugen Diederichs Verlag.
BUBER, M.
1949 *Die Erzählungen der Chassidim*. Zürich: Manesse Verlag.
CAMPBELL, J.
1959 *The masks of god*. I. *Primitive mythology*. New York: Viking Press.
1972 *Myths to live by*. New York: Viking Press.
CASSIRER, E.
1944 *Essay on man*. New Haven: Yale University Press.
FRANKL, V.
1969 "Reductionism and nihilism," in *Beyond reductionism*. Edited by A. Koestler. New York: Macmillan.
HEIDEGGER, M.
1960 *Sein und Zeit*. Tübingen: Niemeyer Verlag.
HEUSCHER, J. E.
1960 Introduction to myths and fairy tales. *Existential Psychiatry* 1:196.
1970 Language, authenticity and human growth. *Confinia Psychiatrica* 8:193.
1971 The concepts of constitution and deconstitution in psychotherapy. *The Human Context* 3:254.
1974 *A psychiatric study of myths and fairy tales*. Springfield: C. C. Thomas.
JACOBI, J.
1949 *Die psychologie von C. G. Jung*. Zürich: Rascher Verlag.
JUNG, C. G.
1954 *Welt der psyche*. Zürich: Rascher Verlag.
Kalevala
1969 *Kalevala*. New York: Everyman's Library. (Translated by W. F. Kirby.)
KERÉNYI, C.
1963 "The psychological aspects of the Kore," in *Essays on a science of mythology*. Edited by C. G. Jung and C. Kerényi. New York: Harper Torchbooks.
KOESTENBAUM, P.
1969 *The vitality of death*. Westport: Greenwood.
KOESTLER, A., *editor*
1969 *Beyond reductionism*. New York: Macmillan.
LENTZ, F.
1971 *Bildsprache der Märchen*. Stuttgart: Verlag Urachhaus.
MAY, R.
1953 *Man's search for himself*. New York: W. W. Norton.
1958 *Existence*. New York: Basic Books.
MEYER, R.
1935 *Die Weisheit der Deutschen Volksmärchen*. Stuttgart: Verlag der Christengemeinschaft.
NATANSON, M.
1966 "Introduction" in *Essays in phenomenology*. The Hague: Martinus Nijhoff.

NEUMANN, E.
1971 *Art and the creative unconscious*. Princeton: Bollingen Series.

Rätoromanische Märchen
1973 *Rätoromanische Märchen*. Düsseldorf-Köln: Eugen Diederichs Verlag.

SKINNER, B. F.
1971 *Beyond freedom and dignity*. New York: Bantam Books.

SPIEGELBERG, HERBERT
1966 "How subjective is phenomenology," in *Essays in phenomenology*. The Hague: Martinus Nijhoff.

TOLKIEN, J. R. R.
1966 *The Tolkien reader*. New York: Ballantine Books.

VON FRANZ, M.-L.
1973 *Interpretation of fairy tales*. Switzerland and New York: Spring Publications.

WEISS, P. A.
1969 "The living system," in *Beyond reductionism*. Edited by A. Koestler. New York: Macmillan.

Dance as Nonverbal Communication

LAURA B. DeLIND

The existence of dance has been noted in every human society for which adequate and competent ethnographic materials have been collected. As an activity it is pan-human. Stated somewhat differently, it emerges as universal behavior.

Despite its omnipresence, dance has been given sporadic and minimal treatment in anthropological investigation. It has been characteristically viewed as a type activity easily set apart from those behaviors presumed to be vital and essential. By implication, we are expected to infer that dance contrasts sharply with those substantial activities which keep humans alive and well within their cultural and ecological environment. It has traditionally been relegated to the domain of the folklorists.

Yet, there remains a curious contradiction. In societies that must maintain a narrow, strategic balance with their environment, which are closely bound to the "soil," where inefficiency and misplaced energies easily prove fatal, frills would be untenable. Still, all people dance.

Anthropologists have sought to define away contradictions such as this, and dance has been subject to many of the same manipulations imposed upon other supposedly discrete forms of human expression. As a result, it has been treated in two general but not unrelated ways.

The first, which I will call the "taxonomic approach," concentrates upon dance, or more accurately upon dances, as things. As things they are categorized with respect to what is felt to be their specific content or form. They are labeled according to purpose (i.e., war, fertility, social, religious), according to internal structure or motor themes (i.e., circular, jumping, linear), according to symbolic or aesthetic qualities (i.e., free, joyful, animistic), or new divisions are created (i.e., body parts used, the number and types of transitions) to further refine the tabulation process (Sachs 1937; Bunzel 1949; Bartenief 1968; Lomax, Bartenief, and Paulay

1968). Such ordering lends itself to old evolutionary moves, for all men do not dance the same dances or move in the same manner. Value judgments are cloaked in standard hierarchical terms. Certain dances are felt to be more complex, more symbolic, more pure and complete than others. They are often correlated with types of societies and levels of socio-technopolitical integration and, by logical extension, with potentials, capabilities, and competences among peoples.

An illustration of the direct articulation made between dance behavior and "innate tendencies" can be found in Sachs' *World history of the dance*. He puzzles over the differences found in dance styles. What would explain, for example, the "striking" differences between Vedda dancing which is "the most difficult contortion and distortion," and "the balanc(ed), measured action" of the Andaman Islanders? The former he feels is "tortured, joyless," the latter "liberated and joyful." He concludes that, "distinction is so great that in the case of two people almost equally primitive, it can hardly be ascribed to levels of culture, but rather to differences in natures" (Sachs 1937:12). Bunzel provides complementary illustration. "Prior to music and language — with which it is yet inseparably connected — dancing is the irrevocable and unique self-expression of man in co-ordinated movements; it leads to highest individual and collective ecstasy dissolved in the moment of tension-and-solution where equilibrium is supreme." A bit later he adds, "Our problem may be stated thus: shall we consider dancing an 'instinct' or an art? *Forms and social contents* of dances will help us to decide such a question" (Bunzel 1949:436, italics added). Inextricably tangled with this notion, and undoubtably responsible for its underpinnings, is the belief that motor or nonverbal behavior is undeniably more innate, affective (i.e., animal-like) than its more rational, instrumental (i.e., humanlike) counterpart, verbal behavior.[1]

The second approach, which I will term the "rationalization approach" treats dance as a type of activity. Here, the activity becomes the thing rather than, as just presented, the specific dance itself, and treatment concerns its psychological function within society. While other behaviors, commonly referred to as daily behaviors, are real or "normal," dance as a type activity is not. It is a release mechanism, an outlet or safety valve for draining the noxious combustibles of daily life. If a people is routinized, it will need a socially acceptable means by which to express suppressed creativity. If people are undifferentiated, there will be adaptive defenses to accommodate unexpressed individuality. If people are insecure, they will find avenues for sublimated aggression and competition. If an antici-

[1] It is not my purpose in this paper to present or elaborate upon the intellectual basis or the historical and modern arguments surrounding this issue. Many of the conflicting opinions will already be familiar to the reader, and the bibliography compiled for the purpose of this paper presents, at least, a general overview of the problems involved.

pated behavior is absent, its negation substantiates its existence. Dance is an activity that is used for such substantiation. The logic, or lack of it, centers around the belief that the individual must maintain an equilibrium — the real state of affairs — by using every means within his multileveled, adaptive mind to insulate him against change (Mead 1961; Katz 1973).

In the first case, prescriptions are propagated by taxonomists with little regard to context or attention to the nature of the behavior they are arranging. Dance becomes a "museum piece," and like any other artifact, it is displayed as a cultural object, replete with its trappings: musical instruments, costumes, or the ritual occasion to which it belongs.

In the second case, dance activity is also prescriptively set apart from the other more usual or typical activities of life. It is seen as the behavior of coordinated bodies that serves as a mechanism through which the needs of the individual and society are more nearly reconciled. Life experience, on the other hand, is not viewed as a bodily phenomenon. Rather, it is approached in terms of shared attitudes, values, meanings, and beliefs *inferred* from behavior. From the start, the two are treated differently and so, not surprisingly, remain different.

There does exist a variation on this "rationalization" theme. And, while its overt functional explanation of dance may appear different, its underlying assumptions are clearly the same. This is the notion that dance reflects those aspects which are most essential to, or representative of, daily life. Dance serves as a means of reinforcing those role behaviors, customs, and beliefs that are crucial to the continued maintenance of the society (Wilson 1958; Kurath and Marti 1964; Bartenief 1968; Lomax, Bartenief, and Pauley 1968). Again, we are dealing with dance as an activity qualitatively set apart from those of daily life. Again, dance is a physical, bodily activity while the rest of social life is considered to be the mental internalization of symbols, expectations, and patterns of social order. Again, we are presented with the definitional game. For if dance activity does not appear to be contradictory, then its function necessarily becomes supportive of real life.

Both the taxonomic and rationalization approaches treat dance as static. Both concern themselves with differences, discontinuities, and dichotomies (i.e., mind/body, verbal/nonverbal, society/individual, daily activity/dance activity). Both have considered dance as a curiosity and have shorn it of its identity as a real dynamic. The categories and divisions have short circuited the real problem which is a behavioral one. Simply, the essential questions and data have been misplaced, misused, and often missed entirely.

But dance need not be relegated to such marginal or static consideration. There is a tenable third approach which proceeds from the position that dance is real, that cultural experience is the organization of learned bodily managements, and that an understanding of behavior requires approaching behavior as behavior. To grapple effectively then with the problem of what dance may be about, we must consider that behavior occurs in bodies. We must not exclude any behavior nor isolate particular forms to be reified and studied in and of themselves. We must not proceed with the belief that dance is exceptional, or that the body is a large unwieldy appendage to which humans are uncomfortably saddled. The human body and its management are very real adaptations to and expressions of the environment — the cultural history and ecological parameters which form the temporal and spatial (i.e., contextual) dimensions of existence. An open approach to the study of dance should concentrate upon bodily behavior within the cultural networks to which it is wedded.

Yet, the same behavior to which we react, with which we interact, from which we so readily abstract, lacks any complete or detailed description. Humans walk in a variety of ways, posture themselves according to the situation, their perception of self and of those about them (Hewes 1955; Sarles 1970; Schilder 1950; Goffman 1959, 1967; Efron 1941; Birdwhistell 1970; LaBarre 1947). Their gestures are principally learned; their bodily functions, regulated; their rhythms, balance, musculature all similarly shaped and managed. In such a manner, humans quite literally come to look and think like those around them. The body is central to perception, for the conceptual models of self and necessarily others are formed through bodily experience (Schilder 1950). Perception in turn, is the result of internalized behavior, housed and realized in terms of a body which has learned its own organization. It is a body which is shaped by a reality, and whose dynamics in turn create reality.

Such a construct is not new. A half century ago, Malinowski was clearly coming to similar conclusions from a different perspective, for he proceeded from concerns of mind, not body. In his discussion of language and meaning, he felt a great need to question traditional notions of context, behavior, and current linguistic assumptions.

> (The whole utterance), becomes only intelligible when it is placed within its *context of situation*, if I may be allowed to coin an expression which indicates on the one hand that the conception of *context* has to be broadened and on the other that the *situation* in which words are uttered can never be passed over as irrelevant to the linguistic expression. We see how the conception of context must be *substantially widened* if it is to furnish us with full utility. In fact, it must burst the bonds of mere linguistics and be carried over into the analysis of the general conditions under which the language is spoken. Thus, starting from the wider idea of context . . . the study of any language, spoken by a people who live under conditions different from our own . . . must be carried out in conjunction with the study of their culture and their environment (Malinowski 1923:306, italics added).

Malinowski's "substantially widened" context must be extended farther than even he anticipated into an expanded concept of language and communication. However, he provides a firm intellectual foundation for making such a move, in spite of his difficulties in finding a useable methodology.

He pointed out "that the conception of meaning as *contained* in utterance is false and futile, that each "statement by a human being has the aim and function of expressing some thought of feeling actual at that moment and in that situation" (Malinowski 1923:307).

It is erroneous to think that each word (or each movement) holds a real essence, a vital kernel of meaning which is brought to life with its every utterance. He criticized the typical philosophical maneuvers which operate upon the belief that "truth is found by spinning out meaning from the word, its assumed receptacle" (Malinowski 1923:308). "Language is little influenced by Thought, but Thought on the contrary, having to borrow from action its tool" (Malinowski 1923:328).

Meaning is found only in action, and linguistic expression is only real within a context of activity and movement. It is the actual, dynamic behavior that shapes reality, and that behavior is both verbal and nonverbal. ". . . language acquires its meaning only through personal participation in (a type) pursuit. It has to be learned, not through reflection, but through action" (Malinowski 1923:311). Language must be viewed "against a background of activity and as a mode of human behavior" (Malinowski 1923:312).

Malinowski's attention to context was proposed as a means to further understand and study language — verbal behavior. While in context, nonverbal behavior was clearly important, it still remained subservient to his linguistic focus and is thereby neglected for any individual or detailed study (Ruesch 1966:209–213; Ruesch and Kees 1956; Ekman 1970:151–158; Ekman and Frisen 1969:49–98, 1971:124–129).

Such arbitrary, but traditionally hardened barriers, should begin to disappear if we approach verbal and nonverbal behavior as aspects of the same phenomenon, as real, meaningful culture and context specific bodily behavior which must be learned, internalized, and reproduced. We have expanded our dealings with behavior beyond Malinowski's linguistic bias, for language is not a "disembodied" phenomenon (Sarles personal communication; Bateson 1951), but rather one whose primary concern must be just that — bodies and their dynamics.

Malinowski's concept of context, the former handmaiden to linguistic concerns, has been transformed into our specific concern. In such a way, it has burst the bonds of mere linguistics to become a cultural phenomenon and a means by which to investigate the nature of human communication; the transmission of body management, perceptions, and values; the adaptive behavior which both creates and handles the cultural world of experience, thus survival. Since the dynamic study of communication concerns itself with the notion that the individual-existential self is about as rare a part of one's being as the concept of new information (Sarles 1970), we must operate within the larger question context: How are bodies managed to fit cultural expectations and how are these expectations internalized and reproduced in culturally recognizable, viable, behavioral ways?

We are dealing with learned bodily managements and the resulting perceptions traceable through behavior. We are dancing in a most investigable way.

With these theoretical rumblings in mind, let us consider two concrete and *ostensibly* unrelated behavioral illustrations.

The canoes glide slowly and noiselessly, punted by men especially good at this task and always used for it. Other experts who know the bottom of the lagoon, with its plant and animal life, are on the look out for fish. One of them sights the quarry. . . . Then, the whole fleet stops and ranges itself. . . every canoe and every man in it performing his appointed task — according to a customary routine. . . . Again, a word of command is passed here and there, a technical expression or explanation which serves to harmonize their behavior towards other men. The whole group acts in a concerted manner, determined by old tribal tradition and perfectly familiar to the actors through lifelong experience. Some men in the canoes cast the wide encircling nets into the water, others plunge and wade through the lagoon, drive the fish into the nets. Others again stand by with small nets ready to catch the fish. An animated scene, full of movement follows and now that the fish are in their power, the fishermen speak loudly and give vent to their feelings (Malinowski 1923:311).

Malinowski has provided a picturesque description of a specific activity in which verbal behavior is significantly infrequent. Rather, we are impressed with bodies which have learned to manage and organize themselves in an activity essential to survival.

These men have learned to behave bodily in well patterned ways which are understood, expected, anticipated, and demanded by other Trobriand fishermen. They can remain still for long periods of time and yet maintain a bodily alert. They must sit in a confined area without fatigue and with a definite idea of the spaces around them and how they must be managed. They can make instantaneous effort transitions from relative nonactivity to high speed and accurate bodily judgments and adjustments. They must be agile enough to maintain shifting balances and strong enough to control canoes, net, and fish. They must have internalized a rhythm which will not intrude upon their surrounds and frighten away fish. Their eyes and eye muscles are managed in such a way as to provide them shade and still penetrate the lagoon waters. Their arms, legs, head, hearing, breathing, eating, and bodily habits are similarly regulated and controlled. But, most significantly, they are men who, while performing different sets of behaviors, can "harmonize," read bodily cues, coordinate movement effectively, and look "right" doing it. They share a similar sense of the world, and they keep it within similar bodies.

Such Trobriand behavior — bodily coordination, and interaction — has become "perfectly familiar to the actors through life-long experience" (Malinowski 1923:311). Trobrianders can only behave as fishermen because they have learned to do so. Their bodily control, rhythms, efforts, and interactive abilities had been developing long before they became fishermen. This learning experience began before birth and is maintained throughout life. Their cultural context has impressed its bodily demands and returned corresponding attitudes, perceptions, bodily and muscular feelings.

Trobriand fishing is highly organized, stylistic, coordinated, and dynamic behavior. It demonstrates successful interaction. It is essential to survival. And, oddly enough, it looks like dance.

Yet, upon closer investigation, the situation is really not odd at all. Dance is bodily behavior and should be described in these same behavioral terms. Conversely, other behaviors can be realized in those terms that describe dance. While Trobriand fishing may look somewhat different from that behavior traditionally defined as Trobriand dance, the emphasis upon the body, its management, motor patterns, gestures, and interrelationships is clear. The similarity between the two is not surprising, for when Trobrianders move, they do so in Trobriandesque ways, and these behaviors come to look and feel "right." They are the "right" ways for Trobrianders to interact, and in a very real sense that is what being a Trobriander is about.[2]

A second and more familiar illustration of learned bodily management can be presented from our own culture and involves the training of a ballet dancer. A beginning dancer will bring to his first lesson an inappropriately managed body. It is the teacher's responsibility to provide the consistent instruction necessary to shape this body into one which will behave in the expected and culturo-aesthetically appropriate ways. Such training takes many years and deliberate practice.

From the first lesson, bodily demands placed on the student provide him with a real, but different (i.e., extended) knowledge of his body. He is told to turn out from the hip, not to roll over on his feet, but to keep them firmly and fully planted on the ground. His weight must be centered over his hips, his knees must be kept over his toes and he must learn the size step and ground spacing which is best suited to his body. He increasingly

[2] This same reasoning transfers easily to ethnic communities. These are countless examples of ethnic stereotyping, and the questions; Are you a ——?, You look like a ——, are really too familiar to need repeating. Yet, the looks, bodily expressions and behaviors which can be so readily characterized, have received scant in depth behavioral (i.e., bodily management) investigation.

becomes aware of new areas of his body. He learns to recognize and manage new sets of muscles, and gradually becomes comfortable behaving in ways which initially seemed quite difficult.

Exercises are practiced on both sides at the barre. The left and right sides of the body (which may at times be used differently) are developed equally, and felt to be equally important. Ideally, there should be no dominant side, and so, no muscular compensations for a weak left or right.[3] The barre must be used only as a light support. Weight must be centered in the body so that it is free to balance at any moment. There can be no "clinging" or "hanging on." The body must move and maintain itself independently in space.

When steps that require traveling across the floor are introduced, new temporal and spatial adjustments must be made. A step must be done in time, but it also must be done in turn. Bodies that already know a great deal about themselves must interact. The student must know, before he begins, how far a particular step will carry him. If he has no room, he must decide whether he can change his direction, or whether he can make his movements smaller, still dancing the sequence properly and in time. He must know about the space he is in, the space to which he intends to move, and how he is going to get there. In addition, he must expect that the other dancers all know the same.[4]

The student then, can anticipate his own movement vis-à-vis other bodies. He can translate a spoken instruction into ballet behavior or a single demonstration into his own body to be performed immediately. He is able to recognize behavior, relate it to his own, and know how it must be performed if it is to look and feel right.

Yet, both this look and feel (which for our illustration is specific to the ballet behavior) have been learned. The student has seen the same bodily movements and combinations of movements performed repeatedly, and he is told to watch as often. "If you don't know the step, watch someone who does. Don't talk about it. Watch, and do it. Don't look in the mirror. You should know if you are doing it right."[5] But how does the dancer know when his body feels right?

[3] The notion of conceptualizing the body in terms of a left and right side should not be dismissed as obvious. My only recollection of my first lesson was my teacher placing a rubber band around my right wrist so that I might keep the two sides apart. I was quite familiar with the two directions. I just had never found it necessary to make the distinction within my own body when I was moving.

[4] It is interesting to note that the instance of bumping other dancers has virtually disappeared by the third year of dance training. It might prove interesting to investigate how the student learns to handle this spacing problem and how the conceptualization of his body changes accordingly.

[5] Frequently, the teacher demands that the class dance with their backs to the mirror. Without this aid, the dancer is forced to concentrate upon his body and rely upon his knowledge and feeling of it to successfully execute the steps. Here again, we are dealing with changes in bodily management and conceptualization.

We have already commented that initially this new bodily management feels anything but right. The student however, learns not only through observing and doing, but through the use of metaphor.[6] He is told to "turn out," "pull himself out of his body," "not sink into his hips," "wrap his foot around his ankle." These commands and corrections initially mean very little. Yet, as he watches, and has his body placed in position by the teacher, he begins to feel what has actually been expected. And finally, when the teacher says, "Good. Keep it like that," the student knows what it is to "pull himself out of his body." He has internalized a dancer's metaphor. He has translated it into his muscles. He knows how to reproduce it. It is real and it begins to feel right.

Yet as this behavior becomes internalized, it shapes both the dancer's body and his perception of himself. He has made certain bodily commitments in order to behave in this manner. If he is a serious student he must take at least five lessons a week, ideally two a day. What adjustments must he make in his schooling, his social life? How will his behavior differ from the behavior of other children his own age? Certainly, he will spend less time running around noisily outdoors with a group of friends, and interestingly enough he will probably no longer want to.

He has learned to sit quietly and wait for his lesson to start. It is not unusual to see a group of students knitting, crocheting, sewing, or reading while they wait for their lesson. These are all, interestingly enough, individual, noiseless behaviors which demand a relatively high degree of concentration and/or bodily dexterity. The relationship they have to the larger behavioral context presents an intriguing problem. He has learned also to sustain a high level of concentration and bodily control during a two-hour class session, and then practice by himself while he waits for his next lesson or rehearsal. He must learn to behave in this manner day after day, for there are few acceptable reasons for missing class.

The student's weight is a constant source of concern, and he is extremely careful of what he eats. Protein, vitamins, carbohydrates, fats, and calories are carefully calculated. He is also careful about when he eats, for to dance on a full stomach is not only uncomfortable, but dangerous. He must not smoke excessively, ice skate, water ski, or participate in any number of behaviors which use the "wrong" (i.e.,

[6] I observed the use of metaphor during a beginning ballet class composed of six and seven-year-olds. The children were standing in the center of the floor working on their arm exercises — a *port de bras*. The instructor's verbal explanation and counting of the sequence was "pumpkin, pumpkin, explode, and down." The children seemed to have little difficulty translating the movements into their own bodies. The many conceptualizations of "pumpkinness" were unfortunately left unexplored. I remember another incident in which metaphor was used in an equally effective, but negative way. My teacher had become annoyed by a student who was not dancing "full out," but ambling thoughtlessly through her adagio. He stopped the class and yelled, "You look like a banana floating in a bowl of vegetable soup." We all knew what he meant, and undoubtedly could have repeated the behavioral qualities on command.

different) sets of muscles. If his hair is short, he will grow it long. If his legs are heavy, he will exercise them unrelentingly. If his nose is too large, he may have it fixed.[7] The dancer then, has developed a clear model of what proper behavior entails, and his time is spent attaining this perfection.

The standard jokes about ballet dancers looking alike contain more true observation than humor. The dancers have learned to use the same muscles, internalized the same metaphors, adopted similar bodily habits and managements. They have learned similar temporal and spatial relationships, anticipate behavior and behave vis-à-vis other bodies in similar manners, shape their bodies in similar ways, share conceptual models and behavioral awarenesses and values. Why should not they look alike? They have learned too much to look any other way.

The way a dancer comes to look is a result of the ways in which he has learned to manage his body. Yet, once learned, these behavioral patterns are often difficult and sometimes impossible to change. An Indian temple dancer for instance, cannot become a successful ballet dancer; the opposite also is true. They have both learned bodily and muscular managements and perceptions that cannot comfortably be translated into the movement demands of the other.[8] The transition, however, from ballet to modern dance, while initially somewhat uncomfortable, is quite common. I recall my utter frustration and anger as a result of my early experiences with modern dance. I had been trained in ballet and had become comfortable and familiar with its bodily demands. I was now in the hopeless situation of being told to improvise — "to be a box." Not only was I incapable of improvisation, having never been required to do so in ballet, I had no idea what it meant to be a box. I stood motionless or, more accurately, paralyzed. I could not move. I just did not know how. While these feelings of complete bodily incomprehension gradually subsided, it took several years before the new movement patterns, bare feet, and different muscular awarenesses became acceptable behavior.[9]

Yet our discussion is not specific to dancers or dance behavior. Our ballet illustration has described nothing more than was presented earlier in the quote by Malinowski. Ballet training and behavior exhibits remarkable similarities to that of Trobriand fishermen. The hours of

[7] Before class, we would wrap plastic around our bodies in order to reduce those areas (i.e., hips, thighs, upper arms) which we felt unsuitable for, or inconsistent with our model of ballet behavior. A more drastic "manipulation" was recently brought to my attention. A dancer had decided to have her breasts surgically removed because she was tired of losing dance parts for anatomical reasons.

[8] Polsky notes a similar phenomenon in the game styles of pool (Polsky 1969:62fn.).

[9] As far as I am aware, there has been no investigation into the question of what bodily managements and conceptualizations preclude or restrict others. Also, the notion of comfort has been given little recognition in behavioral terms. What makes something comfortable or uncomfortable, and where do the boundaries lie? Culture shock too, might profitably be considered in terms of bodily incomprehension.

dance practice week after week, year after year, presents an analogous situation to a lifetime of training within a cultural network. We are talking then about bodies and how they come to be managed to allow for interaction, cooperation, communication — how they provide for and reflect this cultural context.

While we have dealt specifically with dance behavior, our larger concern is Dance. This notion (with a capital "D") becomes the system of learned bodily managements — the dynamic behaviors which result in particular looks and attitudes and serve as the basis for viable interaction. We have expanded our notion of dance in much the same way as we treated Malinowski's language. It is within this expanded concept that all bodily behavior may be considered. Once treated as a similar bodily phenomenon, we may more carefully investigate what particular sets of behaviors are about, how they are learned, and how and where differences exist, be they between specific groups of people or cultures or new dimensions suggested by this behavioral approach. At the same time, our focus upon dance (small "d") behavior has provided the stimulus and insight necessary to handle this larger concern.

Dance behavior, as we have seen, forces us to attend to the body, its movement and management through time and space. It provides a unique vantage point from which to observe behavior within its cultural context and therefore with respect to meaning. It is a cultural exhibition of bodily form management and interaction. It places emphasis upon postures, gestures, efforts, transitions, and the behavioral dynamics that all work to form what we can term the cultural Dance.

Definitionally different behaviors (i.e., a cultural repertoire of learned motor characteristics) may be seen to utilize the same interactional patterns (i.e., behavioral dynamics) that dance displays. The attitude that ". . . dance is composed of those gestures, postures, movements and movement qualities most essential and most characteristic of the activity of everyday life. . ." (Lomax, Bartenief, and Paulay 1968:224), may be viewed in the following manner: Daily activity will find itself composed of the motor themes, bodily managements and relationships most essential and most characteristic of dance (and of Dance). It may well be found that humans Dance their way through life.

The focus on dance behavior then provides an invaluable means of observing systems of bodily interaction, other behavioral correlates, and finally a dynamic perception of self and others which help form the cultural whole. The more precisely the dance movement and its bodily characteristics can be described, the more accurately other cultural correlates can be recognized within different behavioral settings, the contextual factors isolated, and Dance itself understood.

The attitude that motor behavior is ill suited to verbal translation, primal, and thus less essential to, or characteristic of, human interaction, only serves to understate the minimal exploration into behavioral analysis and human communication. While humans do talk, they do a great deal more. They learn bodily to become what they are. Whatever the nature of human nature, it cannot properly be studied for what it is, if we seek only to define it defensively and know it in terms of what it is not. There is no room for theoretical expansion, no means by which new insight or new contexts can develop, and no possibility of approaching a scientific understanding of reality. Once arbitrary dichotomies and dead ends are eliminated, when bodily behavior is no longer initially separated from language, or dance from other cultural activities, we can begin to investigate what they actually are.

REFERENCES

BARTENIEF, IRMGARD
1968 "Research in anthropology: a study of dance styles in primitive cultures," in *Research in dance: problems and possibilities*, 91–104. New York: CORD.

BATESON, GREGORY
1951 "Why do Frenchmen?," in *Impulse*. San Francisco: Impulse.

BIRDWHISTELL, RAY L.
1970 *Kinesics and context*. Philadelphia: University of Pennsylvania Press.

BUNZEL, JOSEPH H.
1949 "Sociology of dance," in *The dance encyclopedia*. Edited by Anatole Chujoy, 435–440. New York: A. S. Barnes.

EFRON, DAVID
1941 *Gesture and environment*. New York: Kings Crown Press.

EKMAN, PAUL
1970 Universal facial expressions of emotion. *California Mental Health Research Digest* 8:151–158.

EKMAN, PAUL, WALLACE V. FRISEN
1969 The repertoire of nonverbal behavior: categories, origins, usage, and coding. *Semiotica* 1:49–98.
1971 Constants across cultures in face and emotion. *Journal of Personality and Social Psychology* 17:124–129.

GOFFMAN, ERVING
1959 *Presentation of self in everyday life*. New York: Doubleday.
1967 *Interaction ritual*. New York: Doubleday.

HEWES, GORDON W.
1955 World distribution of certain postural habits. *American Anthropologist* 57:231–244.

KATZ, RUTH
1973 The egalitarian waltz. *Comparative Studies in Society and History* 15:368–377.

KURATH, GERTRUDE, SAMUEL MARTI
1964 *Dances of Anáhuac: the choreography and music of precortesian dances*. Viking Fund Publications in Anthropology 38. New York: Viking Fund.

LABARRE, WESTON
1947 The cultural basis of emotions and gestures. *Journal of Personality* 16:49–68.
LOMAX, ALAN, IRMGARD BARTENIEF, FORRESTINE PAULAY
1968 "Dance style and culture," in *Folksong style and culture*. Edited by Alan Lomax, 222–247. Washington, D. C.: American Association for the Advancement of Science.
MALINOWSKI, BRONISLAW
1923 "The problem of meaning in primitive languages," in *The meaning of meaning*. Edited by C. K. Ogden and I. A. Richards, 296–336. New York: Harcourt, Brace and World.
MEAD, MARGARET
1961 *Coming of age in Samoa*. New York: Morrow.
POLSKY, NED
1969 *Hustlers, beats, and others*. New York: Doubleday.
RUESCH, JURGEN
1966 "Nonverbal language and therapy," in *Culture and communication*. Edited by Alfred G. Smith, 209–213. New York: Holt, Rinehart and Winston.
RUESCH, JURGEN, WELDON KEES
1956 *Nonverbal communication: notes on the visual perception of human relations*. Berkeley University of California Press.
SACHS, KURT
1937 *World history of the dance*. New York: W. W. Norton.
SARLES, HARVEY B.
1970 "Facial expression and body movement." Unpublished manuscript.
SCHILDER, PAUL
1950 *The image and appearance of the human body, in the constructive energies of the psyche*. New York: International Universities Press.
WILSON, EDMUND
1958 "The Zuni Shalako ceremony," in *Reader in comparative religion, an anthropological approach.* Edited by William A. Lessa and Evon Z. Vogt, 159–169. Evanston: Row, Peterson.

Interrelationships of Individual, Cultural, and Pan-Human Values

C. R. WELTE

1. Twenty years ago the study of values was given much attention at the Wenner-Gren Foundation International Symposium on Anthropology. Prominent anthropologists of the day discussed the two inventory papers (Tax et al. 1953: 322–341), and A. L. Kroeber devoted the last four pages of his "Concluding review" to the subject. Throughout the decade of the 1950's value studies were actively pursued.[1] In England their relation to social structure was emphasized (Firth 1953). In America Kroeber and Clyde Kluckholn wrote extensively on values and culture and tied the two concepts closely together (Kluckhohn et al. 1951; Kroeber 1952, 1953, 1956; Kroeber and Kluckhohn 1952; Kluckhohn 1953, 1958). A massive fieldwork project, the "Comparative study of values in five cultures" of the Laboratory of Social Relations, Harvard University, lasted for six years and generated over fifty publications (Vogt and Albert 1966: 299–305). There seemed to be widespread agreement that value studies were important, and that, as Kroeber had said, "it follows that if we refuse to deal with values, we are refusing to deal with what has most meaning in particular cultures as well as in human culture seen as a whole" (1952:137).

Nevertheless, in the 1960's there was a general turning away from value studies. Firth noted that they "are perhaps less fashionable now" (1964:180). Vogt and Albert, in their summing up of the Harvard "values study" in *People of Rimrock* (1966:1–21), identified many unresolved problems that the study had brought to light and could only hope

[1] Except for American and British studies, there has been little theoretical development of the concept of values as an analytical tool in anthropology. Recent European work has been on the philosophical side and often deals primarily with the metaphysical concepts of the people being studied. For a survey of the work in sociology, and of its limitations, see Hutcheon's recent article (1972:173–177).

that developments in componential analysis might provide solutions. Theodore Graves, in a review of *People of Rimrock* (1967), added charges of circular reasoning and opportunistic stretching of definitions. Manners and Kaplan, in their *Theory in anthropology* (1968), left the subject of values to an article from a sociology handbook which denigrated both values and culture as explanatory concepts (Blake and Davis 1964).

No one had responded adequately to the fundamental need for a classification of values. Commenting on a presentation by Kluckhohn, William L. Kolb said, "I am appalled at the number of systems of classification of values . . . that are running loose today, each claiming . . . to be exhaustive at the same or similar level of abstraction; and each failing to give a full-scale logical ground for its claim" (1961:52). Ethel Albert, in her article "Value systems" in the *International encyclopedia of the social sciences* (1968), found no classification systems that had improved on those of the 1950's. At the end of the decade a philosopher published *Value theory and the behavioral sciences* (Handy 1969). He found the theories wanting and even, at times, incoherent.

The unresolved problems of value studies are still with us. Neither developments in componential analysis nor in other kinds of cognitive studies related to values (Wallace 1962:354–355) have proved to be effective. I think, however, that a clue to a fruitful approach has been given to us by Kroeber and Kluckhohn. They wrote, in their monograph on culture, "culture change seems to be due to the ceaseless feedback between factors of idiosyncratic and universal human motivation, on the one hand, and factors of universal and special situation, on the other. Unfortunately, we lack conceptual instruments for dealing with such systems of organized complexity" (Kroeber and Kluckhohn 1952:111).

In this paper I develop a taxonomic classification of values — based on distinctions between, and interrelationships of, the individual, the cultural, and the universal — and show how this provides a basis for attacking the problems that have arrested value studies. I also analyze the diachronic and synchronic interactions of major classes of values and thus provide a new view of the value factors in the "systems of organized complexity" with which anthropologists must deal. This view is based on the assumption that the calculi of culture describe only one area of the cognitive processes that must be investigated in the study of cultural differentiation and change.[2]

[2] Roger Brown, in his summation of the conference "Transcultural studies in cognition," spoke of the contrasting views of mind as categorical grid or template, and as operating agency or transformer. He said, "We need comparative studies of mind in all its aspects and not of the categorical grid alone. Eventually someone is going to have to nail together the template and the transformer" (1964:252). In value studies the transformer metaphor is apt, particularly if extended to include the interrelationships of rules for behavior in a wider-than-cultural context and the interactions of sets of these rules in decision making.

2. Semantic considerations must be faced at the beginning of any study of values. The widely varying meanings of "culture" are well known. When they are combined with the ambiguities of "universals" and of "values" immediate clarification is called for.

As for "culture," suffice it to say that I use it as equivalent to "social heritage" and place culture in the ideational order of reality. "Universals" cannot be dealt with so easily. In addition to the many meanings of "cultural universals," the association of "universals" with "values" has caused ambiguity because the term "universal values" is often used as equivalent to "universal cultural values." In his "Concluding review" Kroeber entitled the last section "Universal values" but he discussed universal cultural values (Tax et al. 1953: 375–376). Investigations of universal cultural values have been unsatisfactory because of the few universals found, because of the almost contentless level of abstraction of most of them, and because of lack of agreement on what is proved about a value if it is found to exist in all cultures.

In investigating the factors of "universal human motivation," that Kroeber and Kluckhohn referred to in the clue that they gave us, we must make a place for the possibility that there are values that are not part of the social or the genetic heritage; values that are rooted in experience, but which interact with cultural values in motivating the choices men make; values that, if they do exist, radically change the interrelationships that have been assumed to be operative in the field of values. This possibility will be investigated, but to avoid the confusions apparent in the use of "universal values" I shall use the term "pan-human values."

The term "values" itself has proved to be a stumbling block in value studies. As Vogt and Albert comment, "descriptive studies of values usually either offer no definition at all or adopt a verbal definition that does not make effective contact with the data. Any other definition might have been cited without affecting the description. Such 'ritual' use of a definition of values does not contribute to stabilizing the concept" (1966: 6). Kluckhohn had tried to stablize it with a definition that restricted it to concepts of the desirable (as opposed to the desired), but this was the very definition that has been subject to "ritual" use. Furthermore, it has been rejected as too narrow, in the article "The concept of values" (Williams 1968), and, as Albert writes in her ensuing article, "it is doubtful whether a definition of values can be produced that embraces all the meanings assigned to the term and its cognates or that would be acceptable to all investigators" (1968:288). Kluckhohn saw the problem when he reviewed the literature on values and found "values considered as attitudes, motivations, objects, measurable quantities, substantive areas of behavior, affect-laden customs and traditions, and relationships such as those between individuals, groups, objects, events. The only general agreement is that values somehow have to do with normative as opposed to existential propositions" (1951:390).

I think it is clear that an attempt to reconcile the definitions that are in use is not the way to approach the problem. If we were to make a semantic analysis of the domain of values in the usage of social scientists, the results would be startling but hardly useful. "Values" has usefulness only as a collective term to group kinds of values. When a particular kind or class of values is meant, a distinctive term must be used. We need a vocabulary of terms as part of a taxonomic classification.

3. In order to develop a logical classification, the locus of values must be decided upon. We are concerned with cognition, so our approach should be from the point of view of human symbolic processes. An objective symbol may be said to be "valued" and thus may "have value" for a person, but it is on the subjective side of the valuing process that choices are made, so I place values in the realm of cognitive elements with symbolic significance.

Before proceeding we must investigate the various usages of the term "symbol." A symbol can be anything that stands for something else, or has a meaning. The symbol itself may be an idea (notion, conception, mental image, element of cognitive structure) or an objective representation (a word, object, or sign). The meaning may be conventionally assigned, may develop through association, or may lie in the felt importance of an experience to the subject's scheme of life which includes future possibilities as well as present requirements. It is the last of these three kinds of meaning or significance that is applicable to values. They are symbolic in the sense of having this special kind of significance. Values are not words or verbal propositions. The entanglement of values with the verbal level has been one of the causes of difficulties in value studies. The importance of the distinction between values and the terms or labels (conventional symbols) used to express them will become apparent as we progress.

For a formal definition, values must be assigned to a class and then differentiated from other members of the class. In conformity with the locus chosen, let us call the class "elements of underlying cognitive structure." In addition to values, this class includes beliefs (in the broad sense of the existential ideas: those that pertain to what exists or what is considered to be true). Values are distinguished from beliefs by virtue of being affective and symbolic. Thus, values are defined as affective symbolic elements of underlying cognitive structure.

A definition such as this is useful only if we proceed to specify subordinate levels of classification and thus bring down the level of abstraction to where contact can be made with descriptive data. This procedure will form our taxonomy.

The first subordinate level of the taxonomy is that of *categories*. The three categories, *individual values*, *cultural values*, and *panhuman values*, designate the taxa at that level. Differentiation between them is to be made on the basis of the genesis of each category. The next subordinate level is that of *major classes*. Differentiation will be made on a functional basis by means of a cognitive model of the making of a choice. Further subdivision will produce the levels of *classes*, *subclasses*, and *types*. In the case of cultural values, 120 types will be specified for the ordering of descriptive data.

As a clarification of the nature of the categories, I wish to contrast them with *levels of discourse*. Three levels of discourse are sufficient for our purpose. The *universal level* is with reference to all men: mankind as a whole. The *group level* is with reference to all members of a specified society or subsociety: all the bearers of its culture or subculture. The *idiosyncratic level* is with reference to a particular member of a society: one flesh-and-blood culture-bearer. The values in each of the categories can be described at each level of discourse. Ambiguity is avoided by using terms for the categories that are distinct from those used for the levels of discourse. For example, individual values at the universal level of discourse are discussed in terms of the hypothesized processes of the human ego. Furthermore, a discussion of the interaction of individual and cultural values at the group level can be readily distinguished from a discussion of their interaction at the idiosyncratic level. Whether we study a person, a society, or mankind, all three categories of values are applicable.

4. The ontogenesis of each category of values is the basis for differentiating the three categories. The values in each category are rooted in experience but they are learned in different ways. Only the cultural values are part of a social heritage.

The ontogenesis of cultural values is through the special kind of learning called enculturation. During enculturation the beliefs and values of the culture are learned as the result of instruction or example. They make up the social heritage. The beliefs form existential patterns for behavior and the values form normative patterns for behavior. Behavior includes thinking, feeling, and acting. The distinction between the existential and the normative relates to the distinction between fact and value rather than that between real and ideal.

In the case of pan-human values, it is my thesis that their ontogenesis lies in the application of rudimentary powers of symbolization to experiences that all infants must undergo in some measure in order to survive. These are of two kinds: (1) experiencing satisfaction of biogenetic needs, and (2) experiencing activities which provide for satisfaction of those needs. When symbolization is applied to such experiences, values are formed that include the affect due to the relation of the experiences to self-preservation. In Section 7 the content of pan-human values will be discussed and the subordinate major classes (*primate needs*, based on satisfactions of needs shared with the anthropoid apes, and *pan-human standards*, based on experience of essential activities) will be differentiated.

In the case of individual values, we must turn to the psychologists for an hypothesis concerning their ontogenesis. The challenge given by Kroeber, when he wrote "we can hardly forever leave the psychic equipment of man a blank except for a few reflexes, no instincts, and a faculty for symbolization; at least not without a try" (1956:295), has not been taken up by anthropologists. There is certainly no consensus among psychologists, but there are psychologists (sometimes referred to as the "ego psychologists") who have developed a view of man's psychic equipment that provides a basis for an hypothesis concerning individual values. As Gordon Allport wrote, "to such writers the ego-ideal is no longer, as it was with Freud, a passive reflection of the superego, which in turn is conceived as a mere legacy of the parent. The ego through its ideals reaches into the future, becomes an executive, a planner, a fighter" (1960:75).

In the process of maturation, as the ego becomes the executive, the planner, and the fighter, individual values are learned. They develop from the experiences of mediating between the environment, internal tensions, value conflicts, and the struggle to develop and maintain identity. When Ego is able to resolve the conflicts, the values learned in the process become the standards, norms, preferences, and goals of the ego. These *ego values* are guides for future decisions.

The conflicts that Ego cannot resolve, and which it views as threats, may result in the development of *neurotic needs*. These neurotic defenses, obsessions, compulsions, etc., and the related anxieties, can become strong enough to override all other values at times. The ego values and the neurotic needs are the subordinate major classes in the category of individual values. I will not go further into personality structure, but will accept the view of the relative autonomy of the ego that sees Ego as attempting to resolve (among other problems) internal conflicts between values when confronted with a choice to which it must respond.

5. The major classes of values (the taxonomic level, next subordinate to the categories) are the ego values and neurotic needs within the category of individual values; the pan-human standards and primate needs within the category of pan-human values; and the cultural values (the category of cultural values is not subdivided at this level, so the term "cultural values" will designate a major class as well as a category). The genesis or source of each of these major classes has been indicated in the previous section. It will be assumed that, by virtue of their common source, the values in each major class form a set of values that function together when brought to bear on a problem. For instance, we can think of situations in

which cultural values are opposed to primate needs, or in which other combinations of the major classes interact.

Now let us analyze the elements involved in the making of a choice, in order to provide the basis for functional distinctions that will supplement the genetic distinctions we have made. The analysis will also relate the major classes to one another, and distinguish values in general from beliefs, capacities, responses, and objects of interest. Figure 1 represents the steps involved in the making of a single choice by an actor. This synchronic model assumes that the choice is an important one (a considered choice) in which all elements are active, but that questions of whether certain elements are rational or irrational, conscious or unconscious, strong or weak, are not essential to the relationships and distinctions with which we are concerned. The model presents an ideal case, but one that is useful to us.

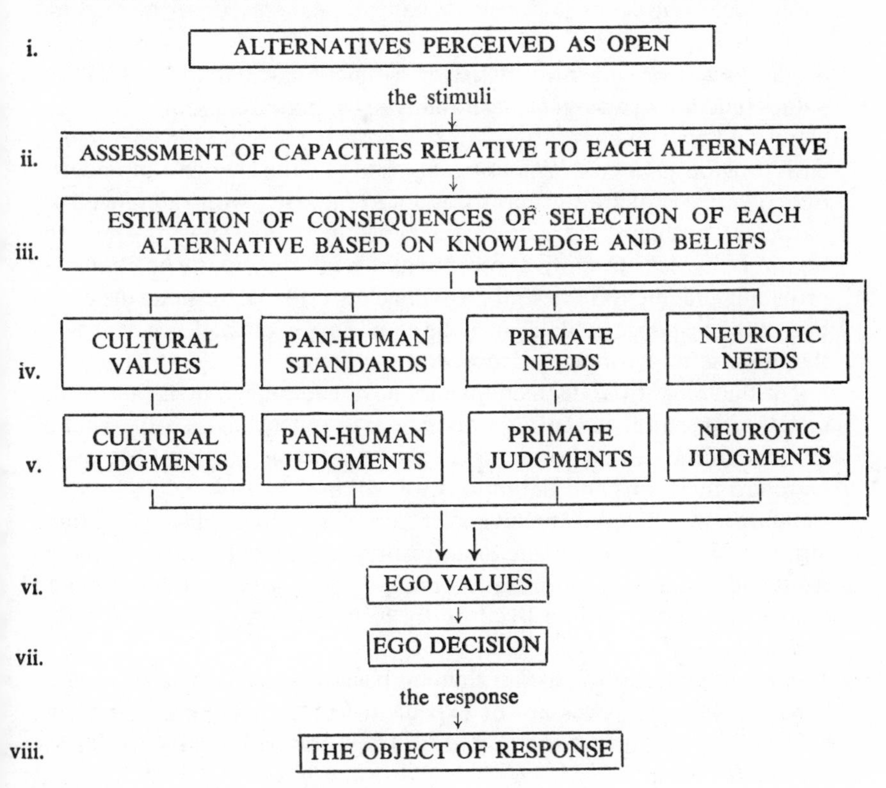

Figure 1. The elements of a choice

The elements of a choice, as numbered in Figure 1, are:

i. Perception of open alternatives: These are the alternatives as perceived by the actor. The processes of perception are prior to this and are not included.

ii. Assessment of personal capacities relative to each alternative: This involves considerations of how well, and at what expenditure of time and effort, the alternatives could be executed.

iii. Estimate of the consequences of selecting each alternative: Beliefs about what is, or can be, or can be expected, are used to make the estimate. Beliefs include what the actor considers to be "knowledge," so that is included in the model.

iv. Consideration of the applicable values in each of four of the major classes with reference to each estimate of consequences.

v. The forming of a judgment or evaluation of the desirability of each consequence in terms of each of the four sets of values.

vi. Application of ego values to the various judgments, and to the estimates of consequences of the alternatives.

vii. Decision by Ego that issues in a response.

viii. The object of interest: Whatever the response is directed toward.

The model provides the basis for a functional definition of values: values function to assess the desirability of estimated consequences when choosing between alternatives. In this process the values in each major class provide *guides for behavior*. That is to say, the elements of underlying cognitive structure group together to indicate, with reference to a particular problem, the pattern for behavior that is valued. Thus, the relation of values to motivation is twofold: (1) they motivate choice by providing judgments of desirability which operate to determine the direction of the response, and (2) once the choice has been made, they provide the impulse to act in accordance with the choice.

To supplement the distinctions that have been made by means of the model, some axiological discriminations are needed. These will permit us to move to the next subordinate level — the taxonomic level of classes. Here are the terms and definitions we will use:

Standards are guides for behavior that are widely applicable and perduring; considered to be justified and justifiable; subject to deep commitment and affect; measures for other values, giving them orientation and direction. They are to be lived up to, to be striven toward, and to be defended.

Norms are guides for behavior that are bases for rules, roles, or ways of behaving that are necessary or expedient for implementing standards, achieving goals, or satisfying needs. They are particularly useful for routine decisions.

Preferences are guides for behavior involving simple likes or dislikes that are taken for granted, or guide choices in the absence of other values. Some preferences are minor variations in modes of fulfilling needs. They are all lightly held or optional.

Goals are guides for behavior related to a planned course of action or

change in a situation that are formulated by focusing applicable standards, norms, preferences, and needs on an area of interest.
Needs are guides for behavior which aim to satisfy biogenetic drives or neurotic anxiety reactions. They are related to deficiency motives, but because of their symbolic nature they function even when the drive is not active.

The foregoing definitions, and the groupings of values that they imply, are arbitrary. In the face of general lack of agreement on terminology we can only adopt the criterion of usefulness and construct a vocabulary that provides an adequate range and whose terms we can distinguish relatively clearly. Using these terms we can specify eleven classes of values and arrive at a level where we can dispense with the term "values" except in its collective usage. Thus we establish that the classes of values are: (1) cultural standards, (2) cultural norms, (3) cultural preferences, (4) cultural goals, (5) ego standards, (6) ego norms, (7) ego preferences, (8) ego goals, (9) neurotic needs, (10) pan-human standards, and (11) primate needs.

6. I not only wish to show the importance of values in the generation of behavior, but also wish to stress that, in addition to cultural values, individual and pan-human values are involved. Nevertheless, I wish to keep in view the importance of cultural values within the cultural patterns for behavior, and that the behavior generated by values in interaction is an important source of culture change. This complex of relationships is shown diagrammatically in Figure 2.

In Figure 2 the large upper left-hand box, labeled "the culture-bearer as actor," is a simplification of Figure 1. Those items that are not part of the social heritage are labeled "the non-cultural." The box represents the actor making a response which results in behavior. This becomes part of the total behavior of the society (labeled "raw behavior"). From perceptions of this raw behavior all culture-bearers abstract patterns *of* behavior that become their views of "what is going on." Through a little understood complex of processes, a consensus is arrived at that results in the selection of some of these patterns of behavior to become patterns *for* behavior (guides) and to be passed on through processes of enculturation. Thus, behavior generated by the interaction of the five major classes of values (as in Figure 1) has become a basis for new cultural patterns.

I have not yet mentioned the two boxes that are enclosed in double frames. These are labeled "processes of cultural development and change" and "culture." They represent scientific abstractions made by anthropologists. In regard to these abstractions, I wish to point out that culture as an explanatory concept must be regarded as patterns *for* behavior (a design for living, a guide). If culture is regarded as patterns *of* behavior (an abstraction from raw behavior), any attempt to use culture to explain behavior is bound to involve circular reasoning. So, the

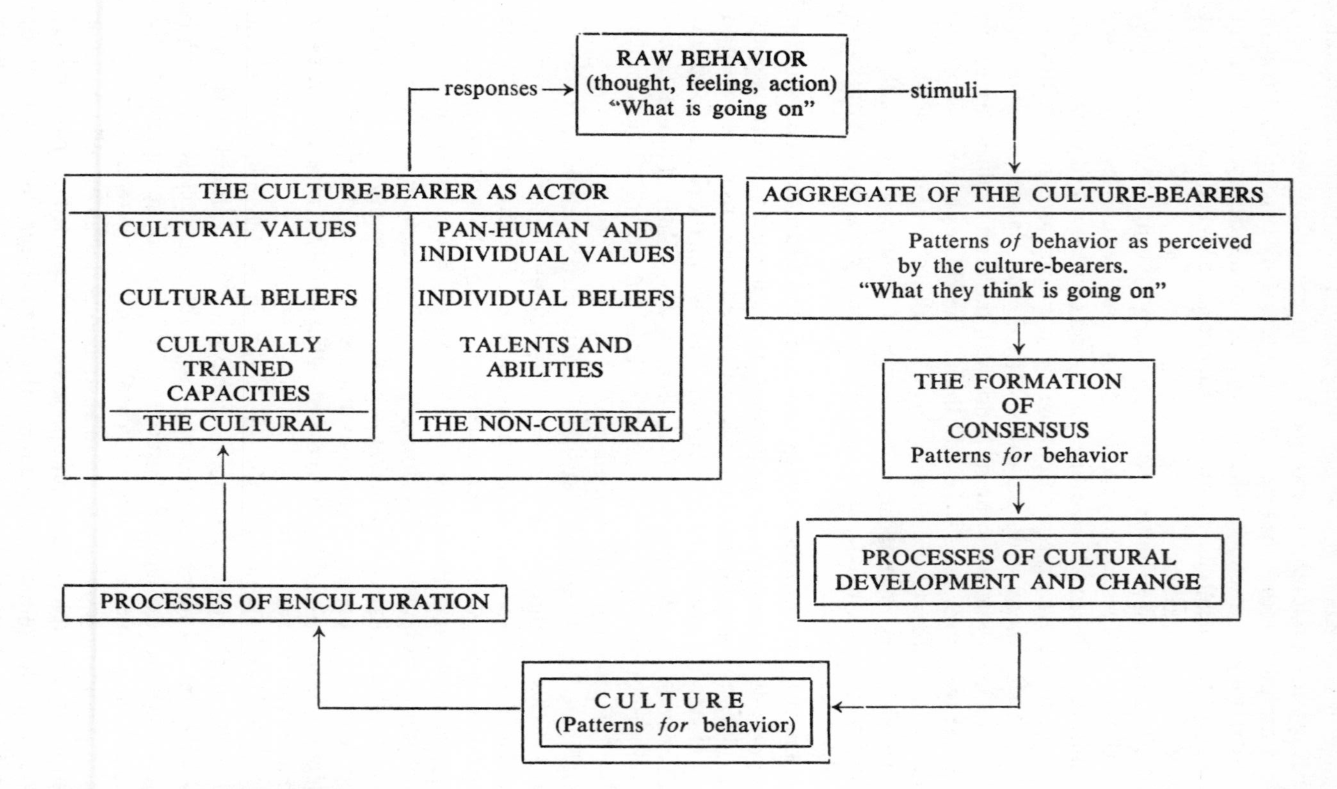

Figure 2. Culture and behavior

anthropologist studies culture as an abstraction, but one viewed as patterns for behavior to be passed on by the processes of enculturation. But, the processes of enculturation are not abstractions. They are going on all the time; most of them can be observed and, in some cases, participated in by an ethnographer. Here is the prime source from which to abstract a view of culture, including cultural values, as patterns for behavior.

In Figure 3, the processes of enculturation are considered in more detail, and culture is divided into traditional culture, operative culture, and subcultures. These divisions of culture are usually sources of conflicting values, and each has its own measure of integration. In studying enculturation all three must be taken into consideration. The diagram shows the three cultures "feeding" their values and beliefs into the enculturation process. Many kinds of selection are involved as they are passed on to the culture-bearers. The internalized culture of each culture-bearer results from the selections made in his particular case.

I use the terms "traditional culture" and "operative culture," not because of particular accuracy or aptness of their connotations, but to avoid other distinctions such as overt/covert, implicit/explicit, and ideal-/real which include considerations of the patterns *of* behavior that I have excluded from my view of culture. The traditional/operative distinction is based on a division within the patterns *for* behavior. I conceive of the traditional culture as patterns for behavior that are old, idealized, formal, venerated. Verbal support may be general, but many of them are followed out only when convenient or in certain situations. The operative culture usually consists of alternative patterns for behavior that are operative when the traditional culture can be circumvented. The nature of the division between the two can be specified only with reference to a particular society. The distinction is not between lip service and practice or between norm and mode, but between two designs for living. The two cultures coexist, and a cultural value system exists in each. In addition there may be one or more subcultures with their own value systems.

Four types of selection are shown in the enculturation processes (Figure 3). There is selection by the culture-bearer when, as a result of temperament, physique, particular experiences, or even the sequence of enculturation, the individual accepts certain items and rejects others. There is selection by enculturators when differences in roles and personalities of the enculturators result in varying emphases. There is selection by "groupings" in which the patterns transmitted depend on the age, sex, class, etc., of the enculturee. Finally there is selection by roles and specialties. The end result is the complexity of the internalized culture shown at the right side of the figure.

The foregoing rather lengthy discussion of matters that are usually taken for granted has been necessary because neglect of essential distinc-

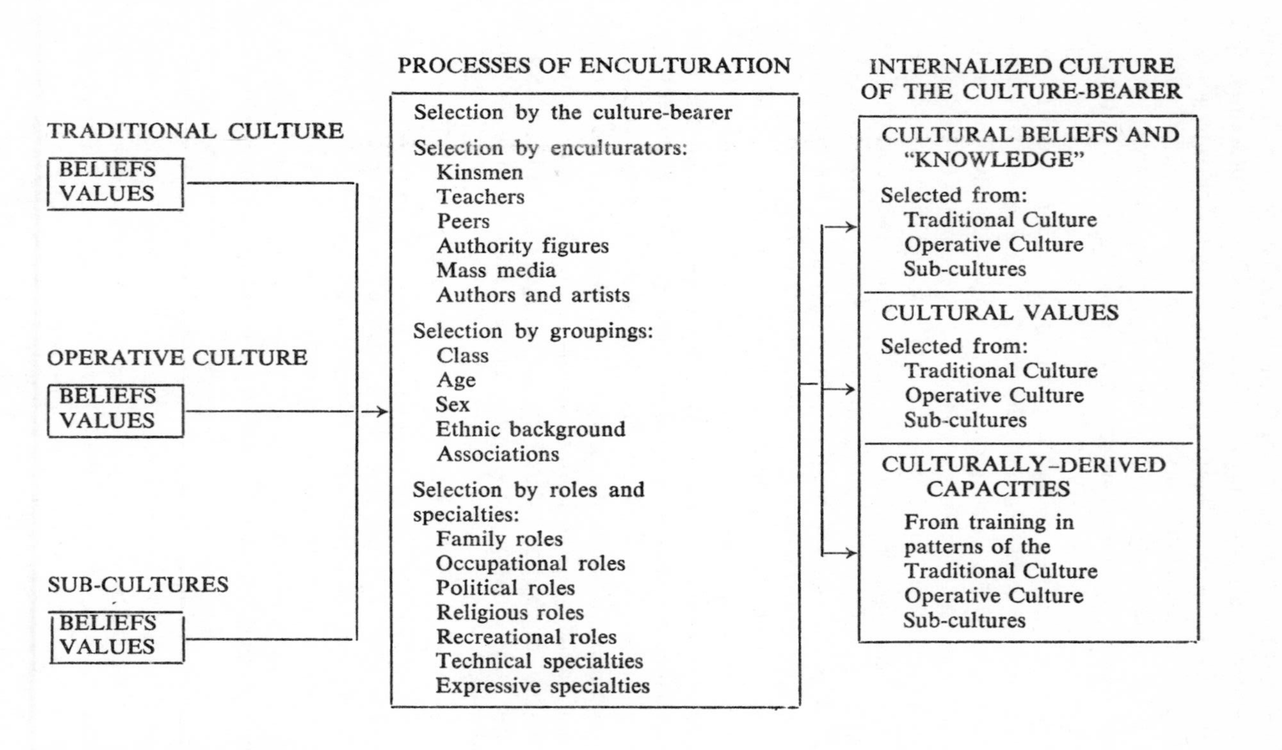

Figure 3. Enculturation

tions and details of processes has resulted in views of culture that tend to discredit value studies. If cultural values are viewed as abstractions, their place in human motivation and processes of culture change will be misunderstood. If they are viewed as elements of underlying cognitive structure, their importance becomes evident, but a full analysis of the cultural processes we have reviewed is required to assess their content and functions.

7. Before proceeding with the consideration of cultural values, we must work out descriptions for the taxa that constitute pan-human standards and primate needs. How may they be described, and what are the observations on which descriptions can be based? These questions must be answered now because cultural values are to be typed, as will be shown in the next section, on the basis of their relationships to the pan-human standards and the primate needs.

As we have seen, the pan-human standards are assumed to have their genesis in symbolizations of infant experiences that have brought satisfaction of genetically transmitted needs. These experiences vary in degree, but not in kind, from one society to another. Erik Erikson and Charlotte Buhler are prominent among those psychologists who consider that the experience of a feeling of "trust" is of fundamental importance. Buhler refers to "that 'trust' with which the infant, almost immediately after it is born, opens its mouth and itself to the world" (1962:158–159). Erikson refers to "a sense of basic trust, which is a pervasive attitude toward oneself and the world derived from the experiences of the first year of life" (1968:96). Buhler emphasizes that trust is revealed in the behavior of all normal infants. Although Erikson is concerned with the development of trustfulness by the reward of trust that prevents it from converting into a feeling of "basic mistrust," some measure of the reward of trust is essential to the survival of the infant, and the mistrust that is experienced is felt as a threat to self-preservation. In addition to trust, conative and exploratory activities on the part of the infant are required for the satisfaction of its needs. All infants that survive will have experienced them.

The essential experiences can be described as trusting (in a mutually open relationship), willing (forming intentions), and exploring or searching (motor activity and sensing). When these experiences of activities and relationships take on affect and symbolic significance, further experiences of the same sort will be sought. Such additional experiences will reinforce the affect and symbolic significance of the elements of underlying cognitive structure, and their meanings will become more generalized, covering the variations in successively wider experiences. The trusting will extend to trusting in other people, to trusting in the physical environment, and to trusting in oneself. The willing or conation will extend toward the taking over of direction of wider areas of one's life and of the consequent decision making. The exploring and searching will extend to use of the full range of one's talents and abilities.[3]

At the universal level the standards can be briefly described as: (1) trust in others, (2) trust in environment, (3) trust in self, (4) use of talents and abilities, and (5) self-direction. Observation of the mothering of infants in a society will give an indication of the degree of early reinforcement of the pan-human standards, and thus an indication of the strength and breadth to which they will develop, but, regardless of the cross-cultural variation in mothering practices, the standards will be formed and will develop to some degree.

The primate needs, in man, take on symbolic significance through symbolization of experiences of actual satisfaction of the genetically transmitted needs. On the symbolic level the needs are part of the value system, however manifestations of the genetically transmitted needs are observable in the behavior of the other primates, particulary the anthropoid apes. Ethological studies are available to provide descriptive material. How they are to be grouped is not critical, but the following designations are useful: (1) physiological requirements, (2) safety–security, (3) affection–belongingness, (4) activity–competence, (5) recognition–esteem.

It must be recognized that the words or phrases that I have used to describe or label the subclasses of the pan-human standards and the primate needs are not the values themselves. The values are the taxa so designated. Words are used to describe the taxa, but understanding their content derives from observations of behavior and analyses of processes of development — not from analyses of the semantics of the descriptions.

In order to avoid undue reliance on verbal labels, and for convenience in tabulations (which will be made in the comparisons with cultural values), I shall use the following letter or letter-number indicators for the subclasses of pan-human values: PHS-A (trust in others), PHS-B (trust in environment), PHS-C (trust in self), PHS-D (use of talents and abilities), PHS-E (self-direction), PN-1 (physiological requirements), PN-2 (safety–security), PN-3 (affection–belongingness), PN-4 (activity–competence), PN-5 (recognition–esteem).

8. We are now ready to complete our classification of cultural values. The first step will be to divide the culture as a whole into three sets of patterns for behavior as shown in Figure 4. The terms "ideational," "social," and "environmental" indicate the nature of the problems or the objectives toward which each of the sets of patterns is directed. Ideational patterns (patterns for ideation) are for dealing with ideas or abstractions;

[3] The assumptions related to the symbolization of essential infant experiences and the generalizing of their symbolic significance as a result of further experience are considered to be warranted by virtue of their adaptive nature. Seeking satisfactions in a successively wider range of experiences leads to the finding of new sources of support for survival.

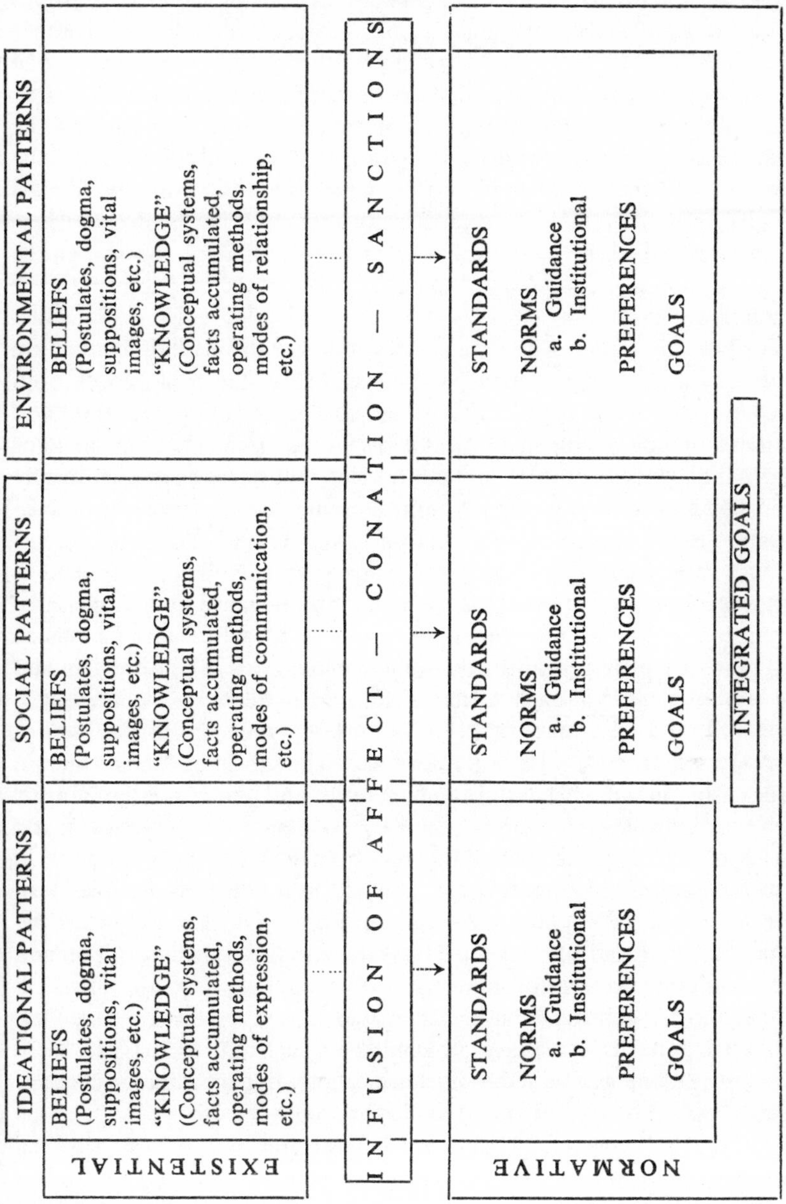

	IDEATIONAL PATTERNS	SOCIAL PATTERNS	ENVIRONMENTAL PATTERNS
EXISTENTIAL	BELIEFS (Postulates, dogma, suppositions, vital images, etc.) "KNOWLEDGE" (Conceptual systems, facts accumulated, operating methods, modes of expression, etc.)	BELIEFS (Postulates, dogma, suppositions, vital images, etc.) "KNOWLEDGE" (Conceptual systems, facts accumulated, operating methods, modes of communication, etc.)	BELIEFS (Postulates, dogma, suppositions, vital images, etc.) "KNOWLEDGE" (Conceptual systems, facts accumulated, operating methods, modes of relationship, etc.)
INFUSION OF AFFECT — CONATION — SANCTIONS	↓	↓	↓
NORMATIVE	STANDARDS NORMS a. Guidance b. Institutional PREFERENCES GOALS	STANDARDS NORMS a. Guidance b. Institutional PREFERENCES GOALS	STANDARDS NORMS a. Guidance b. Institutional PREFERENCES GOALS
		INTEGRATED GOALS	

Figure 4. A terminology for cultural "patterns for behavior"

social patterns are for dealing with other people or with social organizations; environmental patterns are for dealing with the natural or artifactual environment. There have been a number of other threefold divisions of culture proposed (see summaries in Kroeber 1952:154–157; Kroeber and Kluckhohn 1952:98), but all of them either concentrate values in one sector of culture or omit areas that are of significance to values — such as nonexploitative relationships to the natural environment. In Figure 4 the three sectors, or sets of patterns, are arranged in vertical columns with the existential/normative division cutting across them horizontally. Values are given equal prominence in each sector. We now have twelve divisions of cultural values formed by the tripartite division of the standards, norms, preferences, and goals. We have arrived at the taxonomic level of subclasses.

Further subdivision will be necessary to lower the level of abstraction to that of useful descriptions. First we will drop the preferences from consideration. They have been defined as lightly held or optional and their extent may be almost limitless. Now we will use the ten subclasses of pan-human values, set forth in the foregoing section, as a cross-culturally valid basis for defining taxa of cultural values below the level of subclasses. These taxa are at the taxonomic level of types. The subclasses of cultural values are to be divided into types by arranging their component patterns for behavior in accordance with their relevance to the subclasses of pan-human values. Such an arrangement is shown in Figure 5 in which 120 types are provided for by dividing the standards [S], norms [Ng and Ni], and goals [G] in each sector of cultural patterns — ideational [I], social [S], and environmental [E] — into 10 types (20 in the case of norms) in accordance with their relevance to the subclasses of pan-human values listed in the left-hand column. A letter and number designator for each type indicates the sector (I, S, or E), the class (S, Ni, Ng, or G), and the last character of the comparable pan-human value designator (1 to 5; A to E). Separate designators are assigned to norms that provide guidance for conduct [Ng] and those that are primarily oriented toward the maintenance of institutions [Ni]. This is done to facilitate description and in recognition of the importance of institutional support for values. A description of a cultural value system would consist of a description of the normative patterns for behavior that make up each of the 120 types. They can then be compared with the subclasses of pan-human values in order to assess their effect on individual values and on behavior.

Now that we have reached the level of description we are faced with the problems of etics and emics. I think that we are justified in passing directly to an etic frame because of the purpose of the classification, and because we have the hypothesized pan-human values to provide a cross-cultural base. Our purpose is to see how cultural values, pan-human values, and individual values interact. To do so, we need an analysis of

Pan-human values	Sectors of Culture	IDEATIONAL PATTERNS		SOCIAL PATTERNS		ENVIRONMENTAL PATTERNS	
		STANDARDS		*STANDARDS*		*STANDARDS*	
Subclass	PN-1	Type I-S-1		Type S-S-1		Type E-S-1	
,, ,,	PN-2	Type I-S-2		Type S-S-2		Type E-S-2	
,, ,,	PN-3	Type I-S-3		Type S-S-3		Type E-S-3	
,, ,,	PN-4	Type I-S-4		Type S-S-4		Type E-S-4	
,, ,,	PN-5	Type I-S-5		Type S-S-5		Type E-S-5	
,, ,,	PHS-A	Type I-S-A		Type S-S-A		Type E-S-A	
,, ,,	PHS-B	Type I-S-B		Type S-S-B		Type E-S-B	
,, ,,	PHS-C	Type I-S-C		Type S-S-C		Type E-S-C	
,, ,,	PHS-D	Type I-S-D		Type S-S-D		Type E-S-D	
,, ,,	PHS-E	Type I-S-E		Type S-S-E		Type E-S-E	
		NORMS		*NORMS*		*NORMS*	
		Institutional	Guidance	Institutional	Guidance	Institutional	Guidance
Subclass	PN-1	Type I-Ni-1	Type I-Ng-1	Type S-Ni-1	Type S-Ng-1	Type E-Ni-1	Type E-Ng-1
,, ,,	PN-2	Type I-Ni-2	Type I-Ng-2	Type S-Ni-2	Type S-Ng-2	Type E-Ni-2	Type E-Ng-2
,, ,,	PN-3	Type I-Ni-3	Type I-Ng-3	Type S-Ni-3	Type S-Ng-3	Type E-Ni-3	Type E-Ng-3
,, ,,	PN-4	Type I-Ni-4	Type I-Ng-4	Type S-Ni-4	Type S-Ng-4	Type E-Ni-4	Type E-Ng-4
,, ,,	PN-5	Type I-Ni-5	Type I-Ng-5	Type S-Ni-5	Type S-Ng-5	Type E-Ni-5	Type E-Ng-5
,, ,,	PHS-A	Type I-Ni-A	Type I-Ng-A	Type S-Ni-A	Type S-Ng-A	Type E-Ni-A	Type E-Ng-A
,, ,,	PHS-B	Type I-Ni-B	Type I-Ng-B	Type S-Ni-B	Type S-Ng-B	Type E-Ni-B	Type E-Ng-B
,, ,,	PHS-C	Type I-Ni-C	Type I-Ng-C	Type S-Ni-C	Type S-Ng-C	Type E-Ni-C	Type E-Ng-C
,, ,,	PHS-D	Type I-Ni-D	Type I-Ng-D	Type S-Ni-D	Type S-Ng-D	Type E-Ni-D	Type E-Ng-D
,, ,,	PHS-E	Type I-Ni-E	Type I-Ng-E	Type S-Ni-E	Type S-Ng-E	Type E-Ni-E	Type E-Ng-E
		GOALS		*GOALS*		*GOALS*	
Subclass	PN-1	Type I-G-1		Type S-G-1		Type E-G-1	
,, ,,	PN-2	Type I-G-2		Type S-G-2		Type E-G-2	
,, ,,	PN-3	Type I-G-3		Type S-G-3		Type E-G-3	
,, ,,	PN-4	Type I-G-4		Type S-G-4		Type E-G-4	
,, ,,	PN-5	Type I-G-5		Type S-G-5		Type E-G-5	
,, ,,	PHS-A	Type I-G-A		Type S-G-A		Type E-G-A	
,, ,,	PHS-B	Type I-G-B		Type S-G-B		Type E-G-B	
,, ,,	PHS-C	Type I-G-C		Type S-G-C		Type E-G-C	
,, ,,	PHS-D	Type I-G-D		Type S-G-D		Type E-G-D	
,, ,,	PHS-E	Type I-G-E		Type S-G-E		Type E-G-E	

Figure 5. Frame for the classification of cultural values

cultural values that shows their significant relationships to pan-human values (in the sense of fostering or frustrating them). Therefore the differences that are significant in this classification are not the distinctions made by the culture-bearers but the differing relationships of cultural patterns for behavior to pan-human values.

In the domain of values, unconscious processes play a large part, the lexicon is more limited than it appears to be, and there are large unverbalizable areas. Furthermore, some of the distinctions that are important to members of the society are due to elaborations in a particular area of culture — elaborations that are of little relevance to pan-human values. Even if a componential analysis could uncover all the primitive elements of a cultural value system, we would have to group them differently than the culture-bearers group them. We want to know how they impinge on pan-human values, not how they are grouped in the calculus of the culture.[4]

The linguist can describe the sounds of a language he does not know by using the symbols of a phonetic chart. The botanist describes and groups plants regardless of their local names or uses. He employs a taxonomy unrelated to the semantic domain of plants in the language of the people who use them. The ethnographer who is following our system finds in Figure 5, along with the definitions and analyses on which it is based, a frame for classifying and describing types of cultural values in accordance with his interests. The taxa are made up of components of the affective patterns for behavior that he sees being taught, has had explained to him, or constructs from his analyses of motivations of behavior. For example, in any society the great variety of acts that are prescribed through instruction, example, or sanctions are grouped as patterns for behavior, and some of these are verbalized as values. The ethnographer is concerned, not with the patterns or the labels, but with the component acts themselves. These he classifies in accordance with the 120 types of Figure 5. He does not label them but describes them with at least a paragraph for each type.

The number of descriptions required may seem excessive, especially when it is remembered that a full set will be required for the traditional culture and another for the operative culture. However, the normative system of a culture, even within the limits that we have imposed, cannot

[4] As will be pointed out later, the relationships that cause conflicts of interest to us are not to be found in emic categories or in their interaction. We are interested in conflicts that are seen most clearly when the emic categories are disregarded. Furthermore, we do not find the kind of etic frame we need by "passing through" the emic categories. The type of "etic kit of possibilities" that is developed by doing so, and the kind of general theories to account for variance among cultures that Goodenough speaks of (1970:110) are for application to subject matters that are cultural. In such applications the objective is to see each subject matter in the terms of the culture. In contrast, we wish to explore conflicts for which the culture has no terms.

be described adequately at the level of abstraction at which ethnographers are wont to present "the cultural value system." As Kluckhohn said of a differently based, but methodologically similar, descriptive system, "the next step is to organize data and write an ethnography within the framework of the invariant points of reference. The first serious trial will not be easy, but it should be rewarding" (1953:522).

9. Our taxonomic classification system for values is now complete. Starting with the three categories, we have moved to the successive subordinate levels of major classes, classes, subclasses, and, in the case of cultural values, to types. The resulting classifications are shown in Figure 6. At the level of the subclasses we begin to make contact with descriptive data, and here the diagram makes use of the levels of discourse that were set forth in Section 3. The subclasses of the pan-human values were described in Section 7 at the universal level of discourse. That level is considered to be the *base level* for describing pan-human values. This is indicated in the diagram by the use of a double frame to enclose the ten subclasses. The base level for describing cultural values is the group level of discourse. The base level for describing individual values is the idiosyncratic level of discourse. The base levels are those from which descriptions in the same category, but at other levels of discourse, are derived.

The levels of classification and of discourse shown in Figure 6 should also be viewed as levels of abstraction. Values are elements of underlying cognitive structure, and it is these that compose the taxa and are being classified. The idiosyncratic level is the lowest level of abstraction at which we can attach verbal descriptions to taxa whose nature we infer by methods appropriate to the category. Even at that level we arrange our descriptions of cultural values in accordance with the 120 types of the group level of discourse, since that is the base level for description of cultural values. Similarly, the pan-human values are described at the idiosyncratic level with reference to the ten subclasses of their base level. Descriptions at the group level and at the universal level are at higher levels of abstraction. The diagram indicates the type of description at each level of discourse.

In constructing the classification system we started at the level of the categories and worked downward. If we had been classifying objects such as plants we would have started at the level of particular plants and moved to higher levels of abstraction. This is not possible in the case of values — the units of the taxa are not observable — so the classes must be constructed in accordance with the hypothesized system of relationships that we seek to investigate. Once the system is set up, however, the descriptions and designations within it are considered as abstractions from lower levels. In this system the primary divisions of the *summum genus* "values" are not concepts such as "honesty," "justice," or "freedom," but are taxa

Categories	INDIVIDUAL VALUES		CULTURAL VALUES	PAN-HUMAN VALUES	
Major classes	EGO VALUES	NEUROTIC NEEDS	CULTURAL VALUES	PAN-HUMAN STANDARDS	PRIMATE NEEDS
Classes	EGO STANDARDS EGO NORMS EGO PREFERENCES EGO GOALS	NEUROTIC NEEDS	CULTURAL STANDARDS CULTURAL NORMS CULTURAL PREFERENCES CULTURAL GOALS	PAN-HUMAN STANDARDS	PRIMATE NEEDS
Subclasses and descriptions at *universal level* of discourse	Nature of EGO STANDARDS EGO NORMS EGO PREFERENCES EGO GOALS	Nature of OBSESSIONS COMPULSIONS DEFENSES ETC.	The nature of each class (above) within the IDEATIONAL PATTERNS SOCIAL PATTERNS ENVIRONMENTAL PATTERNS (forming 12 subclasses)	PHS-A PHS-B PHS-C PHS-D PHS-E PN-1 PN-2 PN-3 PN-4 PN-5 (The 10 subclasses described)	
Descriptions at *group level* of discourse	Modal EGO STANDARDS EGO NORMS EGO PREFERENCES EGO GOALS	Modal OBSESSIONS COMPULSIONS DEFENSES ETC.	Description of each of the 120 TYPES (Fig. 5) of values of A culture	The modal profile of pan-human values developed in A group	
Descriptions at *idiosyncratic level* of discourse	Descriptions of the ego values and neurotic needs of a particular person		Description of the cultural values internalized by a particular person	The profile of pan-human values developed in a particular person	

Figure 6. The classification and description of values

designated by the category names "individual values," "cultural values," and "pan-human values."

The interrelationships of the three categories are our special concern because therein are found neglected sources of conflicts pertinent to the understanding of intragroup conflicts and culture change. At the universal level of discourse the potential between pan-human and cultural values can be considered by comparing their subclasses shown in adjacent boxes in Figure 6. Then we can investigate the nature of the ego values and neurotic needs that might result, using the hypothesized interactions discussed in Section 4. At this level we get a general view of the potential problems. At the group level the interrelations would be between the cultural values of the society, the modal profile of pan-human values as they developed in that society (the "profile" refers to the relative strengths of the subclasses), and the modal individual values (a part of the modal personality). Here we would get a society-wide view of the conflicts. At the idiosyncratic level one person's profile of pan-human values, his internalized cultural values, and his individual values are in interaction. Here we find the sources of the conflicts as they affect our informants.

Particular note should be taken of the part that individual values play at the group level. They mediate the interaction of cultural and pan-human values and directly influence behavior as well. As a result, behavior may change in ways that neither the pan-human nor the cultural values can account for. The interactions are as shown in Figure 1 and include those discussed in the last paragraph of Section 4. The effect on the behavior observed in the society and on the formation of consensus is outlined in Figure 2. The modal ego values and neurotic needs that are developed in a society eventually contribute to change in cultural values, and are thus partly absorbed by them, but synchronically they are distinct. They may override the cultural values at times and provide unexpected sources of conflict. The action of the pan-human standards should also be noted. They may seem to be a negligible factor since direct manifestations in behavior are rare. This is so because the power of the cultural values and the action of the ego values and neurotic needs combine to submerge them. But, submerged or not, the pan-human standards are involved in internal conflicts, and sometimes are the governing influence on behavior.

The symbolic factors in pan-human, cultural, and individual motivation can be related to one another and to behavior, both synchronically and diachronically, and at all levels of discourse, by means of values.

10. The outline of a theory of values and their interactions that has been presented bears directly on problems of culture change and culture change in its many aspects — micro and macro, voluntary and directed, planned and opportunistic — bears on most of the problems of anthro-

pology. The traditional view of the part played by values in culture change has been rather limited. Felix Keesing wrote, "It has been said metaphorically that the values of a culture, particulary the 'basic' values, act as 'watchdogs' or 'censors' permitting or inhibiting entry and exit of cultural elements. Values, in other words, have a screening effect on stability and change" (1958:400). The dynamic aspects of value conflicts as sources of change have been slighted, and values other than cultural values have largely been ignored. A purely cultural calculus may explain customs, but explanation of change in customs requires a wider calculus that "transcends" culture. That is, one that includes the psychological capacities that provide the potentialities for culture change as Hallowell has explained them (1960:361). It is the prediction of culture change rather than of behavior that is characteristically the anthropological problem.[5]

Pan-human values set up requirements that each culture must fulfill in some measure. All cultures fall short of complete fulfillment. Cultures develop values that tend to frustrate the pan-human requirements and conflicts are produced. As men strive to cope with the resulting shortcomings and frustrations, ego values and neurotic needs are produced that may lead to behavior that intensifies conflict, and this behavior may, through the formation of consensus, lead to the development of cultural values that further aggravate the frustrations. Only with full understanding of the part played by values can one work out measures to break the vicious circle.

When working at the level of the five major classes of values, it is obvious that the synchronic and diachronic interrelationships are complex and introduce many variables. In my view, these five major classes of values act as if they were five "grammars" simultaneously at work in determining behavior. Such complexity precludes a behavioristic analysis and requires the positing of hidden underlying mechanisms. Theories that include such contrasts are useful if they account for a wider range of behavior than do those now in use. The models that I have presented are designed with this idea in view and they appear to have heuristic value. They can generate or suggest hypotheses that have a place in a meaningful whole and can guide empirically based studies. The studies needed include studies of infancy to determine the degree to which pan-human standards are fostered; ethological studies of the anthropoid apes to refine concepts of the primate needs; extensive enculturation studies to

[5] Cultural determinists contend that culture change is to be explained without reference to individuals. My view follows that of Morris Opler: "This idea, that man manipulates and uses his culture for his own ends quite as much as he is subject to it, the cultural determinist resists and ignores. And because he does so, much of the richest data of anthropology passes by him, unrecognized and unappreciated" (1964:525). Hallowell specifically points to the psychological basis for culture change: "The psychological basis of culture lies not only in a capacity for highly complex forms of learning but in a capacity for transcending what is learned; a potentiality of innovation, creativity, reorganization, and change" (1960:361).

apprehend cultural values at their source; psychological studies of ego autonomy, ego strength, and neurotic needs; and studies of symbolization to clarify the processes of symbol formation.[6]

In these studies, research that is less than elegant may have to be accepted. Elegant research is desirable, but those who are doing it now cannot assure us that their detailed studies can ever be combined into a significant whole. René Dubos has given us a microbiologist's view of this problem: "There is a more honorable, though fallacious, reason for the reluctance of the scientific establishment to initiate, encourage, or support the study of the complex problems of human life. It is the strange assumption that knowledge of complex systems will inevitably emerge from studies of much simpler ones" (1968:216).

In setting forth my views on values, many assumptions have been required. It is not my purpose to defend these as the best that can be worked out, but rather to insist upon the need for an integrated set of concepts and assumptions that covers the possibilities envisaged here and that has explanatory and heuristic usefulness. We must be willing to investigate unconventional views whenever they hold promise of shedding new light on problems of man and culture.[7] As Sol Tax wrote in concluding his article "The integration of anthropology," "an essential value in the case of anthropology is its historic reaching for new ideas, tools, and subject matters that may further the study of man" (1956: 326).

11. Culture is man's great creation. It has largely freed him from the bonds of instinctual behavior. But cultural development has been mostly a process of trial and error. Opportunistic solutions have been perpetuated; the values attached to even the best solutions have often outlived their usefulness. These cultural values have become new bonds, and attempts to break free of them have caused dire conflicts. The conflicts themselves have become roots of culture change, and often this has been change toward values of intolerance, dominance, and violence.

Anthropologists are involved in culture change, and not just in the study of it. This has been obvious in applied anthropology, but the inevitability of political implications in almost any kind of fieldwork is now recognized. Our involvement is reflected in the call for a "profound

[6] Studies of symbolization usually concentrate on conventional symbols or on those whose meanings derive from historical associations. These are important in the study of culture, but symbolization in the sense used in this paper — recognition of meaning in relation to a scheme of life — is also a requirement for cultural development. Some in-depth studies of symbolization in psychology are useful but studies that are more broadly based are needed.

[7] Insistence on a concept of culture that views it as man's only guide is one way of shutting off inquiry. An example is the statement by Geertz: "Without the guiding patterns of human culture, man's intellectual life would be . . . buzzing, booming confusion . . . Without the guidance of the public images of sentiment found in ritual, myth, and art we would, quite literally, not know how to feel" (1964:47).

critique," not only of the cultures we are studying, but of the culture we are representing. Cultural relativism has served its purpose by exposing the stumbling block of ethnocentrism; now it should give way to a search for a scientific basis for a profound critique of all cultures. Value studies can provide such a basis, but they should be aimed at cross-cultural comparability and be based on an holistic view of pan-human requirements and capabilities — not just those of "social man," or "economic man," or "man the primate."

A holistic view of human requirements and capabilities implies a model of man. An analysis of values as sources of human motivation and culture change is an essential part of such a model, but many different models should be formulated, and hypotheses derived from them should be tested. We are already confronted with a number of models that, in view of the violence and destructiveness all about us, have been developed to tell us that such is the nature of man. An alternative is to hypothesize that violence and destructiveness occur because the nature of man is being violated by the cultures he internalizes — by cultural values that cause unacceptable internal conflicts.[8] We are being unscientific if we do not investigate both alternatives. We are being shortsighted if we do not realize that, as Leon Eisenberg pointed out in a recent issue of *Science*, "the behavior of men is not independent of the theories of human behavior that men adopt" (1972:123).

REFERENCES

ALBERT, ETHEL M.
1968 "Value system," in *International encyclopedia of the social sciences*, volume sixteen, 287–291. New York: Macmillan.

ALLPORT, GORDON W.
1960 *Personality and social encounter: selected essays*. Boston: Beacon.

BLAKE, JUDITH, KINGSLEY DAVIS
1964 "Norms, values, and sanctions," in *Handbook of modern sociology*. Edited by Robert Faris. Chicago: Rand McNally.

[8] An hypothesis such as this requires a view of the nature of man that admits a "human nature" that is more than "primate nature" and is sufficiently independent of the culture in which it develops to provide a point of reference for evaluation of that culture. Redfield pointed out that the phrase "human nature" has been used in recent psychological and anthropological literature to mean at least three different things. One of those things is that which all men inevitably have as human beings regardless of their particular biological or social heritage (1962:443–444). This usage is applicable to our hypothesis. The hypothesis also requires a view of culture that admits that it can develop in ways that are contrary to some of the needs of man. This is not necessarily in conflict with the view of culture as "adaptive." Cultures can be viewed as adaptive in the sense of providing for man's physiological needs in a given environment, and yet be in conflict with many of the other human requirements that make up human nature.

BROWN, ROGER
1964 "Discussion of the conference," in *Transcultural studies in cognition*. Edited by A. Kimball Romney and Roy Goodwin d'Andrade, 243–253.

BUHLER, CHARLOTTE
1962 *Values in psychotherapy*. New York: Free Press.

DUBOS, RENÉ
1968 *So human an animal*. New York: Scribner's.

EISENBERG, LEON
1972 The *human* nature of *human* nature. *Science* 176:123–128.

ERIKSON, ERIK H.
1968 *Identity: youth and crisis*. New York: Norton.

FIRTH, RAYMOND
1953 The study of values by social anthropologists. *Man* 53:146–153.
1964 *Essays on social organization and values*. London School of Economics, Monographs on Social Anthropology 28. London: Athlone.

GEERTZ, CLIFFORD
1964 "The transition to humanity," in *Horizons of anthropology*. Edited by Sol Tax, 37–48. Chicago: Aldine.

GOODENOUGH, WARD H.
1970 *Description and comparison in cultural anthropology*. Chicago: Aldine.

GRAVES, THEODORE D.
1967 Review of *People of Rimrock* by E. Z. Vogt and E. M. Albert. *American Anthropologist* 69:751–752.

HALLOWELL, A. IRVING
1960 "Self, society, and culture in phylogenetic perspective," in *Evolution after Darwin*, volume two: *The evolution of man*. Edited by Sol Tax, 309–371. Chicago: University of Chicago Press.

HANDY, ROLLO
1969 *Value theory and the behavioral sciences*. Springfield, Illinois: Thomas.

HUTCHEON, PAT DUFFY
1972 Value theory: towards conceptual clarification. *The British Journal of Sociology* 23:172–187.

KEESING, FELIX M.
1958 *Cultural anthropology: the science of custom*. New York: Rinehart.

KLUCKHOHN, CLYDE
1953 "Universal categories of culture," in *Anthropology today*. Edited by A. L. Kroeber, 507–523. Chicago: University of Chicago Press.
1958 The scientific study of values and contemporary civilization. *Proceedings of the American Philosophical Society* 102:469–476.

KLUCKHOHN, CLYDE, *et al.*
1951 "Values and value-orientations in the theory of action: an exploration in definition and classification," in *Toward a general theory of action*. Edited by Talcott Parsons and Edward A. Shils, 388–433. Cambridge: Harvard University Press.

KOLB, WILLIAM L.
1961 "Values, determinism and abstraction," in *Values in America*. Edited by Donald N. Barrett, 47–54. Notre Dame, Ind.: University of Notre Dame Press.

KROEBER, A. L.
1952 *The nature of culture*. Chicago: University of Chicago Press.
1953 "Concluding review," in *An appraisal of* Anthropology today. Edited by Sol Tax et al., 357–376. Chicago: University of Chicago Press.

1956 "History of anthropological thought," in *Current anthropology: a supplement to* Anthropology today. Edited by William L. Thomas, Jr., 293–311. Chicago: University of Chicago Press.

KROEBER, A. L., CLYDE KLUCKHOHN

1952 *Culture: a critical review of concepts and definitions*. Papers of the Peabody Museum of American Archaeology and Ethnology, Harvard University 47–1.

MANNERS, ROBERT A., DAVID KAPLAN, *editors*

1968 *Theory in anthropology: a sourcebook*. Chicago: Aldine.

OPLER, MORRIS E.

1964 The human being in culture theory. *American Anthropologist* 66: 507–528.

REDFIELD, ROBERT

1962 "The universally human and the culturally variable," in *The papers of Robert Redfield*, volume one: *Human nature and the study of society*. Edited by Margaret Park Redfield, 439–453. Chicago: University of Chicago Press.

TAX, SOL

1956 "The integration of anthropology," in *Current anthropology: a supplement to* Anthropology today. Edited by William L. Thomas, Jr., 313–328. Chicago: University of Chicago Press.

TAX, SOL, *et al., editors*

1953 *An appraisal of* Anthropology today. Chicago: University of Chicago Press.

VOGT, EVON Z., ETHEL M. ALBERT, *editors*

1966 *People of Rimrock: a study of values in five cultures*. Cambridge: Harvard University Press.

WALLACE, ANTHONY F. C.

1962 Culture and cognition. *Science* 135:351–357.

WILLIAMS, ROBIN M., JR.

1968 "The concept of values," in *International encyclopedia of the social sciences*, volume sixteen, 283–287. New York: Macmillan.

Meaning in Culture

F. ALLAN HANSON

This paper concerns what the peculiarly *social* sciences — social and cultural anthropology, sociology, history — are all about. It explores the nature of their subject matter and the kind of knowledge we can expect to get about that subject matter. These inquiries will result in the conclusions that sociocultural phenomena and explanations are not reducible to their psychological counterparts, and that to understand sociocultural phenomena is to know their intrinsic meaning. It will also contend that neither of these conclusions entails the view that culture is some sort of superorganic entity or being with purposes of its own.

INDIVIDUAL AND INSTITUTIONAL QUESTIONS

Our first question concerns the nature of the subject matter of social science. Do systems of belief or of social relations, cultures, historical eras, civilizations have peculiar existence in their own right, or are these concepts merely convenient shorthand for the more tangible stuff of individuals engaging in bits of shared and repetitive behavior? This question is often thought to threaten the territorial integrity of many social sciences, for if the latter possibility is the case then all social science is reducible to psychology. Hence, the question has an emotional charge, for such a remapping of disciplinary boundaries is exciting to some scholars and distressing to others. After briefly describing the two sides I will contend that neither is right. There is another, preferable way of talking about what the social sciences study and the division of labor between them which does not get bogged down in perplexing problems about the ontological status of cultural things.

A rationale for those holding that culture has its own special existence was provided in the early nineteenth century by Auguste Comte. This is the view that reality comes in levels. Each level represents an emergent order of existence with laws of its own, irreducible to lower levels. Chemical reactions and combinations, for example, cannot be explained by the laws of physics. The division of scientific labor is thought to reflect layered reality, each level of existence having an autonomous discipline to study its laws. Usually, the levels are identified, from simplest to most complex, as physical, chemical, biological, psychological, and social (Comte 1934). From this perspective cultural institutions, patterns of behavior, and social interaction do indeed represent a level of reality unto themselves, and disciplines like sociology and anthropology carry the mandate to explore it.

Kroeber represented this kind of thinking in anthropology. He wrote: "the mind and the body are but facets of the same organic material or activity; the social substance — or unsubstantial fabric, if one prefers the phrase, — the existence that we call civilization, transcends them utterly for all its forever being rooted in life" (1917:212). Kroeber conceived of social reality — he also termed it "superorganic" — as following a path of evolution essentially independent from organic evolution (1917:210); it seemed even to have some inscrutable but intelligent purpose of its own. Concerning the strikingly regular and repetitive pattern of changes in women's fashions he said (1919:261):

> What it is that causes fashions to drive so long and with ever increasing insistence toward the consummation of their ends, we do not know; but it is clear that the forces are social, and not the fortuitous appearance of personalities gifted with this taste or that faculty. Again the principle of civilizational determinism scores against individualistic randomness.

Many empirically minded social scientists find views like Kroeber's excessive if not downright mystical. They hold that concepts like "culture" and "social system" are only abstractions from the reality of human behavior. To say that culture is a real thing which determines individual behavior and shapes its own development is to be guilty of the fallacy of reifying an abstraction and endowing it with causal influence over the very thing from which it was originally abstracted (Bidney 1944:41–43).

If collective concepts like culture are merely abstractions from individual behavior, then cultural phenomena ought ultimately to be explicable in psychological terms. Such is the view of Melford Spiro, who has argued the "cultural heritage" does not refer to anything that is not already covered by the term "super-ego" (1951:36). George Homans argues that social and cultural phenomena can be explained by the propositions of behavioral psychology (1964; 1967:36–43, 60–64, 73, 103–104), while George P. Murdock analyzes them according to psychological considerations such as the satisfaction of basic needs and drives, habit formation and learning (1965:83–84, 96–101). Recently Murdock has even recommended the abandonment of concepts like culture and social system as "illusory constructs" with no more utility and validity than notions like phlogiston (1971:19).

There is no need to opt for either of the polar positions just described, for there is a better way of looking at the subject matter of social science. As David Kaplan outlined it, this perspective involves the denial of the Comtean position that reality comes in levels or layers. Instead, the division of labor among the sciences stems from variations in our point of view. One way of asking questions about human things produces answers in psychological terms, another way produces them in cultural terms (Kaplan 1965; see also Kaplan and Manners 1972:128–133).

I term this the distinction between individual and institutional questions. It can readily be made clear with an example. One day during my fieldwork on the French Polynesian island of Rapa, I was helping a few men with the heavy job of turning the soil to prepare a taro garden for cultivation. The sun was hot and we were perspiring freely. I picked up a jug of cool water I had brought and asked my comrades if they wanted a drink. They said no. When I then took a drink myself, they looked concerned and one of them told me I should not do that lest I get sick.

Now we can ask the question, why did they refuse the water and caution me against drinking? The question admits of two quite distinct answers, depending on our aim in asking it. If we want to know the Rapans' motives or reasons for acting as they did, we are asking an individual question and the answer in this case would be simply that they wanted to avoid illness both for themselves and for me. On the other hand, we may want to know about the ideas which lead Rapans to believe that drinking cold water when hot and perspiring can produce illness. This is an institutional question. It is not about people at all, but about concepts in their own right. In this case the answer would detail the Rapan system of ideas relating health to body temperature, which in turn is affected by various foods. One implication of these ideas is that drinking cold water when the body is hot and perspiring can adversely affect body temperature and hence endanger health. To summarize the difference between individual and institutional questions: individual questions relate to the motives, intentions, and reasons people have for doing what they do; institutional questions concern concepts, forms of organization, patterns of behavior seen in relation to each other.[1]

What then of the reality of culture? Well, culture exists in the same way that beliefs, values, customs, forms of social and economic organization exist, for culture is the organized total of such things. But that is beside the

[1] My ideas here are very close to Popper's in his lucid discussions (1968, 1969) of the difference between subjective and objective thought (see also Jarvie 1972: Chap. 6).

point, because the peculiarly social sciences do not stake their claim for existence upon having a separate chunk of reality to investigate. They study the same reality that psychology studies. The difference between them is that psychologists ask individual questions about that reality whereas sociologists and their kinsmen in anthropology and history ask institutional questions of it. In asserting this I do not wish to quibble about disciplinary boundaries, such as arguing whether psychological anthropology is a branch of anthropology or psychology. The only point I wish to stress is that sociocultural investigations are not reducible to psychological ones because the institutional questions asked in the former are different from and irreducible to the individual questions asked in the latter.

While there are still plenty of psychological reductionists around today, it is difficult to find a scholar who can fairly be said to reify culture. Later in his career Kroeber explicitly shifted to a view similar to that advocated here (1952:23, 112). And although Leslie White has been accused not only of reifying culture but also of deifying it (Bidney 1950), as I read him, White is driving at a distinction not so much between kinds of substance or levels of reality as between ways of inquiring into human phenomena. There is no clearer statement of the distinction between individual and institutional questions than his:

> When things and events dependent upon symboling are considered and interpreted in terms of their relationship to human organisms, i.e., in a somatic context, they may properly be called human behavior, and the science, psychology. When things and events dependent upon symboling are considered and interpreted in an extrasomatic context, i.e., in terms of their relationships to one another rather than to human organisms, we may call them *culture*, and the science, *culturology* (White 1959:231, original emphasis; see also 1969:77).

MEANING

Thus far we have talked a good deal about questions but very little about answers. Now I want to discuss the kind of knowledge we can expect to get when we inquire into human phenomena. Of special interest will be whether we get the same kind of answers when we ask individual and institutional questions.

It has frequently been argued that there is a fundamental difference between human and natural phenomena. As Collingwood characterized it, natural events have only an "outside," whereas human events have also an "inside" consisting of the thoughts, motives, purposes of the agents (1946:113–114). The "inside" of human events is their intrinsic meaning. Natural events lack this meaning: the moon does not mean to orbit the earth, nor plants to grow, nor glands to secrete, they just do. But meaning is so crucial in human events that they are often unintelligible unless their meaning is known. Jones crawls about peering intently at the ground, because he means to recover a contact lens he dropped. Head-on collisions sometimes occur on the shoulder of the highway, because each driver meant to avoid a crash by swerving off the road. For Max Weber, "the specific task of sociological analysis. . . is the interpretation of action in terms of its subjective meaning" (1947:94; see also 101 and 1949:72–82).

"Meaning" is notorious for its multiplicity of meanings. Wittgenstein termed it an "odd job" word which is called upon for a variety of tasks (1958:43–44); Ogden and Richards (1923) wrote a whole book about what it means. As the foregoing examples make clear, however, the most common meaning of meaning when used with reference to human phenomena is intentional (see Weber 1947:93). The meaning of human act is the agent's intention, purpose, motive, or reason for doing it.

In addition to human acts, meaning of the intentional sort is intrinsic to man-made objects, because an artifact's design crystallizes its maker's purpose. Think of the intentional meaning inherent by design in shoes, parking meters, or scissors, or of the delay and debate devoted to fixing nuances of meaning in the shape of the table at the Vietnam peace negotiations in Paris. Artifacts belong to what Dilthey called "objectifications of life" or the "mind-affected world"(Dilthey 1962:114; see also Weber 1947:93): "everything human beings have created and in which they have embodied their thoughts, feelings and intentions."[2] Dilthey also perceived meaning in the human arrangement and combination of objects. "Every square planted with trees, every room in which seats are arranged, is intelligible to us. . . because human planning, arranging and valuing — common to us all — have assigned its place to every square and every object in the room" (Dilthey 1962:120). Thus, while chairs, tables, and blackboards all have their meanings, a new meaning emerges from their combination to form a classroom. And think how one can turn a seminar room into a lecture room simply by rearranging the furniture.

Dilthey's "mind-affected world" consists of more than artifacts and their configurations. "Its realm extends from the style of life and the forms of social intercourse, to the system of purposes which society has created for itself, to custom, law, state, religion, art, science and philosophy" (1962:120). In other words, that intrinsically meaningful thing which Dilthey variously and expressively referred to as "objectifications of life," "objective mind" and "mind-affected world" is nothing other than "that complex whole" that anthropologists call culture.

At this point a problem develops with the intentional concept of

[2] The passage quoted is from some introductory remarks by H. P. Rickman, editor of the Dilthey volume.

meaning. This notion is adequate when the focus is on human acts or artifacts, for their meaning can be found in the purposes or motives of the people who did or made them. But how can we talk of the intrinsic meaning of cultural institutions, when no one intended them?[3] We can, of course, talk about the intentional meaning of people's *use* of institutions: why they affirm them, conform to them, manipulate them, rebel against them, and so on. These are human acts and their meaning lies in the purposes, motives, reasons of the agents. But we cannot explain the meaning of cultural institutions themselves in this manner, for here there are no intending agents.

Note that in the terminology developed above, this problem occurs precisely at the shift between individual and institutional questions. Individual questions are associated with intentional meaning, in that when we ask why people do the things they do, answers in terms of the intentional meaning of their acts are appropriate. But institutional questions focus on cultural phenomena in their own terms and not on people, so answers in terms of intention, reason and so on are not fitted to them. How, then, can we hold with Dilthey and others that cultures and their institutions are intrinsically meaningful?

At least some of the time, Dilthey tried to get around this problem via the notion that intentional meaning is in fact valid for cultural phenomena because institutions can intend things as well as people can. So in a passage quoted above he spoke of "the system of purposes which society has created for itself" (1962:120; see also 129). And Tuttle (1969:75) attributes to Dilthey the view that "the true causal forces in history are found in the motive-deliberations of the systems and not in the motive-deliberative actions of any 'mere' individuals who compose the system." But this solution is scarcely acceptable for, in reifying culture not just into a thing but into something that thinks, it badly distorts normal use of language. Words like "intend" and "deliberations" relate to the conscious thought or design of thinking agents. To apply such terms to cultural institutions, which do not think and have no consciousness, is in my judgment a confusing, unfortunate, and unwarranted use of those words.

[3] To be sure, a few institutions like the American form of government were explicitly designed and can hence legitimately be treated in terms of their intentional meaning, although a good share of any such discussion would be devoted to how such institutions have developed in unforeseen ways. But the concept of intentional meaning seems entirely alien to the vast majority of customs and institutions — the British constitution, capitalism, marriage, motherhood, honor, patrilineal descent, preferential cross-cousin marriage, Oxford University — because no one ever sat down and designed them.

IMPLICATIONAL MEANING

An alternative solution to the problem, and the one I advocate, is to recognize that there are many different kinds of meaning, and that the meaning intrinsic to cultural institutions is not of the intentional sort. Just what kind of meaning it is will be readily apparent if we consult a few uses of that word in ordinary language.

If someone asks, "What was the meaning of Caesar's crossing the Rubicon?" or "What do you mean by keeping my daughter out until 3:00 A.M.?" the concept of meaning in question is clearly of the intentional sort. But consider some other questions about meaning: "What does the theory of evolution mean?" "What is the meaning of the mother-in-law taboo?" "What does it mean to have good manners?" The answers one is likely to get — dealing with matters like inaccuracy in the biblical account of creation, systems of kinship, marriage and residence, and not blowing one's nose on the tablecloth — do not relate to intentions at all. Instead, they concern *implications* of the things in question. This concept of meaning, which we may term implicational, is quite different from the intentional variety. I suggest that the meaning intrinsic to cultural institutions of all sorts — scientific theories, religious creeds and practices, social organization, ethics, and so on — is of the implicational type. Every cultural thing, like the Rapan prohibition against drinking cold water when hot and perspiring, is linked by implication to other cultural things, like a general hot-cold theory of health and disease, and therein lies its meaning.

A focus on implications produces a view of culture as a logical system analogous perhaps to a geometry or a scientific theory, albeit a rough and complex one with internal contradictions and tensions as well as reinforcements. Those authors who have noticed the implicational usage of "meaning" tend to characterize it in logical terms: "every proposition has systematic or logical meaning, so that its full meaning consists in all the propositions which it logically implies and which are required to define its terms" (Nagel 1934:146).[4] From this it is clear that context is critical to the idea of implicational meaning. The meaning of a whole is in its parts and their organization; the meaning of a part is in its logical articulation with other parts to form a whole.[5]

Now let me summarize the main points of my argument thus far. Human phenomena are intrinsically meaningful, and they are best understood and explained by making their meaning intelligible. However, one can ask different kinds of questions about human phenomena, and the

[4] See also Langer (1957:53–56); Rickman (1967:95) and White (1969:xxv).

[5] See also Winch (1958:108). Since dissatisfaction with some of Dilthey's formulations led us to the discussion of implicational meaning, it should be noted that he also recognized meaning in the part-whole relationship (Tuttle 1969:15–16, 80–87).

answers involve different kinds of meaning. Individual questions are about the needs, motives, desires, aims, purposes of people; their answers are in terms of intentional meaning. Institutional questions are not about people at all. They inquire into ideas, beliefs, customs, forms of social organization as such, and their answers demonstrate implicational meaning. I want to stress that institutional questions are not reducible to individual ones, nor vice versa. They move at different levels, asking different kinds of questions and receiving different kinds of answers. Hence, they neither conflict nor compete.[6] Either approach can be pursued independently of the other; together they provide a comprehensive picture of human phenomena. Having made these distinctions, from this point on our discussion will be concerned with the logic of institutional questions.

THE LOGIC OF QUESTION AND ANSWER

One important methodological issue in the study of institutional questions is the manner or format in which the meaning of cultural institutions should be described and analyzed. Since implicational meaning is essentially a matter of logical relationships, I think an excellent model for our purpose can be found in Collingwood's views on logic and metaphysics.[7] These views developed out of Collingwood's reaction to the idea that the questions of Western philosophy are timeless: that ancient, medieval, and modern philosophers have merely offered different answers to the same questions. He held that the questions themselves have changed and therefore that distortion and misunderstanding can be avoided only if one examines the views of a given thinker in terms of the questions he asked. His concern with seeing propositions in the context of questions led Collingwood to formulate a "logic of question and answer." In this logic,

> . . .no two propositions, I saw, can contradict one another unless they are answers to the same question. . . . The same principle applied to the idea of truth. If the meaning of a proposition is relative to the question it answers, its truth must be relative to the same thing. Meaning, agreement and contradiction, truth and

[6] This is often not recognized. Homans, for example, holds that institutional questions can be reduced to individual ones. But this results in the myopic view that the only thing to explain about a social norm is why people conform to it (Homans 1967: 60–64). Together with a collaborator he made a similar error of assuming that systems of unilateral cross-cousin marriage are fully explained if we know what motives and sentiments induce people to follow their rules (Homans and Schneider 1955; see also Needham 1962).

[7] It may seem curious to look to Collingwood for guidance in the study of institutional questions, because he has been criticized by both Hodges (1944:103) and Popper (1968:29) for reducing sociocultural phenomena to psychological ones. Such a criticism may apply to Collingwood's stress on the reenactment of the thought of past individuals for historical understanding (1946), but as the sequel will demonstrate it is far from the mark with regard to his concept of metaphysics.

falsehood, none of these belonged to propositions in their own right, propositions by themselves; they belonged only to propositions as answers to questions: each proposition answering a question strictly correlative to itself (1939:33).

Moreover, a question and its answer taken as a unit has its proper place in a "question-and-answer complex" such that each answer gives rise to the next question in an ordered chain of thought.[8] Questions of truth and meaning should not be asked of particular answers, but of the question-and-answer complex taken as a whole (Collingwood 1939:37–39). Nor does the process stop here. The truth and meaning of question-and-answer complexes should be determined in the context of the metaphysical beliefs of the culture or historical period: what people "believe about the world's general nature; such beliefs being the presuppositions of all their 'physics,' that is, their inquiries into its detail" (1939:66).

The most general metaphysical beliefs are "absolute presuppositions"; with them the questioning activity comes to an end. Rather, this is where the process of question and answer begins. Absolute presuppositions are not themselves answers to any questions; they are the ultimate assumptions which give rise to all questions (Collingwood 1940:31–33). Because they follow from no other questions or suppositions, absolute presuppositions are arbitrary. Probably this is what Wittgenstein had in mind when he wrote "At the foundation of well-founded belief lies belief that is not founded" (quoted in Needham 1972:71).[9]

Clearly the hallmark of the logic of question and answer is its emphasis on context. One should understand the beliefs, concepts and theories current in a particular age or society in the context of the presuppositions assumed and the questions asked at that time and place. Hence, it is not surprising that for Collingwood metaphysics is properly a historical study. Its mandate is not to speculate about the nature of reality, but to describe systems of thought which have actually existed, to determine what their absolute presuppositions were and to demonstrate how these generate the sequences of questions and answers which were pursued (Collingwood 1940:49–57, 63). Although Collingwood's ideas refer primarily to philosophical and scientific thought, it is easy to expand them to provide a paradigm for the institutional approach to social science. The affinities between the implicational concept of meaning and the logic of question and answer are clear, for both emphasize that the meaning of an idea or belief lies in its context, in its relation to other ideas and beliefs. In fact, the question-and-answer sequences that radiate out from absolute presupposition are nothing other than chains of implication. If one stipulates

[8] I believe this "question-and-answer complex" is what Collingwood elsewhere termed "reflective thought" (1946:307–315). It is simply a series of logical implications.

[9] Readers who wish to explore Collingwood's logic of question and answer and the concept of absolute presuppositions further might begin with a criticism by Donagan (1962:66–93) and a defense by Rynin (1964).

that in addition to verbal entities like beliefs and concepts these chains include customs, rites, forms of social organization, artifact design, manufacturing techniques and so forth, then Collingwood's logic of question and answer becomes an ideal model for the description and analysis of cultural institutions.

One example of the kind of analysis I have in mind is Miller's study (1955) of how absolute presuppositions held by Europeans and Fox Indians as to whether or not the cosmos is ordered hierarchically have all kinds of fascinating implications for concepts of the afterlife, the form and stability of government, the nature of collective action, relations among kinsmen, and so on. Again, the Rapan practice of carefully specifying *where* everything occurred when telling a folktale or simply recounting the events of the day, their tendency to measure the value of a man by how much land he owns, and their custom of burying the placenta under the threshold of the family home are all implications of the absolute presupposition that location is a source of order and permanence in the world (Hanson 1970:46–48).

CULTURAL DETERMINISM

Nearly every author whom we have cited in connection with an institutional approach to social science — Dilthey, Kroeber, White — has somewhere reasoned to the effect that culture works out its own purposes in history, follows its own laws, determines its own development. In discussing the world of objective knowledge (arguments, theories, systems of thought considered in their own right), Karl Popper acknowledges that these things are human creations, but characterizes that fact as "overrated." He continues:

> But it is to be stressed that this world exists to a large extent autonomously; that it generates its own problems . . . and that its impact on any of us, even on the most original of creative thinkers, vastly exceeds the impact which any of us can make upon it (1969:272).

It is easy to conclude from propositions like these that culture is some sort of an entity — perhaps even an intelligent entity — in itself. Although I reject any such reification of culture, I think there is a great deal of truth in the thesis of cultural determinism. In this concluding section I want to suggest that the concepts developed above enable us to talk about how culture participates in its own development without implying a superorganic thing or purposive agent in itself. As with so much else in this discussion, the distinction between individual and institutional questions is crucial to my argument at this point. One important approach to social change is via individual questions, where one seeks the motives, intentions, reasons, and rewards which lead people to behave in new or different ways (see Murdock 1965:149–150). But an equally important approach is via institutional questions, where one analyzes changing institutions in themselves, in their relations to each other. My discussion here is concerned with social change at the institutional level.

I have contended that the intrinsic meaning of culture is implicational in nature, relating to the ways in which a culture's component institutions presuppose and imply each other. These implicational relations are dynamic in nature, modifying and developing over time. This means that a culture's present state has a good deal to do with its future state, because the institutions of any particular time have implications that become manifest in the institutions of a later time. Consider for example how a science, working within a particular paradigm or theoretical framework, develops and changes as the unforeseen implications of the theory are worked out. Or consider the vast array of implications of the invention of agriculture for population size and density, sedentary settlement patterns, occupational specialization, the growth of towns and cities. This dynamic quality of institutional implications is, I suggest, what Popper and the others are driving at when they talk about cultural phenomena generating their own problems, following their own laws or determining their own development.[10]

Because such changes are logical or implicational in nature, the evolution of a culture is intelligible after the fact. That is, one can discern the seeds of one period in an earlier period. Doing this, of course, is one of the main activities of historians and other scholars interested in social change. These points also imply the possibility of predicting the future development of a culture. There are, however, so many variables internal to the culture and in its environment that the likelihood of predicting successfully is probably about the same as that of successfully predicting the future form of a biological species — something which also develops out of its present state.

I have stressed that a culture is not a completely harmonious system. Change may result from conflicts, strain, or disequilibrium among institutions (Collingwood 1940:74–75). For example, conflicting views on the age of the earth and the origin of the species on the parts of Christianity and science led to an abandonment of literal biblical interpretation of these subjects by most branches of Christianity. Again, urban life and large, close-knit kin groups seem to be incompatible, so that where the former increases, the latter diminish.

Finally, changes in a culture's environment can render institutions ill-adapted and hence set the stage for their modification. In such a case

[10] In talking about the dynamics of cultural implications I am very close to Popper's concept of unintended consequences (1966:96–97).

the organizational principles (or "absolute presuppositions") may remain intact while the changed institutions may be analyzed as different and better-adapted implications drawn from those presuppositions. For example, an ordering principle among the Tiwi of North Australia is that a man's prestige and influence depend upon how many women he can gather about him. Formerly, ambitious men strove to marry many wives for this purpose. With conversion to Christianity, polygyny was replaced by monogyny. But the ordering principle persisted, in new guise: today men of importance keep a covey of *consanguineal* kinswomen (sisters, daughters, etc.) about them (Hart and Pilling 1960:107–111). I have argued that change in the social and political organization of Rapa during the nineteenth century is also a case of new manifestations derived from the same underlying principles of organization (Hanson 1970:200–206). As the French so insightfully put it, "*le plus ça change, le plus c'est la même chose*."

REFERENCES

BIDNEY, DAVID

1944 On the concept of culture and some cultural fallacies. *American Anthropologist* 46:30–44.

1950 Review of *The science of culture* by L. A. White. *American Anthropologist* 52:518–519.

COLLINGWOOD, R. G.

1939 *An autobiography*. London: Oxford University Press.

1940 *An essay on metaphysics*. Oxford: Clarendon.

1946 *The idea of history*. Oxford: Clarendon.

COMTE, AUGUSTE

1934 *Cours de philosophie positive*, volume one (sixth edition). Paris: Alfred Costes.

DILTHEY, WILHELM

1962 *Pattern and meaning in history*. Edited by H. P. Rickman. New York: Harper and Row.

DONAGAN, ALAN

1962 *The later philosophy of R. G. Collingwood*. Oxford: Clarendon.

HANSON, F. ALLAN

1970 *Rapan lifeways: society and history on a Polynesian island*. Boston: Little, Brown.

HART, C. W. M., ARNOLD R. PILLING

1960 *The Tiwi of North Australia*. New York: Holt, Rinehart and Winston.

HODGES, H. A.

1944 *Wilhelm Dilthey: an introduction*. London: Routledge and Kegan Paul.

HOMANS, GEORGE C.

1964 Bringing men back in. *American Sociological Review* 29:809–818.

1967 *The nature of social science*. New York: Harcourt, Brace and World.

HOMANS, GEORGE C., DAVID M. SCHNEIDER

1955 *Marriage, authority, and final causes*. Glencoe: Free Press.

JARVIE, I. C.
1972 *Concepts and society*. London: Routledge and Kegan Paul.

KAPLAN, DAVID
1965 The superorganic: science or metaphysics? *American Anthropologist* 67:958–974.

KAPLAN, DAVID, ROBERT A. MANNERS
1972 *Culture theory*. Englewood Cliffs, New Jersey: Prentice-Hall.

KROEBER, A. L.
1917 The superorganic. *American Anthropologist* 19:163–213.
1919 On the principle of order in civilization as exemplified by changes of fashion. *American Anthropologist* 21:235–263.
1952 *The nature of culture*. Chicago: University of Chicago Press.

LANGER, SUSANNE K.
1957 *Philosophy in a new key* (third edition). Cambridge, Massachusetts: Harvard University Press.

MILLER, WALTER B.
1955 Two concepts of authority. *American Anthropologist* 57:271–289.

MURDOCK, GEORGE PETER
1965 *Culture and society*. Pittsburgh: University of Pittsburgh Press.
1971 Anthropology's mythology. *Proceedings of the Royal Anthropological Institute* (1971): 17–24.

NAGEL, ERNEST
1934 Verifiability, truth, and verification. *Journal of Philosophy* 31: 141–148.

NEEDHAM, RODNEY
1962 *Structure and sentiment*. Chicago: University of Chicago Press.
1972 *Belief, language, and experience*. Chicago: University of Chicago Press.

OGDEN, C. K., I. A. RICHARDS
1923 *The meaning of meaning*. New York: Harcourt, Brace.

POPPER, KARL R.
1966 *The open society and its enemies* (fifth edition). Princeton: Princeton University Press.
1968 "On the theory of the objective mind," in *Proceedings of the Fourteenth International Congress of Philosophy*, volume one, 25–53. Vienna.
1969 "Epistemology without a knowing subject," in *Philosophy today*, volume two. Edited by Jerry H. Gill, 225–277. London: Macmillan.

RICKMAN, H. P.
1967 *Understanding and the human studies*. London: Heinemann.

RYNIN, DAVID
1964 Donagan on Collingwood: absolute presuppositions, truth, and metaphysics. *Review of Metaphysics* 18:301–333.

SPIRO, MELFORD E.
1951 Culture and personality: the natural history of a false dichotomy. *Psychiatry* 14:19–46.

TUTTLE, HOWARD NELSON
1969 *Wilhelm Dilthey's philosophy of historical understanding: a critical analysis*. Leiden: E. J. Brill.

WEBER, MAX
1947 *The theory of social and economic organization*. Translated and edited by Talcott Parsons and A. M. Henderson. New York: Oxford University Press.

1949 *The methodology of the social sciences*. Translated and edited by Edward A. Shils and Henry A. Finch. Glencoe: Free Press.

WHITE, LESLIE A.

1959 The concept of culture. *American Anthropologist* 61:227–251.

1969 *The science of culture* (second edition). New York: Farrar, Straus and Giroux.

WINCH, PETER

1958 *The idea of a social science*. London: Routledge and Kegan Paul.

WITTGENSTEIN, LUDWIG

1958 *The blue and brown books*. New York: Harper and Brothers.

The Concrete Dialectic of Self-with-Society

CALVIN O. SCHRAG

The problem of the relation of the self to society has been with us for some time, having undergone multiple expressions throughout the development of Western philosophical and scientific thought. Plato, in one of his better known writings, *The republic*, works out the relationship between the individual soul and society in such a manner that a direct analogy obtains. The three parts of the individual soul (reason, spirit, and appetite) are directly analogous to the three classes in society (guardians, auxiliaries, and tradesmen). The self is simply society writ small; and conversely, society is the self writ large. Subsequent philosophers, from Aristotle to Sartre, have addressed the same issue, providing a rather colorful variety of answers. Our principal intention in the current undertaking is not to rehearse the manifold positions on the issue that have been taken by the various representatives of the philosophical tradition, but rather to reassess the problem itself so that a fresh approach to the issue might become visible. We will propose a way of looking at the phenomena of self and society in such a manner that the recurring abstract formulation of the issue as a problem is undermined and a more originative level of experience of self within a social world is brought to light. In our discussion and analysis we will attempt a return to a primordial and concrete dialectic of self-with-society that antedates the constructionism of both formal philosophy and institutionalized science. This return to an *arche*, an origin, will then provide the proper point of departure not only for any ontology of the self and society but also for the manifold projects of the several human sciences. More specifically, our format will be designed so as to enable us to address the following topics in a sequential manner: (1) redefining the framework of inquiry; (2) the return to originative experience; (3) towards an ontology of self-with-society; and (4) implications for the human sciences.

REDEFINING THE FRAMEWORK OF INQUIRY

Every philosophy and every science proceeds from an inquiry framework that is layered with epistemological and metaphysical presuppositions, either tacit or explicit. The inquiry framework for Western philosophical reflection became sedimented early in its development and was articulated in the form of a substance-attribute metaphysics and a subject-predicate grammar of thought. It was thus that substantiality in metaphysics and objectivistic thinking in epistemology became normative for philosophical inquiry. Consequently reason itself was destined to be understood in terms of the objectivistic and classificatory operations of the mind wherewith the various entities that populate the world could be sorted out and defined. This concept of rationality found a congenial home in Galilean science and set the stage for a subsequent mathematization of nature and praxis. The book of nature was understood to be written in the script of mathematical notation requiring an objectifying rationality for its discernment. Thus, the advent of Galilean science solidified a prejudice that had its roots in the thought of the ancients — the prejudice of pure theory and objective reason. As a consequence both philosophy and science lapsed into a situation of crisis from which neither has yet fully recovered.[1]

This prejudice of pure theory and objective reason continued to inform the developments of modern philosophy and was given a new expression in the philosophy of Descartes. Descartes followed Galileo in making mathematics the touchstone for philosophical method. This led him to picture the world of nature as an abstract continuum of extended substances (*res extensa*) which is in some kind of commerce with the world of mind, pictured as a mosaic of thinking substances (*res cogitans*). The consequences of this inquiry standpoint and resultant metaphysical construction for the developing theories of self and society were profound. The self was construed as a self-identical and unitary mental entity inhabiting a changing and composite physical body, and society was defined as a serialization and aggregation of insular units of mind. Within such a scheme of things both self and society suffered the determination of an abstract metaphysics of substantiality, and the significations attaching to the concrete experiences of self interacting with society remained occluded.

Admittedly, the standpoint of inquiry that informed deliberations on matters of self and society underwent modification with the emergence of transcendental idealism (Kant), but the prejudice of pure theory and objective reason remained in force. In transcendental philosophy the

[1] For a trenchant examination of the consequences of Galilean science for the Western ideal of reason see Husserl (1970).

Cartesian thinking substance is transformed into a transcendental ego, which supplies the categories for understanding the physical and social world. Although transcendental philosophy avoids the Cartesian substantializing of consciousness, it gravitates into an abstractionism of its own — an abstract intellectualism, which has recourse to the *a priori* categories of a universal, transcendental consciousness. Within this transcendental perspective the concrete genesis of meaning in the daily associations and public affairs of interacting selves remains suppressed. Indeed, the self, as universal and transcendental consciousness, is from the start cut off from the social world as the practical field of its concerns and involvements. Both the natural and social world achieve signification only as "objects" *for* a constituting consciousness. Clearly such a framework of inquiry, with its resultant abstract intellectualism, will not put us into proximity with the phenomena of self and society in their concrete interdependent formation process.

The developments within empiricism, in many cases the result of conscious efforts to correct the abstract intellectualism of the idealist tradition, contributed little to the needed radical reorientation of the framework of inquiry. Classical empiricism for the most part simply illustrated the traditional prejudice of objectivism in a new way. Reason, in empiricism, became a *controlling reason*, geared in the direction of measurement and quantification. This empiricist reconstruction of reason was quickly embraced by the developing human sciences, which became increasingly preoccupied with feedback control procedures and techniques. The narrowing of the range of reason by empiricism, whereby reason became *technical, controlling reason*, was accompanied by a corresponding narrowing of the meaning of *fact*. The so-called facts which the empiricist aspires to analyze, classify, and quantify are discrete and isolable physical and behavioral properties. Both the facts of nature and social facts are construed as discrete properties and relations, and as such they are dislodged from their contextual, configurative field of interest in which they are first announced. In the end, facts are reified as abstract entities, amenable to the manipulation of technical reason. Thus, we observe in the narrowing of reason and the corresponding restriction of the fabric of fact the emergence of an *abstract* empiricism which in the final analysis is not all that different from the abstract intellectualism of the idealist tradition.[2]

The consequence of such an abstract empiricism for an understanding of the process of self and social formation is that the self is pictured as a solipsistic cluster of physical and mental properties, severed from all intrinsic, meaning-laden connections with nature and society. Abstracted

[2] For a discussion of the common prejudice of empiricism and intellectualism as it figures in rendering an account of the phenomenon of perception the reader is referred to Merleau-Ponty (1962).

from its concrete experiential field, the self is reified as a bundle of discrete properties and relations. Society also, within such an inquiry framework, suffers the painful abstraction from the concrete, lived-through experiences that comprise the everyday, public world. It becomes an aggregate of isolated selves defined within a matrix of objective relations. The concrete connective tissues of self-with-society are torn asunder.

We now wish to propose a suspension of the above versions of the traditional prejudice of pure theory and objectifying reason and undertake an investigation of the lived-through odyssey of self and society as an on-going process of individual and cultural formation. This investigation will proceed in such a manner that the experiential roots in the consummate reciprocity of self-with-society remain transparent, avoiding the closure both of an abstract intellectualism and of an abstract empiricism.

THE RETURN TO ORIGINATIVE EXPERIENCE

Our suspension of the traditional framework of inquiry in philosophy and the human sciences, imbued with the prejudice of pure theory and objectivist thought, has a double aspect. On the epistemological side, it involves a suspension of the restrictions on the range of reason, which is either mathematized (as in Galilean science and intellectualism) or technologized as a feedback control procedure geared toward measurement and manipulation (as in empiricism). On the metaphysical side it involves a suspension of substantiality (classical rationalism and Cartesianism) and elementarism (empiricism). We wish to put out of play the worldview of classical rationalism in which self, nature, society, and God are all represented as substances, as well as the worldview of modern empiricism in which reality is pulverized into elemental, discrete, and atomistic constituents. Suspending these epistemological and metaphysical prejudices, we will work our way back to a more originative experience of an interdependent upsurge of self and society.

Our proposal for a return to originative experience, it should be noted, is not without its historical antecedents. In particular, mention should be made of Edmund Husserl's use of his celebrated phenomenological reduction to reclaim the concrete, functioning intentionality of the experienced lifeworld; of Merleau-Ponty's existential phenomenology of perception, which recovers the originative posture of seeing the world from the perspective of incarnated consciousness; and of William James' efforts to establish a "*radical* empiricism" that would undercut the presuppositions of elementarism and a unidimensional consciousness that informed the doctrinal content of traditional empiricism.[3] In our present project we will at times critically appropriate some of the procedures and insights of Husserl, Merleau-Ponty, and James in addressing a problem that has seemingly produced an impasse for current investigations in philosophical anthropology and the human sciences.

The general requirement that we return "to the phenomena themselves", first set forth by Husserl, retains its validity in the present context. However, we are of the mind that the Husserlian understanding of "phenomenon" needs to be radicalized. Phenomenon, in our usage, is no longer simply "object-for-consciousness," nor is it restricted to the "object-as-meant" in separation from its existential concretion. We are proposing a more primordial and more originative notion of phenomenon — a *prephilosophical* signification of phenomenon that is liberated both from the idealistic legacy which continued to influence the thought of Husserl and from the scientistic connotations of an empiricistic phenomenalism.

With respect to the topic at hand the phenomenon that we wish to uncover is that of self-and-social formation in its primordial posture. A phenomenon is *that which shows itself*. The showing that is at issue in the present context is the showing of self and society in a consummate formation process, as a global interdependence. Our use of hyphens in the grammatical construction, self-and-social-formation, is a linguistic indicator of the global and unitary character of this originative showing, indicating that what is at issue is a continuous and multidimensional lived-through experience. On this level, self and society have not yet come to the parting of the ways; they are dialectically constitutive one of the other. Subjectivity is always already intersubjectivity, and sociality is always the thought and praxis of individual subjects. As the unitary character of this phenomenon needs to be emphasized so also must its concreteness. The formation process at issue is a *process* precisely because of the *concresence* or "growing together" of self and society in a reciprocating development. Thus we have chosen our title. "The Concrete Dialectic of Self-with-Society." The adjective in the title serves a dual purpose. It indicates that the dialectic is an intercalating of self and society within a process of concrescence, and it distinguishes our notion of dialectics from the formal and abstract dialectic of pure thought that was exhibited in some of the later developments of Hegelianism.

To articulate with more precision the concrete dialectical interplay that determines the global phenomenon of self-with-society, we have substituted the preposition, "with," for the more common but very general connective, "and." What shows itself in the originative phenomenon is

[3] The contributions of Husserl, Merleau-Ponty, and James to a new approach to the structure and dynamics of experience are discussed in more detail in Schrag (1969).

not a composite or juxtaposition of two numerically distinct entities existing side by side, but rather two interdependent aspects of a unitary global presence. The originative experience of selfness always incorporates a dimension of social awareness. The self can neither constitute itself nor become aware of itself in isolation from other selves. Awareness of self is achieved only by moving through situations of being *with* other selves. To undergo an experience of self is to acknowledge the presence of other selves and to be acknowledged by other selves in one's presence. Self and other self are thus dialectical moments within a consummate reciprocity of mutual acknowledgement.

The originative experience of self-with-society takes shape against the background of a commonly understood and shared world. Neither self nor society are worldless. They show themselves within a horizon or field of praxis in which multiple meanings are inscribed through the speaking and listening, arguing and agreeing, deliberating and planning, working and playing that comprises an intersubjective world. This world is already understood, though vaguely, by the self as it moves from concrete situation to concrete situation. This intersubjective world, which functions throughout as the connective tissue that unites self with other self, is initially disclosed in a prephilosophical and prescientific manner as an ever-changing and ambiguous field of concern and practical engagements. This dimly apprehended horizon or field is always present along with self and society, and hence will need to be rendered explicit and thematic if we are to come to terms with the originative experience of self-with-society.

TOWARDS AN ONTOLOGY OF SELF-WITH-SOCIETY

Ontological analysis and interpretation is the procedure whereby prephilosophical and prescientific experience is made thematic and rendered explicit. It constitutes an effort to comprehend the structures of the concrete. An ontological analysis of self-with-society, if it is to retain its roots in concrete experience and not succumb to the abstractionism of either traditional intellectualism or traditional empiricism, will need to move out from and return to the origin, namely, the lived-through intersubjective world. The world of self-with-society will need to be elucidated in such a manner that its original contours are not disfigured.

William James, that genius of concrete elucidation and description, already spoke of a "world of pure experience," which he found to be more vital and vibrant than the anemic world of atomistic sensations and disjunctive relations constructed by traditional empiricism. Husserl's celebrated notion of the concrete *Lebenswelt* is by now well known by interpreters of contemporary European thought, as is also Heidegger's existential ontology in which the primordial posture of man is seen as that of "being-in-the-world" (*In-der-Welt-sein*). The French phenomenologist, Maurice Merleau-Ponty, saw fit to appropriate Heidegger's existential-ontological notion of world in an effort to trace the genesis of meaning back to a *monde perçu* as the ever-present horizon for human thought and action. Our current effort will proceed with the composite contributions of James, Husserl, Heidegger, and Merleau-Ponty in mind. Appropriating some of their methods and insights in our ontological analysis, we will attempt to explicate the global phenomenon of world in such a way as to provide a fresh approach to the traditional problem of self and society, hopefully liberating this problem from its "problematic" formulation.

The global phenomenon of world affords different perspectives or profiles. The profile that is particularly germaine for an ontological clarification of the experience of self-with-society is the presence of world as *intersubjective* world. A principal ontological feature of this intersubjective world is the being of language. Language plays a privileged role in the disclosure of the dialectical experience of "being-with." The dialectic of self-with-society is announced principally in the being of language. It is important to recognize, particularly in this context, that the being of language includes both structure and event. The structural aspect of the being of language has to do with the spoken tongue as a particular, institutionalized linguistic system (English, French, German, Spanish, etc.) reflecting specific phonemic, grammatical, and syntactical patterns. The being of language as event has to do with the concrete speech act in the utterances of propositions, declarations, commands, and exclamations. Language is both structure and event, an institutionalized system and a creative act. The aspect of structure and system makes language readily available as an object for linguistic science; the aspect of event and creative act requires that language be approached in such a manner that its preobjective and precategorial establishment of meaning in the full bodied act of speaking remains transparent. Neither of these two sides of the being of language can be reduced to the other. They constitute, if you will, the cross section of the continuing operation of language as performance. If one side is reduced to the other then the original fabric of the phenomenon of language is disfigured.

The interweaving of structure and event in the performance of speech shows itself on the originative level in the concrete dialogic encounter. A dialogue presupposes a common spoken language. If there is to be any genuine transaction of meaning between the participants in a dialogue they must share at least one common spoken tongue and be able to put into play the delivered grammatical syntactical forms that characterize it. However, a dialogue is not a sedimented formal structure; it is a creative activity which vitalizes a structure in its very performance. Dialogue as event illustrates language in action, borne by the dialectical interplay of the word as already spoken in the institutionalized spoken tongue and the word as being spoken in the creative speech act.[4] Structure and event, system and activity, conspire in the genesis and formation of meaning.

The performance of language in the dialogic encounter provides a decisive disclosure of the phenomenon of being-with-others. It reveals the sociality of the self as a *character indelibilis*. In the dialogic interchange the other is present with the self as the self becomes present to itself. The self becomes present to itself by taking over the rejoinders (affirmations and negations) of the other in a mutual adventure of meaning formation. Thoughts pass from the self to the other and from the other to the self in an ascending process of mutual discovery. In the concrete interchange the thoughts of the participants are reciprocally appropriated, interwoven in such a manner that the separate contribution of each remains indistinguishable. The adventure of meaning displayed in the dialogic transaction is a shared activity and a joint project. Only in a post-reflective stance, when one does a post-mortem on the dialogue as performance, can the episodical histories of two separate selves be discerned. In such a post-reflective stance every codification of separate meanings contributed singularly by the two selves remains highly artificial and ambiguous. It is thus that attentiveness to the accomplishment of shared meaning in a dialogic performance will show how the performance itself is an originative disclosure of self-with-society in a world of intersubjective experience. It also will provide a sheet-anchor against an abstractive theory of the self on the one hand and an abstractive theory of society on the other.

Not only, however, does the performance of language in the dialogic transaction decisively disclose the bond of "being-with," the shared activity that constitutes the very telos of a dialogue, it is also the point of origin for the experience of the other self as other. The "being-with" at issue in the life of a dialogue is not a relation of absorption. There is unity but neither numerical nor existential oneness. In its concrete unfolding as a shared activity in which a reciprocity of meaning establishment is at work, there is also a display of duality. If there were no such display of duality, then the dialogue would be translated into a species of monologue. Dialogue requires a connective tissue between self and other that does not dissolve into a relation of absorption. It is thus that the life of dialogue allows for an individualization of self, a self-formative process, if you will, that is equiprimordial with its dialectical opposite, a process of

[4] Merleau-Ponty (1962) has submitted the illuminating distinction between *parole parlante* and *parole parlée* in his discussion of the body as expression and speech. *Parole parlante* is roughly equivalent to what we understand by the word as being spoken in the creative speech act, and *parole parlée* corresponds to our rendition of the word as already spoken in the institutionalized spoken tongue.

cultural or social formation. It is here that we are able to discern the background for the birth and development of the basic ontological polarity that structures the intersubjective world — the polarity of personal individualization and cultural participation.

Up to this point we have not yet spoken of the self as individual or of a cultural substructure that delivers a variety of patterns of thought and praxis through some manner of group participation. We have not been able to do so thus far because the unitary phenomenon of the self-with-society experience does not have supplied with it already fashioned notions of the individual and of culture. These are, if you will, later arrivals, emerging out of the concrete experience of otherness within the dialogic transaction. So it now becomes mandatory to provide some clarification of these emergent ontological polarities.

As always we need to proceed in such a manner that the sirens of abstract intellectualism and abstract empiricism remain unheeded. Harboring the prejudice of pure theory and objectivist thought, traditional rationalism and traditional empiricism work with the concepts of an abstract individual and abstract culture and are then saddled with the Herculean task of uniting in some way that which is conceptually disparate. Abstract intellectualism has attempted to explain individuality either by invoking a metaphysical principle of individuation of form by matter (as in traditional realism) or through recourse to a self-identical, monadic ego (as in modern idealism). Abstract empiricism has attempted to account for individuality through a collogation of successive mental properties and body functions (as in contemporary behaviorism and positivism). In intellectualism and empiricism, individuality is prejudged either as a metaphysical category, a logical condition, or a collogation of discrete properties. Individuality is no longer seen in connection with its genesis in a concrete intersubjective world, but is reified as a bloodless abstraction. Whitehead had already perceptively named what is going on here. He called it "the fallacy of misplaced concreteness" (Whitehead 1926).

Proceeding from the originative matrix of intersubjective world experience our ontology marks out a different route to an understanding of the self as individual. We look for the criterion of individuality not in an abstract epistemological condition but in the domain of *human action*. Individuality, we submit, is not a given datum or condition. It is an *achievement* or *acquisition*, attained through an arduous struggle for self-affirmation. As an acquisition it, of course, needs to be acquired; but it can also be lost or sacrificed. It is not given once and for all. Ontologically comprehended, individuality is a modality of the presence of self-with-society. It is a possible mode of being in an intersubjective world. In its ontic expression this individuality, achieved through self-affirmation, assumes various postures. In political life it can be expressed in a declaration of inalienable rights and the freedom of self-determination. In the economic sphere it can appear in the guise of regulations against interference in the organization of forces of production. In religious life it often makes its appearance in the legitimation of the inner voice of conscience. Within the wider institutional existence of man it comes forcefully to the fore in the efforts to achieve personal identity amidst the encroachment of the various agencies of conformity and depersonalization. In all of these ontic manifestations we see the process of self-affirmation at work, whereby the self, contextualized as self-with-society, seeks to win its personal freedom and existential identity and thus achieve an individual mode of existing in an intersubjective world.

However, as our investigation of the structure and dynamics of the dialogic transaction has already shown, the self never becomes a self in isolation. The self requires the attitudes and responses of other selves to achieve self-understanding and actualization. It is thus necessary that we acknowledge a process of social formation through cultural participation that compliments the process of self formation as an achievement of individuality. Ontologically understood, cultural participation comprises another possible modality of the presence of the originative intersubjective world, complementing that of personal individualization. And as the struggle for individuality through self-affirmation undergoes multiple ontic expressions, so does cultural participation. Politically we see the expression of cultural participation in the democratic principle of the mandate of the people. Economically it is manifested in the prescription of social controls for the production and distribution of wealth. In the religious life of man the modality of cultural participation governs the establishment of ecclesiastical sanctions. Within the wider context of public affairs we see the structural element of participation in the various appeals to communal goals as a corrective to the caprice of nondirected individualism.

We have attempted to show how individualization and participation, the constitution of the personal self and the formation of culture, are correlative structural elements of the intersubjective world. They are the ontological fibers, understood as possible modes of being, that structure the concrete self-with-society experience. Personal individualization and cultural participation are modalities of being that unfold within the ongoing development of the experienced life-world, set forth in a privileged way in the concrete dialogic transaction. It is within these modalities that the originative experience of self-with-society undergoes its struggle for actualization and fulfillment.[5]

[5] For an illuminating discussion of a concrete illustration of the ontological polarity of personal individualization and cultural participation the reader is referred to Haydu (1970). Through careful attention to the biographical details of Rousseau's life and the salient

Along the way we discovered that the individuality of self formation and the cultural participation of social formation are neither metaphysical nor epistemological givens. They are postures of thought and action that vary from situation to situation. They are, if you will, stages of achievement and loss of individuality and participation. Individualization and participation mark out polar tendencies or directions within the intersubjective world. When the tendency toward one element of the polarity becomes dominant the full expression of the other is thwarted. An overly accentuated drive toward individualization entails a suppression of participation, and an excessive concern with participation threatens the expression of individuality.

The drive toward a self-affirmation of one's individuality can become so intensified that it shadows the presence of the other. When this occurs not only is there a loss of cultural participation, there is also a concomitant threat to selfhood. The self that loses the other stands to lose itself as self, given the requirement for the acknowledgment by the other as a condition for the achievement of self-consciousness. On this point we do well to heed Kierkegaard's caution that the self that seeks to constitute itself as infinite loses itself as finite self. As Kierkegaard has shown, the exercise of self-affirmation can take on a demonic form. The self that rebels against all limiting and conditioning factors within its finitude and seeks to ground its own being as unlimited and infinite succumbs to the despair of defiantly willing to be itself and thus losing itself. The ultimate result is despair because the self that is envisioned in this demonic self-affirmation can never be achieved. Stated paradoxically, it is achieved only at the expense of the loss of the self as concretely actualized within the context of relations of dependence on environment, society, and God.[6]

Kierkegaard's profound analysis of demonic self-constitution provides us with an extreme boundary situation of loss of self and world. His delineation of the extreme enables us better to understand the more ordinary types and degrees of radical self-affirmation. Historically, tendencies in this direction appear in the attitudinal frames of romanticism and existentialism, particularly in their postures as movements of revolt against essentialism in philosphy and conformism in man's institutional

---

features of the Romantic Period, Haydu demonstrates the dynamic interrelatedness of individual transformations and cultural patterning.

[6] Kierkegaard writes: "By the aid of this infinite form the self despairingly wills to dispose of itself or to create itself, to make itself the self it wills to be, distinguishing in the concrete self what it will and what it will not accept. The man's concrete self, or his concretion, has in fact necessity and limitations, it is this perfectly definite thing, with these faculties, dispositions, etc. But by the aid of the infinite form, the negative self, he wills first to undertake, to refashion the whole thing, in order to get out of it in this way a self such as he wants to have, produced by the aid of the infinite form of the negative self — and it is thus he wills to be himself. That is to say, he is not willing to begin with the beginning but 'in the beginning.' He is not willing to attire himself in himself, nor to see his task in the self given him; by the aid of being the infinite form he wills to construct it himself" (Kierkegaard 1951:107–109).

life. Romanticism sought to extricate the self from all participation in the manifold cultural forms of society; and existentialism, at least in its more extreme individualistic expression, surfaced as a vehement attack against any and all forces that threaten personal freedom. In the contemporary life of man one is daily made aware of the many garden varieties of individualism that seek to elevate the self above all cultural content. In politics, economics, religion, morality, and art one can find pockets of perfervid individualism that tend to sever the self from society in such a manner that a "mirage self," without reality and without concretion, is constructed.

But there is also the opposing tendency — the tendency toward an excessive and intensified allegiance to a particular cultural form that threatens creative self-actualization. This tendency is discernible in the recurring collectivist proclivities that make their way into the institutional fabric of society. When a collectivist state submerges personal freedom and ascribes more reality to the group than to the individual we observe the disruption of the dialectic of self-and-social-formation that is at work in the intersubjective world. When a religion or a religious sect centralizes its authority in such a manner that the individual is subordinated to tradition and doctrine we see another example of cultural participation threatening personal individualization. Again, this tendency is noticeable in the widespread dominance of the bureaucratic-technological frame that has settled in on contemporary society. In all this we can observe the surrender of the individual self to the imposed constraints of a collective mind, which at once threatens creative self-actualization and meaningful participation in the cultural forms of man's social existence.[7]

In the above analysis we have sought to work out the fundamental polar structure of concrete, originative, intersubjective world experience. The requirement for any ontological analysis, if it is not to cut itself off from the intersubjective world, is to move out from lived experience and return to this experience for its justification. Our analysis of the polarities of personal individualization and cultural participation has followed this route, proceeding from the concrete dialectic of self-with-society as displayed in the dialogic transaction.

In the concluding section we will delineate some implications of our preceeding elucidation and analysis for the projects of the several human sciences.

[7] Paul Tillich (1952) has approached the correlative phenomenon of individualization and participation from another perspective. Of particular interest in Tillich's analysis is his imaginative use of the notions of "the courage to be as oneself" and "the courage to be as a part" in an elucidation of the phenomena of self-and-social-formation (see particularly chapters four and five).

IMPLICATIONS FOR THE HUMAN SCIENCES

The question as to the task and goal of the human sciences is one that needs to be asked time and again. It is a particularly vital question to be asked at a time in which the human sciences have lapsed into a situation of crisis. There are some grounds for concern as to whether the current sciences of man have a clear conception of their tasks and goals.[8] In this concluding section we wish to rephrase this question within the context of our above elucidation of originative experience and our ontological analysis of individualization and cultural participation as they arise within a lived-through intersubjective world. Although our above elucidation and ontological analysis is admittedly relevant for an inquiry into the tasks and goals of all the human sciences, it is particularly relevant for the disciplines of psychology and sociology for they, more explicitly than the others, treat the correlated phenomena of self-with-society as a specific topic of inquiry.

The central task of a human science, most broadly defined, is to provide knowledge about *humanitas*. It would thus seem that a paradox resides in the very heart and soul of its project. It is a scientific undertaking that has a humanistic subject matter. As a science it must proceed, at least in part, in accordance with the canons of objectification and quantification; yet, its subject matter is the *human subject*, immersed in manifold concrete life situations, who already understands his existence in a prescientific and preobjective manner. How can the scientific and the humanistic fraternize in the very doing of a human science without sacrificing the one upon the altar of the other? There may already be an oversupply of literature dealing with the continuing debate on the scientific vs. the humanistic culture, which became so popular over the last two decades. In any case, there is no need here to rehearse the many ramifications of this debate. Rather we wish to mark out a new approach to the issue in light of our previous analysis.

If indeed the several human sciences are to afford knowledge of *humanitas* in its various posturings, a regulative principle that guides their various methodological and conceptual designs will need to be installed at the start. We shall refer to this principle as the *principle of reflexivity upon origins* — the origin in this case being the precategorical experience of self in dialectical commerce with society. Only if this principle of reflexivity is installed at the beginning of a human science can one avoid the fallacy of

[8] For a brief statement of the crisis that has settled in on the scene of contemporary scientific investigation into the behavior of man see Schrag (1975). For a more extensive treatment of the issue in the recent past see Scheler (1961) and Cassirer (1944). More recently the French philosopher, Paul Ricoeur, has addressed the sad state of affairs within the republic of the human sciences and has issued a rather harsh indictment: "The sciences of man are dispersed into separate disciplines and literally do not know what they are talking about" (Ricoeur 1967:390).

misplaced concreteness in the guise of reifying abstractions and free-floating theory construction that have so plagued the human sciences throughout their rather brief history. It is thus that any human science, if it indeed is to retain a self-consciousness of its project and goal, will need to be fully cognizant of this requirement of reflexivity upon its origin in the formulation of its methodological designs and the construction of its conceptual schemes.

It would be the special task of a comprehensive philosophy of the human sciences to investigate the full impact that such an awareness of origin would have on the designs of the various disciplines in the science of man. In our much more narrowly defined present essay we are able to address only a few of the considerations: (1) reflexivity and model construction; (2) reflexivity and functionalist analysis; and (3) reflexivity and interpretive understanding.

The cognitive attitude of the human sciences, in their effort to attain knowledge of humans and society, is that of detachment and objective analysis. The thought and praxis of everyday life with its attendant concerns is approached in such a manner that it stands over against the investigator and thus becomes objective. The various constellations of concern then are sorted out so as to mark off various regions within the intersubjective world. Aware of an indigenous ambiguity and artificiality in marking off such regions, the human sciences nonetheless need to effect a division of labor in which different regions are assigned to different investigators. It is thus that the global intersubjectivity of everyday thought and action is sectioned off into regions of concern that are either principally economic, political, psychological, sociological, religious, linguistic, etc. And the various special sciences are given birth. Developing with this division of labor and specification of regions in man's sociohistorical life there has occurred an inevitable proliferation of portraits or models of man. Already in the nineteenth century the confluence of Marxist and utilitarian interests produced the portrait of *homo oeconomicus*. More recently we have been enjoined to take note of the emergence of *homo politicus* (Lauwell 1948), *homo sociologicus* (Dahrendorf 1972), *homo symbolicus* (Cassirer 1944), *homo ludens* (Huizinga 1962), *homo significans* (Barthes 1972), and "psychological man" (Rieff 1959).

The merit of these models of man is that they make possible a thematic delineation of specific regions of concern within the wider cultural life of man so that the analyzable features of the behavior of man can be studied. They provide a conceptual perspective for viewing the phenomenon. The particular model at issue will designate certain questions to be asked and certain procedures to be followed. In doing so it will also exclude certain questions and procedures which do not bear directly on the point of view that is delineated by the model. Some of the above models may lend themselves more to quantifying methodological procedures (statistical analysis, use of questionnaires, case studies, etc.) than do others. Hence, the use of a particular model will require an epistemic consciousness of what counts as significant within the designed perspective. In the use of such models or portraits of man by the several human sciences we observe a legitimate employment of objectifying procedures. The above serialized models illustrate various forms of self-objectification. A particular posture of human experience is objectified by the human scientist in such a manner that it is made to stand over against a detached, investigating consciousness. Such objectification, we maintain, is not only permissible but it is necessary for the very project of the doing of science. A science, whether a human science or a natural science, needs to proceed in accordance with the canons of objective and demonstrable knowledge. As we have argued elsewhere, the current crisis in the human sciences is not the result of the use of objectification, quantification, and measurement. The use of such procedures retains its legitimacy and natural justification in all scientific endeavors. So it is not objectification per se that is at issue in our proposed critique of the sciences of man. What is at issue is the recurring tendency on the part of the human sciences to transform a methodological procedure into a metaphysical principle. This happens when a particular model of man is no longer seen as the conceptual construct that it is, but is rather proffered as the determination of the reality under investigation. The *humanitas* which is always the subject matter of any particular science of man overflows every conceptualization of partitive profiles. Consequently there should be installed in the heart of every human science a double requirement — a continuing epistemological and methodological refinement of its analytical procedures and reflexivity upon the originative experience of a concretely developing, intersubjective world. It is thus that our principle of reflexivity upon origins, which was delivered through our elucidation and ontological analysis of the concrete self-with-society experience, provides the human sciences with a regulative ideal. This regulative ideal supplies the requisite insurance against the loss of the concrete dialectic of self-and-social-formation. It provides a safeguard against the fallacy of misplaced concreteness, in which a reified conceptual construct is mistaken for the concrete texture of experienced reality.

Another area in which our principle of reflexivity upon origins has relevance for the investigations of the human sciences is in the domain of functionalist analysis. A science of human behavior is always in some sense an analysis of the functions performed by man as a social animal. Man is understood in terms of his social functions. It is common in the literature to speak of these functions as roles played by the individual in his various postures of cultural participation. At any given point of development in his lifetime an individual may at the same time play different roles and perform different functions. Answering to a multiplicity of social demands and institutional requirements the individual functions simultaneously or successively as father or mother, brother or sister, uncle or aunt, employer or employee, teacher or student, citizen or alien, and so on. These roles can then be analyzed vis-à-vis the degree to which they exemplify biological, blood relationships and the degree to which they exemplify cultural predicates. It is through such analyses that the problem of nature and culture is articulated by the human sciences, and as even the initiate into the sciences of man readily becomes aware, the existing literature is replete with proposals for solving this problem.

Now the relevance of ontological analysis is not that of providing another scientific hypothesis toward the solution of this problem, nor that of carrying the functionalist analysis of man and society further. Its relevance is registered on another level of questioning, whereby a reflexive turn to the origin of the plurality of roles and functions is effected. Such a reflexive turn again makes visible a world of intersubjective experience which is older than the classificatory analysis of functions and roles and provides a corrective to the tendency in functionalist analysis to reduce the being of man to his functions. The vibrant, dynamic, and on-going phenomenon of self-with-society is not exhausted in any given collagation of functions. It perpetually overflows these functions, which are in the end conceptual constructs that emerge from a more primordial source of self-understanding and social formation.

A third area in which the principle of reflexivity upon origins bears directly on the projects of the several human sciences pertains to the issue of interpretive understanding. In the construction of various models and images of man and in the definition of social roles and functions the sciences of man are aiming at an explanation of human behavior. This human behavior is already imbued with a prescientific self-comprehension. The socialized self, whose thought and action the human scientist seeks to explain, already understands itself in its daily, mundane engagements and concerns. It already endows its varied activities and projects with meaning. An interpretive self-understanding is hence always already at work in the on-going process of everyday life. Unlike the objects of inquiry in the natural sciences, which do not ascribe significance to their behavior, the human subjects under investigation by the human sciences endow their concrete life-world with manifold meanings, and hence already understand themselves in and through their thought, speech, and action.

This hermeneutic of everyday life, operative on the level of the self's originative dealings within a natural and social world, again calls for a consciousness of the need for a reflexivity upon origins. The models that are constructed, the classifications that are devised, and the quantified data that are collated should in the end reflect in some way the prescientific and precategorical meaning establishment that is at work on the level of originative experience. If there is no such reflexivity upon origins, then the concretely experienced dialectic of self-with-society is occluded, and the stage is set for the entrance of either an abstract empiricism or an abstract intellectualism or — as is more likely the case — a specious combination of the two.

It is thus that the human sciences themselves will need to come to terms with the issue of interpretive understanding — as Max Weber had already so clearly perceived in his analytic of *Verstehen*. *Verstehen* will need to become a feature of the methodological consciousness of any science of man. More precisely, insofar as the human scientist is himself a participant in the on-going process of self-and-social formation, interpretive understanding will need to become self-reflexive. As a methodological technique it will need to remain reflexive upon an on-going process of interpretation and understanding that antedates the methodological project. *Verstehen* as methodology is then properly understood as an interpretation of interpretations. The point of paramount importance at this juncture is that methodological design follows originative understanding. The interpretive understanding esconced within the concrete dialectic of self-with-society should direct and inform the method. The experiential forms of the life-world should govern the method, rather than having the method predefine the experiential forms. Here the employment of the principle of reflexivity upon origins attains relevance on the level of methodological analysis and construction, regulating the investigations of the human sciences in such a manner that a methodological concealment of the originative experience is avoided.

REFERENCES

BARTHES, ROLAND
1972 *Critical essays*. Evanston, Ill.: Northwestern University Press. (Translated by R. Howard.)

CASSIRER, ERNST
1944 *An essay on man*. New Haven: Yale University Press.

DAHRENDORF, RALF
1972 *Homo sociologicus*. Westdeutscher Verlag Opladen.

HAYDU, GEORGE G.
1970 Interrelated transformations of Rousseau's life and of Western culture of his time. *American Journal of Psychoanalysis* 30:161–168.

HUIZINGA, JOHAN
1962 *Homo ludens*. Boston: Beacon.

HUSSERL, EDMUND
1970 *The crisis of European sciences and transcendental philosophy*. Evanston, Ill.: Northwestern University Press. (Translated by David Carr.)

KIERKEGAARD, SØREN

1951 *Sickness unto death*. Princeton: Princeton University Press. (Translated by Walter Lowie.)

LASSWELL, HAROLD

1948 *Power and personality*. New York: W. W. Norton.

MERLEAU-PONTY, MAURICE

1962 *The phenomenology of perception*. New York: Humanities Press. (Originally published in 1945 by Librairie Gallimard in French. Translated by Colin Smith.)

RICOEUR, PAUL

1967 *Readings in existential phenomenology*. Edited by N. Lawrence and D. O'Connor. Englewood Cliffs, N. J.: Prentice-Hall.

RIEFF, PHILIP

1959 *Freud: the mind of the moralist*. New York: Doubleday.

SCHELER, MAX

1961 *Man's place in nature*. New York: The Noonday Press. (Translated by Hans Meyerhoff.)

SCHRAG, CALVIN O.

1969 *Experience and being*. Evanston, Ill.: Northwestern University Press.

1975 The crisis of the human sciences. *Man and World: An International Philosophical Review* 8 (2).

TILLICH, PAUL

1952 *The courage to be*. New Haven: Yale University Press.

WHITEHEAD, ALFRED NORTH

1926 *Process and reality*. New York: Humanities Press.

Those Who Make Music with Their Chains

GEORGE MILLS

> Reality is not a fixed entity. It is a contingent interlocking of moving events. And events do not just happen to us. We are an integral part of every event. We enter into the shape of events, even as we long for an absolute in which to rest. It may be just this longing for an absolute in which our concepts might *not* have to be responsible for our percepts, and so indirectly our reality, that explains the hostility of our ordinary intellect to these shadowy modes of mind.
>
> JOSEPH CHILTON PEARCE, *The crack in the cosmic egg*, p. 3–4

PERSONAL NOTE

I received my graduate training in a department of social relations, having been besprinkled with sociology, clinical psychology, and social psychology, but totally immersed in social anthropology. When as a student I grew weary of the fat tomes of the social relations library, I used to walk across the hall to the philosophy library where I would run one finger down the rows of books like a stick down a picket fence until I came to the slimmest book I could find. I would then take down that book and dip into it. I found some fine things that way, including a book I quote from later, George Santayana's *Platonism and the spiritual life*.

There I found more than brevity. I found or renewed my acquaintance with Epictetus, who calmly told his torturers that if they cranked the rack one more notch they would break his leg, and who, when they did, and the leg cracked, just as calmly said, "See, I told you so"; with Democritus, the laughing philosopher, as well as with Han-shan, the solitary, another laugher, whose laughter swirled gaily away with the wind of Cold Mountain; with Lin Yutang's discussion of the old rogue, a bit of Chinese sociology that still seems delightful; with Chuang Tzu's views on the usefulness of the useless; with the pantheistic declaration for which Giordano Bruno was led into a public square of Rome and burned to death; with the heresy for which Spinoza was made to suffer complete ostracism by the Jewish community in Holland; and with many more. I could not get it out of my head that these were all free men, in some sense heroes, capable of taking life into their own hands, of making up their own minds; capable, above all, of paying a tremendous price for the privilege. It took me some time to see that refusing to pay that price could itself become a tremendously costly undertaking. I think I would have put myself out as an indentured servant to any one of these men had he walked into the library and been willing to take me on, the last possibility being as remote as the first. At that time I seemed to have no choice but to shuffle back across the hall and once more take up the tome where I had left off. I knew that, despite its fatness, I would find no listing for freedom, heroism, or delight in its index, and the realization saddened me. I wanted to stand up and bang on the oak table and catch everyone's attention and then make this speech: "As I understand our doctrine, we are all creatures of the society into which we happen to be born — speak its language, eat its food, fear what it is afraid of, rejoice in what exalts it, and finally die the death it has prepared for us. Because in all this we have no choice but to become whatever society chooses to make of us, some of you prefer to say that we are the puppets or the prisoners of society, making brief, mechanical appearances on stage or languishing out our lives behind remorseless bars." Without hesitation in my fantasy, I go on to say, "If I agree, it is because I want no more to be a victim of illusion than you do. What bothers me is that I have heard rumors that there are prisoners in other corridors of this iron cage who have managed to pierce the walls of their cells with small holes that allow sunlight to fall in patches on the floor and that they dance around these patches of light, cheering and singing and clanking their chains in unison." I pause dramatically at this point to let my words sink in, and then I continue. "Imagine that! They know how to make music with their chains. I want to know more about those extraordinary people. Why is it that, with all the peoples we have scribbled about, no one has done a monograph on 'Those Who Make Music With Their Chains'?"

If my approach to this topic seems a trifle personal, I submit that every commitment, even to science, if it is not automatic, is personal. I have done my best to make what I say honestly accord with my own experience, its quandaries as well as its conclusions.

WAY OF LIFE

The first question is: How are persons bound into society? This is the usual sociological way of putting the question. That it heavily biases the answer in the direction of society, as if society were the independent variable and person the dependent variable, will come clear in a moment. Nevertheless, I now let the question stand in its familiar form.

The customary answers to the question are: The person is bound into society, first, by the playing of social roles; second, by his possession of social values; and, third, by the instrumentalities of selfhood. The new born human infant is of malleable genetic stuff. If a WASP infant is put into an Oriental family and brought up there, he will not only speak an Oriental language and live by Oriental customs, but also, to the extent that motor habits are learned, he will look more Oriental than he would have if he had stayed on in the WASP family. The male child in being treated as a male rather than as a female learns to play the proper sex role and is rewarded when successful and punished when a failure. Reward and punishment by other persons is supplemented by his own rapidly developing feeling that one kind of behavior is appropriate for him while another is not, and when these value judgments become sufficiently significant, they issue as judgments concerning the worthiness or unworthiness of the self. As G. H. Mead has shown, the judgments by others and the self-reflexive judgments are closely bound up, one with the other. Mary comes home from her delightful play in the mud to discover that mother reacts with horror. Mary finds herself in a state of conflict. On the one hand, she has responded positively to the mud, but on the other hand, her attachment to her mother has suddenly caused a negative note to be thrust into the proceedings. Mead says that, in order to sort out the confusion, Mary begins to take the part of her mother and to look back upon her own experience with her mother's eyes. Mary herself is now heard saying, "Mary, Mary, you got all muddy." The conscious self, through the reactions of those people who are important to it, comes to see itself as others see it and to be capable of the sort of duplicitous judgments that Mary passes on herself. To put this another way, society constantly hovers on the edge of neurosis because society and neurosis are both cut out of the same stuff of selfhood.

An agent of society is a person who concludes that his fulfillment as person, at least in a particular situation, derives from his acting so as to support the expectations he shares with others of his social milieu. A judge is an agent of society when he acts so as to bring a convicted criminal to his death. He is also an agent of society when he acts so as to kill off in himself all that he fancies is incompatible with his playing the role of judge.

We can now add a fourth answer to the question: How are persons bound into society? In addition to social roles, selfhood, and values, there is experience. Penfield's experiments (1952) exemplify what I mean by experience. Upon stimulating a human brain electrically, Penfield found that he activated memory events that were less like surrogates (symbols, images) of the original experiences the person was remembering than they were like total recalls of the original experiences. It is as if the river of Heraclitus not only constantly flows in and through each of us but is also wholly and perpetually contained in each one of us. I shall explain in a moment why experience is to be included with social roles, selfhood, and values. This concept of experience is tremendously important, as we shall see, for a revised view of the relations of person and society.

One source of confusion in all of this is a failure to discriminate between the two models we use for thinking about the connections between person and society. We are most familiar with the body model. Mary's body is a discrete entity as compared with the body of her mother. When we think in terms of the body model, we refer to Mary as self and mother as other (or vice versa), as if one self were coterminous with Mary's body and the other with her mother's physical presence. The whole point of Mead's treatment is that the relation of self and other does not come into existence until the limits of body have been transcended. Unless Mary needs or loves and, in some sense, is her mother, Mary will remain indifferent to her mother's strictures. Unless mother needs or loves and, in some sense, is Mary, mother will remain indifferent to Mary's muddiness. In addition to the body model of two discrete entities there is the model of interpenetrating identities. Two discrete entities may bring about exchanges with one another while undergoing minimal internal change, as when two nations, for the sake of expediency, undertake diplomatic relations with one another. In the relation of interpenetrating identities, however, the two entities together create a third entity of which both become parts or components, and the attention of both is fixed, not only upon each other as discrete entities, but also upon this third entity which both together have brought into existence and for which both are responsible.

When Mary says to herself, "Mary, Mary, what have you done?" she is indeed playing the part of mother, but she is also preparing herself for the part of a new Mary which her mother's surprising response has brought to Mary's attention. And in rehearsing both of these parts at once, she is contemplating the possible vicissitudes of that third entity, Mary-and-her-mother, in which both Mary and her mother now somehow find themselves involved. It is certainly no foregone conclusion that Mary will knuckle under to mother. She may do so. But she may also learn how to appear to knuckle under while continuing to dabble in the mud. Or she may decide, even at a tender age, that she must be both true to herself and honest with her mother because both of these are implied both in being true to herself and honest with her mother. If she does knuckle under, she may discover later that psychologically this was an exorbitant choice requiring much hard work to redo. In some ways, the struggles of Mary and her mother may become struggles to see whether it is Mary or her mother who is to be master of this new entity, Mary-and-her-mother.

There is nothing mystical about this model of interpenetrating identities. A whole is not the same as the sum of its parts. Four dots can be arranged in such a way as to be connected by either a circle or a square. The existence of a connection, and its nature, should one come into existence, is not given by the dots themselves but is supplied from without. Once the connecting pattern has been established, it affects all components equally. The example of Mary and her mother is different in that the pattern of their relation is not imposed from without but rather emerges from choices made by the two components. Once established, that pattern is no more reducible to the separate natures of the two entities, Mary and her mother, than the circle or square is reducible to the four dots.

All four ways by which the person is bound into society follow the model of interpenetrating identities rather than the model of discrete entities. I already have spelled this out in terms of selfhood. Social roles are often accompanied by the concept of social statuses, the statuses being the abstract niches in a social system abstractly conceived, while the roles are those abstract niches as occupied by particular persons. A status is less a prescription for behavior than a range of possible behaviors appropriate for that niche, this openness being necessary to accommodate the variability of the persons making up a social system. It is only as a particular person fills a particular status that the role takes on definition sufficient to elicit the definite responses of other persons filling other statuses. It is not simply that a status shapes persons, but also that persons, through their handling of roles, have a profound influence upon the system of statuses.

Much the same is true of values. It is not merely that shared values bestow identity upon persons by defining this as acceptable, that as not, but it is also that persons give life to shared values by accepting their rule or cast doubt upon them by challenging it. This is in accordance with the fictional quality of human life. Our social realities are at bottom conspiracies. A dollar bill is worth what value we are willing to impute to it; the value of the gold which supposedly backs up the dollar bill is as fictional as the value of the dollar bill itself. The value of the gold that backs up social values is conspiratorial in two ways: The agreement is not only to accept these values as norms, but also to ignore the arbitrary quality of our agreement (arbitrary in view of the valuational alternatives). At least implicitly, there is an agreement to find the roots of social life, not in agreement itself, but in some reality that, conceived as larger than and external to persons, is thought of as demanding unquestioning obedience from persons. Thus, culture becomes a god whose commandments are cut deep into the golden tablet of every psyche. I am not denying that society has roots in nature, but I am objecting to the classic way of conceiving this rootedness which puts out our eyes and chops off our hands at the wrists and our feet at the ankles. Procrustean obedience has one terrible consequence: When culture turns against persons, as it is capable of doing, Procrustean people are helpless before a doom which they have brought upon themselves.

In experience, we find again this same fusion of two worlds to form a third. It is my experience in the sense that it is the flickering of my nervous system which no one else, locked up in his own cage of arteries, may directly share; and yet my experience is not only of my discrete self but also of other entities in the world, or so we assume. Experience is situational. Penfield's electric probe rouses trees, skies, faces living and faces long dead, as well as feelings, all of these made equal in the same patterns of vividness. Experience makes other a part of myself and forces upon me the task of doing my part — of variable scope — in constructing the meaning of both self and other.

Notice the delightful balance that obtains. If statuses are properly called social statuses in that it is shared expectations which provide the stable component and persons who contribute the ephemeral and idiosyncratic component, then experience, in the same way, is properly referred to as personal experience, however social it may be, simply because society has no giant nervous system of its own and cannot become the vehicle of experience in the same way that persons can. This balance is sufficient to make us skeptical of any simpleminded scheme that sets up society as the independent variable, person as the dependent variable. Man is not only a social animal, as the classic definition has it. He is also a personal animal. If we forget either half of this double definition, we make it impossible to understand the remaining half.

While agents of society are best regarded, not as a special way of binding persons into society, but rather as servitors of roles, values, and selfhood, they manifest the same doubleness of reference the sociologist catches when he uses status as the companion term for role. The status is a set of social expectations awaiting a person; the role is a person who has accepted the challenge of the expectations. Human life is impossible unless both status and person are present in a condition of hyphenation. Similarly, if human experience is made up of both personal and social components, then the way of life we build upon this fact will not work unless some of us satisfy the requirements of being personal by being social (agents of society), while others satisfy the requirements of being social by being personal (meditation).

The easiest way to picture the relationship of interpenetrating identities is the yin-yang, the familiar circle containing two spermlike entities nestled closely together, one light and one dark. The light half is centered on a dark dot, the dark half on a light dot, signifying that the two halves interpenetrate in such a way that midnight begins at noon and so on, whatever pair of "opposites" is under consideration. The enclosing circle represents the larger whole which yin and yang taken together comprise: yin-yang. When we look at the "opposites" as counterparts of one another, we speak of yin and yang. When we look at them as interpenetrating to form a third system, we speak of yin-yang. I have discussed the interpenetrative nature of selfhood, social roles, values, agents of society, and experience in order to make clear that person and society are related to one another as yin and yang, i.e., that they are counterparts which comprise a more inclusive order of reality, the yin-yang.

To look more closely at the relation of person and society, let me change the picture. Instead of the yin-yang diagram, imagine two circles drawn on a sheet of paper so that they intersect or overlap. One circle represents person, the other society. Their overlapping creates three areas. If person is red and society is yellow, then the area of overlap will be orange. This means that in addition to the area of interpenetration of the two (orange), there is a red area (purely personal except as the red is continuous with the orange) and a yellow area (purely social except as the yellow is continuous with the orange). As an example of red, I suggest the work of Gregor Mendel. In 1866 Mendel published his discoveries concerning the basic quantitative laws of heredity in a scientific journal. His fellow scientists, with one exception, paid no attention to his work, and he died perhaps believing himself to be a failure. Around the turn of the century, Mendel's conclusions were rediscovered and his previous publication of them was brought to light. Kroeber (1948:366–367) discusses the question whether Mendel's discoveries were social or personal discoveries, and he concludes that since they took place around the middle of the nineteenth century, they were personal discoveries and did not become social discoveries until the turn of the century.

Similarly, the yellow area designates aspects of society that transcend the person. The number of people in a group affects the nature of the group apart from any other influences at work. Another example of yellow is that in a state, the law, including legally approved methods for changing the law, may be above every person in the society, including the head of state. Orange represents the area of interpenetration of social and personal factors. Most science, unlike that of Mendel's original discoveries, results from personal manipulation of collective tools.

Each of the component systems, person and society, has its own characteristics and works according to its own principles. The person is basically a psychobiological entity. Society is an aggregation of persons. Unlike a person, an aggregation of persons is never found isolated in reality.

Something beyond a mere aggregation of persons is necessary to bind those persons together and give them a sense of common identity and common destiny. Even though an aggregation of persons is never found as such in reality, there are good reasons for isolating it conceptually under the heading of society.

One technical term for that which binds an aggregation of persons together is culture. In 1871, E. B. Tylor established the term in English, defining it as "that complex whole which includes knowledge, belief, art, law, morals, custom, and any other capabilities and habits acquired by man as a member of society" (Kroeber and Kluckhohn 1952:81). Culture is a set of shared expectations, and it is shared expectations, elaborated as roles, values, and views of self, with agents of society to administer them, which bind together an aggregation of people having a certain genetic makeup.

The phrasing, "capabilities and habits acquired by man as a member of society," is ambiguous. Its most obvious meaning is: acquired from the common stock of expectations available to members of that society. If this is taken to be the meaning, then we must regard Mendel's original contributions as lying outside of his own culture. A less obvious meaning makes membership in society a necessary but not necessarily sufficient condition for the acquisition of capabilities. On such a definition, Mendel's contributions become part of the stock of his culture even as they go unrecognized. This second view places personal experience on a par with culture as a possible source of capabilities and habits.

That which society and person have in common, making it possible for them to interpenetrate and form a larger system, is expectations. Human beings cannot cooperate with one another unless they define situations in common and know more or less what to expect from one another. To the extent that expectations are shared, they are social. The expectations by which a human being lives are always to some extent social but never completely so. No prefabricated set of expectations can provide an answer for every question, a response appropriate to every predicament. A person must often make his own choice from among several appropriate possibilities; in some instances he falls back upon what seem to be novel improvisations. The pivot of expectations is experience whose social and personal aspects are inextricably bound up with each other. Experience is of a nervous system and therefore of a person; hence experience is also inextricable with selfhood. This means that experience is the focus of three kinds of work. First, the agents of society work upon experience to shape it and keep it shaped to the socially approved system of roles and values. Second, the psyche, the predispositions of the living creature, work upon experience for the attainment of its own ends. The work of psyche may be consonant with the work of society, but it need not be. The third form of work is that which selfhood performs upon experience. If the work of society and psyche are harmonious, selfhood lies fallow. When the work of society and psyche are at odds, self may bestir itself to take experience in hand and do what it can to cope with the painful division of the person that psyche and society have brought about.

Persons are unique. Even identical twins differ in some experiences and see the world from their own points in space. This idiosyncratic aspect cannot be eliminated. This means that each person is a particular shade of red, interpenetrating with a range of yellowness to produce a special shade of orange. For any society, we may generalize one picture of red and yellow interpenetrating one another and use this in comparison with a similar picture abstracted from a different society. This kind of thing, done with sophistication, is the stuff of anthropology. This generalized picture, however, may not prove appropriate once we shift perspective and take a look at red-yellow as they appear to any person in that society, i.e., one of the numerous hues of red. Indeed, the anthropological generalization may apply to no one of the persons making up the society; there may be no single, personal representative of that which results from this anthropological averaging out. Much confusion has resulted from our failure to make distinctions at just this point. We talk about culture as the mode that characterizes society — the aggregation of persons. The concept of culture is needed because similar aggregations of genetic stuff, similar societies, may conduct their lives by quite different modes. In addition to the distinction between society and culture, we need to be clear as to when we are talking about the red-yellow system from yellow's perspective and when from red's, when we perceive the inclusive system from a generalized social perspective and when from a particular personal perspective. If culture suffices as the terminological companion for society, we need some additional designator of the overall system as a companion for person. I suggest the concept of way of life. You and I may be members of the same society, fluent in the same culture, and yet have quite different ways of life. These terms do not and cannot represent rigid distinctions but rather allow for shifts of perspective within a system that is complex but not without its claims to coherence.

We ought also to maintain a clear distinction between Culture One — the substantial agreements making it possible for the persons of a society to live together — and Culture Two — the anthropologist's abstraction from Culture One. We should make a similar distinction between Way of Life One — the personal actuality — and Way of Life Two — the psychologist's abstraction or the person's abstraction when psychologizing his own experience. It is tempting to turn Ones and Twos into verbs by saying that Oneing is always living life. Twoing is always talking about living, psychologizing it, philosophizing about it, anthropologizing it. For two reasons, to do this would be a mistake. First, philosophizing may have an effect upon experience; so far as it does, it becomes a part of living.

Second, talking about living may itself become a mode of living; teacherly careers are carved out of such monologues.

The scientist bungles whenever he assumes that his abstractions, his culture patterns and configurations, are actually at work in the experience of persons. For this reason alone, it is important to keep in mind the distinction between Ones and Twos.

I think of experience, vividly available for recall, as the key to all that a human being is. It is also the common pivot around which the two systems, person and society, revolve. While these systems are different kinds of systems, each having its own nature, principles of operation, goals, etc., they are also interdependent and mutually interpenetrating in such a way that each forms a part of the other. Locked together as they are, they comprise a third system having characteristics that are not reducible to those of either of the component systems. In order to keep the terminology straight, I am calling this third system culture when looked at from the viewpoint of the social component and way of life when looked at from the viewpoint of the personal component. Culture and way of life take two forms: the substantive forms by which persons, collectively or individually, live their lives; and the descriptive or theoretical forms produced by the professional interpreters of human behavior. Thus, experience leads us not only into nonpersonal aspects of society, on the one hand, and into nonsocial aspects of the person, on the other, but also into the workings of culture and way of life.

One concomitant of this arrangement is reflexivity: selfhood. Stimulus-response psychology, though itself a magnificent manifestation of human self-awareness, seems totally inadequate for understanding the immense potentialities hidden in our human version of reflexivity.

MEANING AND BALANCE

The arrangement as described so far is static. Momentum is usually provided by means of a concept of need or drive. When these concepts are grounded in physiology, various gadgets may be used for measuring the intensity of needs and drives. However rigorous a tissue-depletion conception of need may be, its inadequacy is apparent from what has already been said. Persons, even in their nonsocial aspects, are nothing without society. Because society is not a psychobiological organism, tissue depletion has no relevance to it. I am not saying that tissue depletion is unimportant or that it does not provide motivation under certain circumstances. I am saying that tissue depletion is inadequate for understanding what motivates a way of life that involves both social and psychobiological components.

Some would cut the idea of need loose from the idea of tissue depletion about the need to be good citizen. Then it is obvious that some people have a need to be bad citizens or to be good citizens in some context which, from the viewpoint of the predominant mode, makes them appear to be bad citizens. By the time one has multiplied the list of needs to cover all of the vicissitudes of experience, the list has become so fragmented and self-contradictory as to be useless. Whatever impetus needs do provide in the human scheme, they are inadequate for solving the problem of what keeps the more inclusive system going.

The question of motivation must start from the indivisible unity on the yin-yang pattern, of two entities that can be referred to as either person and society or as culture and way of life. I shall use the term "meaning" to designate a motivational perspective taken on the more inclusive scheme.

Many of the untranslatable concepts dear to the hearts of cultures and institutions are expressions of meaning: progress, honor, fair play, the Navaho view of harmony, the Zen concept of satori, etc. Meaning, as the nucleus of feelings of patriotism or ethnicity, may dye innumerable details of a way of life. Such concepts, as I have mentioned, exemplify meaning from the perspective of culture.

The idea of progress does not become motivational until there are persons for it to motivate. However far sociologists may push the analogy of society and organism, the sociocultural system will never be a person. It is persons who act, persons who must be motivated to act, persons who are motivated on the basis of the kind of meaning their lives have. Persons do have needs, physiological balances that become upset and must be restored. But persons also respond to more abstract imperatives associated with the sociocultural component of the way of life. Men go to prison for stealing because they are starving; they also offer themselves up on the battlefield to keep the world safe for democracy or whatever.

When we speak of agents of society, the phrase often takes on too formal a meaning, as if agenthood were a characteristic of some roles but not of others. Rather, being an agent of society is being a spokesman for the shared component of a way of life; it is a perspective which is not only formalized in certain special roles, such as president, judge, priest, teacher, but also permeates every role. Mary would not be in debate with her mother were she not also in debate with her self as agent of society. Each of us, as both creature and creator of his own way of life, is the agent for the social component, however he conceives it, entering into that way of life. At the same time, each of us is an agent of the other component of his way of life, of himself as person. If we are going to use the phrase "agent of society," we ought to find its counterpart, agent of the self, equally appropriate.

Two sets of claims are made upon a person. Thus, we find Mary torn between the claims her mother makes upon her and the claims her own experience makes upon her. Giving life meaning is an attempt to reconcile these two sets of claims. Sometimes this reconciliation calls for a sacrifice of needs. When government agents offered to teach the Kiowa how to replace their wizened corn with a more delectable variety, the government agents met with unaccountable resistance. Pressed for an explanation, one old Kiowa said that the Kiowa had always grown the wizened corn, that wizened corn was Kiowa corn, and that giving up wizened corn seemed to presage the end of Kiowa culture. At the other extreme, the Marquis de Sade's way of life called for a certain indifference to the public outrage which his philosophy excited.

Meaning, then, is the relection in the mirror of the self of the quality of a way of life. Because much of the psyche does not rise into consciousness, meaning may go on record at a level of the psyche that is below conscious awareness. For the neurotic, life has a painful meaning, and the pain is bewildering because at least part of its cause is beyond the neurotic's ken. When a way of life is successful in integrating the totality of a person's experience, it takes on more positive meaning. As the sources of positive meaning become habitual, the issue of meaning itself disappears. As Chuang Tzu says, "When the shoe fits, the foot is forgotten." So is the shoe. When the way of life is right for a person, reflections on both way of life and meaning are forgotten. This is so, not because the two sets of claims, social and personal, have ceased to assert themselves within the experience of the person but because those claims have been reconciled. That we are in trouble today is suggested as much by the quantity of verbiage spilled out on the subject of life's meaning, social values, and the like, as by the prevalence of suicide and other self-negating practices.

The concept of meaning is useful for several reasons.

First, scientists keep hoping that there is some objective way of comparing and classifying ways of life. A list of needs seemed for a time to provide such an objective tool. The trouble is that the needs of the person and the requirements of a society, as they actually function, derive from the meaning of the way of life being studied and can be arrived at only through experience of the way of life. As Dorothy Lee points out, when the anthropologist classifies a Hopi journey into the Grand Canyon as an economic activity designed to increase the salt supply, he overlooks the larger, "metaphysical" meanings this journey has for the Hopi and is guilty of the worst ethnocentrism (Lee 1959:73).

Second, even when the means for satisfying the needs of a person are intact, a breakdown of meaning may occur. Thus, the successful owner of a paint factory, whose life others may envy, one day gets up from his desk and walks out of his office forever — to become a writer or try some other role that promises to restore meaning to his experience.

Third, as a complex organization, a way of life is precarious, especially in situations where alternative ways of life, alternative kinds of experience, have become threateningly visible. As the burden of anxiety increases, so does the precariousness of meaning. One day the point is reached where many people feel that a few additional inches in the length of male hair constitute a challenge to their whole way of life.

Fourth, the unpredictability of social events and the irony of social action are rooted in states of personal experience. Even with full, therapeutic knowledge of a psychological sort, one cannot predict what will emerge from a person's colloquy with himself over the meaning of his life. These colloquies are the growing points of both person and society.

Finally, the concept of meaning is crucial for understanding the problem of balance as it characterizes a way of life. By balance I mean that the ways of life that a scheme comprises must be such as to meet the demands of both person and society sufficiently to keep intact the system that embraces both. It may be that the classic arrangement skews the system in favor of the sociocultural component. A couple of paragraphs from Alfred Kroeber will make clear what I mean.

> . . . only a fraction of all the men congenitally equipped for genius ever actualize as such. Only a fraction are ever found out, or allowed the rank, by history. This fraction is the same as the proportion that the number of generations recognized as fruitful and genius-studded, in all lands, bears to the number of barren, geniusless generations. This proportion can hardly be reckoned as greater than one in four, and may be as little as one in ten, if we take into account all the regions and eras of the world in which it is customary not to recognize any geniuses as having occurred.
>
> There is a point of impressive significance here. Human biological heredity runs good enough to produce, once in every so many hundred thousand or million births, an individual so highly gifted as to be capable of becoming one of the lights of our species, a benefactor or a creator whose work will live in history; and yet the nature of our culture manages to neutralize or frustrate from seventy-five to perhaps ninety out of every hundred such great geniuses, or to depress them into mere second-rank talents or transient leaders of soon-forgotten days. Ideally considered, this is a tremendous waste from the point of view of those concerned with human achievement (Kroeber 1948:339).

In a footnote, Kroeber has this to say:

> This over-all wastage of 75 per cent or more of the finest congenital talents born in the human species may seem to constitute a blasting indictment of human culture. But without culture, the waste would be a complete 100 per cent. Culture is admittedly still an imperfect instrument.

This means that balance is first of all what we experience it to be. Human lives throughout history have maintained sufficient meaning, if Kroeber is correct, to maintain the species despite what, from an external viewpoint, may seem to be an exorbitant sacrifice of human potentiality. That we have become aware of the sacrifice does not mean that our forebears experienced their lives as sacrificial. It means rather that knowledge of these matters is a vital component of meaning, and therefore of the state of acceptable balance that prevails at any moment as between the personal and social components of a way of life.

Maintenance of personal meaning ought to be one of the concerns of the social system. Those for whom life has lost its meaning have lost their incentive to be good citizens as well as good persons. There is a point at which the agents of society, no matter how strictly they conceive their function, must become agents of persons or they cease to act wisely as agents of society. One of the difficulties is that as a way of life undergoes threat, we tend to accelerate the splitting apart of components by developing, in the name of health, obsessive devotion to one half or the other. The agents of society increase the oppressiveness of measures designed to coerce us into collective salvation, while persons become convinced that participation in open and fulfilling undertakings with other human beings is no longer possible.

SHIFTING THE CENTER OF MEANING

Becoming one's own person involves a shift in the center of meaning. Growing up tends initially to locate the center of meaning in culture; the infant's way of life is rudimentary; the child's, tentative. We in the U.S. are apt to think of adulthood as the final plateau up to whose heights infancy and childhood slope. If all goes well, the barbarous infant is inducted into the culture in such a way that he more or less automatically finds a satisfactory place within the social system and so falls into a meaningful way of life. Automatically means with a minimum of conscious appreciation and manipulation. Of course he must make choices of his own, but he is helped in this by preexisting patterns of expectation for one in his role and station. I conform even as I praise myself for my individuality, so that the more strictly I conform, the more certain I am that I am an individual. Under such circumstances, it is easy to confuse coherence of culture with coherence of self, a mistake I do not discover until I find myself in circumstances that destroy cultural coherence and so lay bare the incoherence of self. Travel in a foreign country can be such a set of circumstances. At the same time, it is a simple matter to protect myself by remaining within a tube of Hilton culture whose valves ingeniously allow money to flow out while preventing ghastly revelations from flowing in.

When automaticity breaks down, various eventualities become possible, one of which includes an increase in conscious awareness of what enters into a way of life and willingness to assume responsibility for my own. As this happens, a significant shift in the centering of my life takes place from the social to the personal component. When this shift proceeds far enough, I can say of myself that I have become my own person, and understand what I mean more clearly than when I formerly praised myself for my individuality.

This initial overweighting of the social is not occasioned by a dark conspiracy among the grand inquisitors who manage our institutions but rather results from the prematureness of human birth. When monkeys are born, they are sufficiently developed to cling to their mothers. If human birth were postponed to a point of comparable development, the infant's head would be too large to come down the birth canal. Pushing back the time of birth renders the infant helpless and prolongs the ministrations of adults, without whose care the infant would either die or fail to develop what we regard as normal human capabilities. The examples of illegitimate children locked away from the world in closets attest to this. In terms of the yin-yang scheme, this prolonged helplessness represents a tremendous invasion of the psyche by the social component before the psyche is sufficiently developed to question or resist. Because it is this invasion that gives the psyche much of its form, the psyche grows up predisposed to follow those drummers who beat out the familiar marches without asking where it is being marched to or what other tunes are available. This is the inculcation of social values. In the past, social science has dropped the matter here. Russians never pop up among Zuni Indians nor do U.S. citizens spring to life from the wombs of Tiv women; Ifugaos always produce Ifugaos; Kwakiutls, Kwakiutls. What other evidence for you need to agree that society plus culture is primary and determinative? The key to the whole works?

And yet there is equally good evidence that the issue is not this simple. There are the sports, like Gregor Mendel, whose socialization somehow involves a process of personalization enabling them to work in ways that though incomprehensible to their contemporaries, prove finally to be socially meaningful.

There are also the small group experiments of Solomon Asch (1955). Although Asch set these up to look like experiments in perception, in my terms they are tests of where in the yin-yang scheme a person's center falls. The test of perception was a simple one. With no complicating factors, people rarely made an error. The complicating factor that Asch introduced was a prearranged consensus dedicated to the proposition that an erroneous perceptual choice was the true one. The person being tested was thus caught between the testimony of his own senses and the insidious attraction of a consensus urging him to negate his senses. Asch found that only one-quarter of the subjects were completely independent and never agreed with the erroneous judgments of the majority.

It may be said that the forces present in a classroom experiment are not the forces present in the socialization of a person. I agree. Nonetheless, Asch's experiments provide an important analogy with what takes place every day on a larger scale. There is a similar doubleness of claims made upon the person, a similar precariousness of meaning and doubt about how one ought to respond, and there is a similar disparity between those who go along with the consensus and those who prefer to remain true to their own experience. There is also a similarity in the psychological cost of one kind of choice or the other — in any event, a choice is made, not choosing being itself a kind of choice.

In terms of the yin-yang scheme, the conflicts analogous with those contrived by Asch are of four kinds: between different ways of life; between aspects of the social system; between personal and social expectations; and between aspects of the person.

Politically, the world has shrunk, if not to one global village, then to several villages whose inhabitants warily eye one another's cultural obsessions. Social classes have their own ways of life which may fall into conflict. The archival apparatus of today — museums, galleries, libraries — contain records of human possibilities which may at any moment be reactivated, as when the counterculture organizes itself in ways inspired by American Indian tribes. It is no wonder that life becomes precarious when extinct ways of life jostle the already jostled thriving ways of life.

Conflicting role expectations, as when students, administrators, and parents make different demands upon teachers, exemplify conflict between aspects of a social system, while Asch's experiments indicate what I mean by conflict between personal and social expectations.

Finally, each of us must accept the limitations of an existence which is mostly one thing at a time. If I decide to do this now, I cannot also do that. If I decide to become an opera singer, I shall never know what it would be like to become a teacher of the classics, a career toward which I also have some leaning.

Any one of these kinds of conflict may create a rift in the automaticity of a way of life with a resultant eruption of awareness and new meaning.

At this point I seem to have plunged the whole argument into confusion. On the one hand, I talk about conflict between social and personal expectations, and, on the other, I say that all experience, in which such conflict must take place, is personal. If all conflict is of experience, and all experience is of the person, then all conflict is of the person. If that is true, how is conflict between social and personal expectations possible? This is a key point that explains why being a member of society can be such a neurotically insidious experience; it is important to be clear about it.

We talk about social controls as if that which controls my behavior, all of the rewards and punishments, were external to me. Indeed, in one sense, this is so. If I am walking down Death Row attended by a priest, wishing the electric chair away will not cause it to disappear. At the same time, threat of the loss of anything — possessions, social approval, even life — is effective only as long as I value possessions, social approval, or life more than I value self-approval or as long as I make the valuing of one of these integral with my valuing of my self. Shift the center of my life, change its meaning, and I take the hemlock as calmly as Socrates did or I ride to the gallows seated on my own coffin as John Brown did, content to have done my work well.

The distinction that must be made is between the externality of the sanctions and the internality of that which makes them effective. The electric chair remains external and inevitable but may loom no larger in my experience than the family rocker. Its presence I can do nothing about; that presence is a threat only if my experience is such as to make it a threat. To the extent that I manage my experience, I have a choice as to whether or not the electric chair, as threat to my existence, is a threat to my integrity.

That most people behave and most theorists theorize as if external sanctions were externally effective does not make them so; there remain those, brought up in the same scheme, who prove this false. The conclusion I reach is unequivocal: All of the forces of a culture, however conflicting, so far as they touch a person, are of the experience of that person who, at least potentially, has a capacity for determining which parts of his experience will play a vital part in his identity and which will remain indifferent. When it seems as if others control me, it is not that others control me, but that I permit them to control me; that I have not realized the ways in which I control myself. The customary distinction between self and other based on the physical model makes it easy to maintain the illusion of external control. Ignorance makes me a conspirator in my own misery. If I am caught in a half-submerged neurotic conflict, the energy of immense frustrations being fed back so as to exacerbate the causes of the conflict, and if I see this conflict as exclusively due to stupid parents or a rotten social system, then there may be no way out except by ordering a revolver for dessert. Even my efforts to change circumstances, being neurotic, will enhance the neurosis and the neurotic circumstances. Our leaders often seem caught in this bind. They do not deal with their personal problems because they have projected them upward as public issues. They do not deal with the public issues because they are masks for personal, subterranean conflicts.

Experience is a paradox; all we have is a paradox. On the one hand, we make the assumption that life is not my dream or nightmare, that the world exists independently of me and does in some sense enter into my experience, and that therefore I may be in error about it. On the other hand, my experience is the only thing-in-itself which I shall ever know, all other things-in-themselves coming to me filtered through my own thing-in-itself. I may have no control over the ten thousand things, but I do have some control over the one thing — the senses, feelings, intuitions, thoughts, memories, anticipations, evaluations, and imaginings that pour into my experience, of which my world is composed. This may not seem much, yet it makes the difference between heaven and hell. It is fun to imagine the consequences of a social policy that knows under whose doormat the keys to heaven and hell are to be found and, by means of its schools, lets this secret out.

And so, while ignorance makes me a conspirator in my own misery, misery may yet deliver me from ignorance. Human beings are strange objects whose wholeness does not come into being until they have been broken. This paradox is easily resolved. What I experience as being broken is being opened up; as if our routine socialization were accomplished at producing boxes, handsome, gracious, efficient-looking on the outside, but nailed shut around contents that remain a disturbing mystery. Every unopened box, as the Greeks knew, is a Pandora's box. Sometimes the story ends here. At other times, the box is opened. It is the nature of such boxes, the key having long been misplaced, that the opening is violent.

In this vale of soul-making, as Keats called it, there is one more conflict to be considered — that between the limitations of the particular culture, without which no human organism is transformed into a human being, and the transcendent background of humanity, nature, and of the universe that supports all cultures, all ways of life, no matter how antipathetic to one another they may be. If culture arouses patriotism, the universe may arouse a piety that takes precedence over patriotism. Santayana refers to this capacity for piety as spirit and says that spirit is a treacherous inmate of the animal soul. He might have added of the social system as well. In *Platonism and the spiritual life* he writes that spirit,

> . . . has slipped in . . . from beyond the gates: and its home is the desert. This foreignness is moral, not genealogical: spirit is bred in the psyche because the psyche, in living, is obliged to adjust herself to alien things: she does so in her own interest: but in taking cognizance of other things, in moulding a part of her dream to follow their alien fortunes, she becomes intelligent, she creates spirit; and this spirit overleaps the pragmatic function of physical sensibility — it is the very act of overleaping it — and so proves itself a rank outsider, a child rebellious to the household, an Ishmael ranging alone, a dweller in the infinite (Santayana 1927: 66–67).

The life of spirit, Santayana summarizes, is to be attached to no values. One critic makes this correction: The life of spirit is to be attached only to the one value of being attached to no values. This is to misunderstand what spirit is up to. Spirit has no values because it has all values at once or all values that it is able to encompass. Which is the same as saying that it has them all in the special light of eternity.

> This eternal aspect of things is also their immediate aspect, the dimension in which they are not things but pure essences; for if belief and anxiety be banished from the experience of any object, only its pure essence remains present to the mind.

And this aspect of things, which is immediate psychologically, ontologically is ultimate, since evidently the existence of anything is a temporary accident, while its essence is an indelible variation of necessary Being, an eternal form. The spirit lives in this continual sense of the ultimate in the immediate (Santayana 1927:66–67).

Spirit represents the triumph of selfhood. Culture guards its rights against the rights of the person by making attachment to what is outside of the person a condition of that person's attachment to his self. Since what is outside of the person is beyond the person's control, his self, his sense of his own worth, is also beyond his control. Loss of status is synonymous with loss of soul. When we in the U.S. lose our souls, we often try to regain them by pouring fresh energies into the very devices that brought about the loss; we overcompensate. Despair and breakdown may finally ensue, and out of them may rise that darkest of all questions: Who am I? It is a dark question because the easy answers have been ruled out. Was it not the easy answers, despite the tacit promises of culture, which brought on disaster? Under the aegis of this question, I may come to see how I have made myself into a machine that destroys itself. I may find ways to simplify and order my attachments so that eagerness for the things of life flows back. This acquisition by the self of the power to fix meaning is the maturation of the self. With maturation, the self can no longer be understood through a reduction to drives, on the one hand, or roles, on the other. It lives by rules of its own discovery, which are the rules of meaning. All that has come before is transformed, made subordinate to its sway; it has the power to say no to any prompting of psyche, any ukase of society. It wins through to the power to affirm what lies beyond the boundaries of both. Self may come to value itself with such profound joy that none of its circumstances, including the circumstance of life itself, is its equal, and so the loss of any one of these, including the loss of life, poses no threat.

With the appearance of self as spirit, the center of the way of life undergoes a radical displacement. It is to be found neither in culture nor in the way of life, having leaped entirely outside of the yin-yang scheme. This means that we must make one more addition to the yin-yang scheme before it is complete.

THE LEAP OF MIND

One curious and important fact I have glossed over. The medium of experience is mind. Either everything in the universe is mind or everything comes to us through mind, to the extent that all we ever experience of the material world, if such exists, are its mental counterparts. Whatever dimensions human life includes, those of which we have any awareness cross in experience; whatever diverse modes reality assumes, they come home to us translated into the mode of experience. Mind becomes the occasion of heroism, the centering of the person upon himself as architect of his experience. We are imprisoned in mind. Every effort to escape reveals new ramifications of mind, until the conclusion is urged on us that such a splendid prison can hardly be a prison at all.

A Hopi Indian analogy will make this clear. The Hopis are said to conceive of a complex universe made up of variegated forms, each of which is dependent upon all of the others, no one of them taking precedence over the others. The Hopis are but one of these forms, their chief contribution to the workings of the cosmos being their annual cycle of ceremonial dances which, among other things, is a celebration of this arrangement. The clearest expression of Hopi metaphysics comes in the masking, part of the ceremonial dances. The masks may cover the entire face, the most human aspect of the human being. At the same time that the mask mutes the Hopi's humanity, it provides a blank space upon which to paint the symbols of that metaphysical language into which the different identities of things — plants, animals, natural elements, human and imaginary forms — are gathered. All things are made equal by this symbolic language, as all things were given equal existence by the universe itself. In much the same way, mind, or more concretely, experience, may be seen as the symbolic equivalent of the Hopi cosmos, all things coming to us translated into its condition of vivid equality. That our application of value distinctions to the contents of experience, however socially necessary it may be, is secondary is suggested by what Santayana calls the life of spirit, the return to an ontologically primary condition of having no values.

Experience thus becomes the focus of one more set of double possibilities. On the one hand, mind is given to man, not created by man. On the other hand, man has power over mind for the conceiving of universes, selves, societies, inventions, theories. We can, with the Hopis, emphasize the givenness and moderate our own humanity to a position of equality with the other forms, or we can, as we tend to do in the U.S., exalt ourselves and take pride in man's "mastery over nature." The joke is that science masters nature only by learning how to solicit its cooperation. One of the firmest conclusions of science is that, whatever our Faustian fantasies may be, we remain a tribe of Hopi Indians who carry on their own most important tribal dances in the laboratory.

It will probably remain true that mind cannot know mind any more than teeth can bite themselves. It is reasonable to assume that somewhere I reach a limit that is built into the nature of things apart from my conception of the nature of things. With all that is given to us, meaning is not given. Prefabricated meanings, such as culture provides, are always in jeopardy. And the key choice in the creation of meaning is the meaning I impute to mind itself, medium of all human experience. While other aspects of society and the person may be understood, mind, as the vehicle of understanding, transcends understanding, and so I must impute meaning even to that which strains my capacity for meaning.

The one principle that seems clear is that my scope as human being is limited not only by the actual scope of mind, which is beyond my comprehension, but even more by the definitional scope that I attribute to mind. My being inevitably shares in the largeness or narrowness I bestow upon mind. Genuine experiment with the potentialities of man must begin with the largest possible definition of mind.

Mind is that everyday part of my identity that it is easy to take for granted; it is helped in this by being the transcendent part, the terrifying part. By transcend I mean to go beyond without leaving behind. Mind is my self but it is also a transcendent presence trailing off into obscurity, a presence always present under the form of its absence. Consider the heart, beating on for years without a moment's cessation, perfectly altering its rhythm to fit our activities, and doing this whether or not we give it any thought; that it should be alive and still retain the incredible patience of stones is one of those marvelous conjunctions that make human being possible. It is that same transcendent mind that speaks to me in the elemental language of dreams. I have had one dream that I regard as clairvoyant, providing me with knowledge which I had no way of acquiring through the usual avenues of information. The agency of that dream, which I am calling mind, is without doubt my self in the sense that it was I and not someone else who slept and had that dream, and yet I am not wholly it, of it, in the sense that I cannot comprehend how it came by the knowledge vouchsafed in the dream. I am all the questions mind raises, questions that have no answers except as I give them answers, inevitably becoming whatever those answers make of me.

Here again is the leap into other. We have met this leap three times before: first, as Mary identifies herself with mother and mother with Mary; second, as the person identifies himself with the anonymous, shared expectations that constitute culture; and, third, as a person's experience incorporates within itself aspects of other as well as of the person. Because no human being exists without the help of parents, culture, and the cosmos, it is possible to think of these as three forms of piety.

With experience, a child learns to see around his parents; to perceive them as human beings struggling with their lives exactly as he struggles with his own; and to realize that the looming, Sophoclean qualities were bestowed upon those frail creatures by his even frailer self during its infantile helplessness. And so too with the obsessions of the society into which he happened to be born. Anthropology may be of help in demonstrating that while human problems are repetitive, our ways of working them out vary from society to society, all normal, all mad. Thus, society also comes to have backsides; I may view its ways with that mild satire that includes myself in its embrace. Yet mind cannot be walked around in the same way that I walk around my statuesque parents and society. Even the mystics who purport to have gone further in this direction than anyone else talk about being dissolved in the mystery, which falls short of glimpsing the great backsides of God.

The implications for heroism are considerable. If mind is both self and other, who is the hero, self or other? Mind is at once the instrument of fate as well as the agency through which I meet the vicissitudes of fate. The hero is the one who takes his fate into his own hands, never certain but what his hands are themselves hostages of fate. A quality of matter-of-factness enters in, like that of one of those ancient priests who elevates the host as if he were changing a flat tire. Even the misery and pain which were a turning point in my life, as I look back on them, seem little more than ideas whose brilliance momentarily dazzled and blinded the ego.

Mind has depth as well as conscious surface, and about all the conscious surface can do is hold itself in readiness, as the surface of water holds itself in readiness for the bobbing up of objects released from the bottom. There is no courage like that with which the depth dares the surface. Gregor Mendel tasted the courage, even as he was baffled by it.

In its depths, mind has no doubts about itself. In my depths, I may be troubled by the surface storms which this complex scheme generates, but I am in no doubt about my self. In my depths, the question of my worth is never raised. The task of conscious mind is, by being aware, to calm the superficial turbulence and finally join itself to that profound identity which is beyond all doubt. It is beyond doubt because, from the beginning, it has formed part of the whole. The depth is wise with a wisdom that transcends both birth and death. Therefore it will not do to talk of suicide and Santayana's life of spirit as if they were the same. In suicide, the conscious self rejects its ancient imbeddedness in the more inclusive, wiser whole. With spirit the surface learns to move in harmony with the depths, much as the ancient Taoist learned to ride the whirlwind.

A mind that is capable of transcending all of its conditioned states must also remain unconditioned throughout all those changes, otherwise transcendence would not be possible. This unconditioned state is like a child, outside of every reality adults construct, capable of perceiving what no one else has dared to recognize, that all along the Emperor has been naked. It is this unconditioned state which fuses with, or reaches through, our conditioned states to play upon our chains as if they were musical instruments.

THE NAVAHO HERO

The central point is that we are dealing ultimately with a universe which, however distinguishable its aspects, remains a single, comprehensive entity. The distinguishable aspects of this single entity are person, society, and physical world, and I pictured the relation among these as three overlapping or interlocking circles. These circles are intended to make clear that while each component, viewed separately, has its own characteristics and works according to its own principles, no one is self-contained, sufficient by itself. Because each requires the others for its existence and support, all become members of one another to form the totality of the universe. Thus, persons are psychobiological creatures; society is an organization of persons; the physical world is the ground of both person and society. Cosmos sometimes is used as synonymous with the physical world, but I am using the term to include person and society as well.

If this is accepted, it becomes clear that the workings of any one component cannot be fully understood without knowledge of the totality, each part being dependent in some sense upon every other part, everything being both cause and effect of everything else.

Another epistemological difficulty is that either everything that exists comes to a human being through mind or everything that exists is mind. I am using mind to include sensing, involuntary, and subliminal processes, as well as conscious thinking. If I reach out and touch the table, I may conclude that "the table is hard," but in fact I have not experienced the hardness of the table, only a sense of pressure which I label an experience of hardness while attributing the quality of hardness to the table. What I am certain of at any moment is only the iridescent play of mind, the sensations, bodily awarenesses, longings, impulses that constitute my being at that moment. It makes little practical difference whether a human being is mind or everything comes to him through mind, for in either event, a person, being mind, can do little more than experience its manifestations; he cannot know mind objectively any more than teeth can bite themselves. The only tool we have for knowing mind is the mind that constitutes our being, and this identity of subject and object makes objective knowledge impossible. Our one certainty is immersion in mind. This statement makes sense only as we think of mind, not in terms of some particular definition or conceptualization of mind, but rather in terms of that predefinitional flux of qualities, that iridescence, which is human life.

The mystic concludes that knowledge is impossible and turns to other undertakings. The scientist agrees that he will never arrive at the absolute and ultimate truth but finds this no hindrance. If knowledge of the whole is out of the question, some knowledge of the parts is possible, and such provisional knowledge is, for the scientist, well worth the effort required to obtain it.

Yet we do fashion a world out of being. We do this by making assumptions. I assume that other people, trees, and stones exist independently of my experience of them. I do not know this to be so, I cannot know it, but I can assume that it is so and behave as if it were. I experience directly the continuous stream of iridescence that is my life, but I do not experience the independent existence of other entities. Every human world is precariously perched upon the foundation of its own "as ifs." Every human world is a virtual world, it being effective whether or not what they stand for are actualities. What happens all too often in science as in daily life is that we allow mind to set up a truncated conception of mind and experience, we learn to take the truncation for granted or see it as inherent in the nature of things, and so lock the prison door upon ourselves, even as we still clutch the forgotten key in our hand. The mind, at least in potentiality, transcends every virtual world it creates and, in its full capability, is never the prisoner of any of its worlds.

Rather than appreciate the virtualness of the world we live in, we are apt to take it for granted, to identify it with reality, and to reject all discrepant views. As infants we are vulnerable both to the social structuring of assumptions and to the assumption that reality is on our side, not the side of those who differ. This vulnerability may diminish the tendency to participate in the largerness of mind, to see through the taken-for-granted nature of the world, but it does not obliterate the capacity to do so. If mind transcends every virtual world it sets up, and if the person is the bearer of mind, then the person harbors a transcendent capacity: the capacity to detect genuine discrepancies between his own nature and that of his culture and the capacity to redo the assumptions on which his personal world, his way of life, is based. In this way the person may attain a greater degree of fulfillment. Some people, like Mendel, fall as if by accident into a larger world and are themselves bewildered by it, while others, like Spinoza, consciously frame, even in the face of disease, suffering, old age, and death, new experience of serenity, spaciousness, and light.

Historically, those who experience the largerness of mind have often seemed to jeopardize those who do not. Persecution and the threat of death are familiar instruments of the guardians of the taken-for-granted world. That some persons persevere in self fulfillment at great cost to themselves attests to the genuineness of their discovery of the more inclusive aspects of experience.

If Kroeber is correct, past societies in working out their own principles have used mind in such a way as to minimize the potentialities of persons. At the same time, there is no necessary opposition between the shared expectations necessary for a culture and the fulfillment of persons. Both are functions of mind. Both may be reconciled in new and more harmonious social and personal arrangements. Knowledge of past limitations is the first step in lifting those limitations, if not for others, then at least for myself.

The hero of Navaho mythology provides a good example of transcendence. For the Navaho, what lies beyond the known world is alien and fearsome. The Navaho hero is the one who journeys into this alien and forbidding world to subdue its powers, by trickery if need be, and bring them back as gifts for his people. The Navaho hero often begins by being cast out. He may gamble away his family's property until the family, in disgust, throws him out. Thus, his adventures beyond the confines of the known begin. We may say that no person consciously sets out to become a hero, but rather, when the time is ripe, tricks himself into the pain and misery which make heroism about the only alternative to suicide. In one Navaho story, the protagonist, having departed a scapegrace and returned a hero, announces that he can no longer stand the stink of his people's hogans. His goal now is to live with the Holy People he met in the course of his wanderings. When the precarious truce between the limitations of conscious mind and the vastness of mind not yet risen to consciousness has been broken, it may be that no new truce is possible, only mutual understanding and union of the two — which may be what Santayana means when he speaks of the life of spirit as being attached to no values. I did not request this vastness, but having acknowledged and accepted it, I must live it out, however arduous and perplexing that task may be.

REFERENCES

ASCH, SOLOMON E.

1955 Opinions and social pressure. *Scientific American* 193:31–35.

KROEBER, ALFRED

1948 *Anthropology: race, language, culture, psychology, prehistory*, revised edition. New York: Harcourt, Brace.

KROEBER, ALFRED, CLYDE KLUCKHOHN

1952 *Culture, a critical review of concepts and definitions*. New York: Vintage Books.

LEE, DOROTHY

1959 *Freedom and culture*. Englewood Cliffs, N. J.: Prentice Hall.

PEARCE, JOSEPH CHILTON

1973 *The crack in the cosmic egg*. New York: Pocket Books.

PENFIELD, W.

1952 Memory mechanisms. *Archives of Neurology and Psychiatry* 67: 178–198.

SANTAYANA, GEORGE

1927 *Platonism and the spiritual life*. New York: Charles Scribner's Sons.

Panel Papers

INTRODUCTION

The present volume originated from a section of the IXth International Congress of Anthropological and Ethnological Sciences, 1973, entitled "Experience forms." What follows was part of the proceedings of the Congress. Some of the preceding papers were also contributions to it. As interest continued in the point of view expressed by them, additional articles were included to render a more comprehensive illustration of what human sciences are doing and achieving in the field of experience forms. It is hoped that what we have tried to accomplish here will be helpful in the continual effort to find reliable regularities in the life of human beings and in the ambient world he–she creates.

THE SEMIOTIC ECOLOGY AND EXPERIENCE FORMS
by *George G. Haydu*

When we are interested in the fullest happening that a person transacts in his human world, we come to the description of experience. The term *experience entity* tries to come to grips with the additional observations that show that experience is an individual, discrete, unitary structure with distinct organic form characteristics. In the study of human behavior in a particular culture this idiomatic pattern has become an increasingly significant feature. The functions and transformations of the configurational nature of these patterns do not have to be opposed to common human needs; they are the particular definitions and satisfactions (if all goes well) of these needs (Hallowell 1955; Haydu 1958; Malinowski 1945). From a physiological point of view it was difficult, even a few years ago, to conceive of such experience entities (psychemes), their intrapsychic genesis, their orderly encoding, their lawful reenergizing and transaction. Since the work of Penfield (1952), however, experimental evidence has become increasingly forthcoming. He could activate through mild electric stimulation in the temporal lobe of the brain complete and discrete experience entities. An entity had a unitary patterned quality. It was an event-integral of the people in it, ambient sounds and noises, and many other components. It had its individual intimacy, a great immediate aliveness. These psychemes are not symbols or images of something. They are somewhat like the complete recall of an experience.

Since the above studies took place, the individuality and plasticity of the central nervous system has been thoroughly established (Fetz and Finocchio 1971; Horn, Rose, and Bateson 1973; Møllgaard et al. 1971; Pettigrew and Freeman 1973). We are now able to conceive how discrete patterned wholes can form, reverberate, scan, and evoke others of kindred resonances (Ungar 1970). These are not just drives or concepts coursing in huge nerve assemblies. By scanning and reverberating resonances, needs are able to find their tools, instrumentalities, and skills that can eventuate in action. This action in turn is registered back constantly; it is fed back. The whole contexture of the event constitutes an experience entity that is then encoded (Haydu 1972; Thatcher and John 1977; Powers 1973). Thus an experience form is the result of a structural inner pattern. It is very much a multicomponential affair; the form arises from the very structure of the event. The sequential continuum of these psychemes constitutes one's experiential life.

An experience form is a patterned integration. Some features of it can be separated, abstracted from its quasi-holographic nature (Haydu 1962a). We can concentrate on their aspects of separating and grouping of features; that is when we think of concepts and things (Gregg 1967; Simon and Kostovsky 1963). We can study or find the aspects of their individual affect tone; that is when we compare emotional qualities (Davitz 1964; Haydu 1962b). We can see how psychemes carry information about particular parts of the world; that is when we think of learning (Bower and Theios 1964; Estes 1964; Restle 1970) and remembering (Ellis 1973; Levin et al. 1973). But in all the above instances, as also in studying perception (Stevens 1972) or imagery (Paivio 1971), the locus and individuality of the total experience and the genesis of the global experience form leave their indelible marks even on the most abstract or quantitative feature we can derive.

Human beings live in a world of interpersonal give-and-take that is the outcome of a huge collective enterprise (Cassirer 1944; Herskovits 1948). The most primitive need is only partly a programmed inborn one; it must find its cultural definition and accommodations. Any need, any instrumentality, comes to us through the interpretive and sense-making action of our semiotic ecology. We are steeped in this system of significations and consequences. It contains the practical signs and meanings of all things that matter. It is culture at its most pertinent (Sebeok, Hayes, and Bateson 1964). Yet the system of signs and significations must become experience forms of individuals in order to come to life or even to be maintained as something alive. This becomes especially crucial when change must occur, when new experience entities must arise to transact the business of life in a new manner (Haydu 1958). The particular double encoding (Bohannan 1973) that must occur — one in the individual and the other in the system of the semiotic ecology that is the cultural world — takes place through the interpersonal give-and-take of people. The semiotic environing system does not change automatically or by easy substitutions. It is not the same as a nonliving Platonic realm or the eternal objects of Whitehead. The culture forms can change only when the living old ones confront some obstruction in individuals — whose every activity is behavior (Arensberg 1972) and who resolve this predicament in new experience entities through their interpersonal give-and-take (Haydu 1970).

The above considerations make the expression of experience entities a vital cultural function. Even in maintaining old forms, such expression is essential. But when transformations must take place for very pragmatic reasons, the expression and communication of experience entities is even more vital. Bodily stance and kinetics, the dance, plastic and pintoreal shapes and other forms can convey the new transfigurations. Even the metaphor of the poem or poetic ritual or ritual worship is not just a symbol standing capriciously in place of something else (Fernandez 1974); it stands for a new experience form that was a created part of a particular person's psyche. If we try to satisfy Skinnerian objections (which I am not at all certain we should), the psyche could be defined as the encoded system of wants and their reinforcers. That system has its history, its transformations, its particular unity and individual consistency (Fuller 1967; Renner 1972). And yet, the experience entities of the participants in a particular culture have so much in common that social and cultural conclusions can be drawn from the semiotic study of the members (Osgood 1964). Thus, individuality of the psychemes and the shared nature of the semiotic ecology are not contradictory. Both are essential aspects of cultural dynamism (Haydu 1973a) and can be neglected only at the peril of great deficiencies of understanding.

There has been a great amount of discussion how one can penetrate and understand an alien culture. The discipline of history writing had met this problem before (Dilthey 1913; Windelband 1904). How can you understand an eighteenth-century Englishman or a fifteenth-century Croatian? You look at the artifacts first, the different documents, the remains of a previous way of life. Theorists of history devised complemental analysis. It consists of letting the artifact or a document tell its own story through the discovery of cross referentiable features. This method surfaces again in culture studies as holistic analysis (Angyal 1941; Oxnard 1973). We have even seen lately mathematical and algorhythmic varieties of these that can be applied in the study of a particular semiotic ecology (Nelson 1973; Thom 1972). It is necessary to realize that the nomothetic discursive way of study emphasizes and restricts the total experience to its conceptual components and their relations to similar (or dissimilar) concepts. On the other hand, an ideographic or musical way of studying the same experience emphasizes and expands the individual inner resonances and their near or distant harmonics. The same point of view can be expressed in the two different ways (Haydu 1958; Haydu 1973b) each conveying its different intent, yet springing from the same experiential life. So a great deal depends on what questions we are asking in analytic studies. One kind of methodology may be quite appropriate for a particular type of study and yet totally unproductive, indeed obfuscating, in another. This has to be consistently borne in mind in the study of semiotic and experience forms which are always amenable to analyses of the greatest variety.

REFERENCES

ANGYAL, A.

1941 *Foundations for a science of personality*. New York: Commonwealth Fund.

ARENSBERG, C. M.

1972 Culture as behavior: structure and emergence. *Annual Review of Anthropology* 2:1–26.

BOHANNAN, P.

1973 Rethinking culture: a project for current anthropologists. *Current Anthropology* 14:357–372.

BOWER, G. H., J. THEIOS
1964 "A learning model for discrete performance levels," in *Studies in Mathematical Psychology*. Edited by R. C. Atkinson. Stanford: Stanford University Press.
CASSIRER, E.
1944 *An essay on man*. New Haven: Yale University Press.
DAVITZ, J. R., *editor*
1964 *The communication of emotional meaning*. New York: McGraw-Hill.
DILTHEY, W.
1913 *Gesammelte Schriften*. Leipzig: Teubner.
ELLIS, H. D.
1973 Proactive effects of interpolated anchors. *Journal of Experimental Psychology* 98:233–238.
ESTES, W. K.
1964 All-or-none processes in learning and retention. *American Psychologist* 19:16–25.
FERNANDEZ, J.
1974 The mission of metaphor in expressive culture. *Current Anthropology* 15:119–145.
FETZ, E. E., D. V. FINOCCHIO
1971 Operant conditioning of specific patterns of neural and muscular activity. *Science* 174:431–435.
FULLER, J. L.
1967 Experiential deprivation and later behavior. *Science* 158:1645–1652.
GREGG, L. W.
1967 "Internal representation of sequential concepts," in *Concepts and the structure of memory*. Edited by B. Kleinmuts. New York: Wiley.
HALLOWELL, A. I.
1955 *Culture and experience*. Philadelphia: University of Pennsylvania Press.
HAYDU, G. G.
1958 *The architecture of sanity*. New York: Julian.
1962a Event-experience patterns and conceptual structures: an experimental approach. *American Journal of Psychotherapy* 15:619–629.
1962b Manic-depressive rhythm: its pharmacological modification and the nature of the self structure. *Annals of the New York Academy of Sciences* 98:1126–1138.
1970 Interrelated transformations of Rousseau's life and of Western culture of his time. *American Journal of Psychoanalysis* 30:161–168.
1972 Cerebral organization and the integration of experience. *Annals of the New York Academy of Sciences* 193:217–232.
1973a Psychotherapy, enculturation and indoctrination. *Journal of Life Sciences* 3:25–27.
1973b *Statements and avowals*. Boston: Branden Press.
HERSKOVITS, M. J.
1948 *Man and his works: the science of cultural anthropology*. New York: Alfred A. Knopf.
HORN, G., S. P. R. ROSE, P. P. G. BATESON
1973 Experience and the plasticity in the internal nervous system. *Science* 181:506–514.
LEVIN, J. R., R. E. DAVIDSON, P. WOLFF, M. CITRON
1973 A comparison of induced imagery and sentence strategies in children's

paired-associate learning. *Journal of Educational Psychology* 64:306–309.

MALINOWSKI, B.
1945 *The dynamics of culture change*. New Haven: Yale University Press.

MØLLGAARD, K., M. C. DIAMOND, E. L. BENNETT, M. R. ROSENZWEIG, B. LINDNER
1971 Quantitative synaptic changes with differential experience in rat brain. *International Journal of Neuroscience* 2:113–128.

NELSON, K.
1973 "Holistic analysis," in *Perspectives in ethnology*. Edited by P. P. G. Bateson and P. H. Klopfer. New York: Plenum.

OSGOOD, C. E.
1964 Semantic differential technique in the comparative study of cultures. *American Anthropologist* 66:171–201.

OXNARD, C.
1973 *Form and pattern in human evolution*. Chicago: University of Chicago Press.

PAIVIO, A.
1971 *Imagery and verbal processes*. New York: Holt, Rinehart and Winston.

PENFIELD, W.
1952 Memory mechanisms. *Archives of Neurology and Psychiatry* 67:178–198.

PETTIGREW, J. D., R. D. FREEMAN
1973 Visual experience without lines: effect on cortical neurons. *Science* 182:599–601.

POWERS, N. T.
1973 Feedback: beyond behaviorism. *Science* 179:351–356.

RENNER, K. E.
1972 Coherent self-direction and values. *Annals of the New York Academy of Sciences* 193:175–184.

RESTLE, F.
1970 Theory of serial pattern learning: structural trees. *Psychological Review* 77:481–495.

SEBEOK, T. A., A. S. HAYES, M. C. BATESON, *editors*
1964 *Approaches to semiotics*. The Hague: Mouton.

SIMON, H. A., K. KOTOVSKY
1963 Human acquisition of concepts for sequential patterns. *Psychological Review* 70:534–546.

STEVENS, S. S.
1972 Neural quantum in sensory discrimination. *Science* 177:749–762.

THATCHER, R. W., E. R. JOHN
1977 *Foundations of cognitive processes*. Hillsdale, N. J.: Erlbaum.

THOM, R.
1972 *Stabilité structurelle et morphogénèse*. Reading: Benjamin.

UNGAR, G.
1970 *Molecular mechanisms in learning and memory*. New York: Plenum.

WINDLEBAND, W.
1904 *Geschichte und naturwissenschaft*. Strassburg: Heitz.

PANEL DISCUSSION by *F. Allan Hanson*

Dr. Haydu has very neatly and efficiently crystalized a notion of experience forms which is not only psychological but also to some degree neurological. I am sorry to say that in some ways my paper serves to muddle up that clear focus because my research on the philosophical side of anthropological questions leads me to add another dimension to the picture.

One of the primary problems that faces the whole enterprise of anthropology is the nature of cultural things. Does culture really exist in its own right, or is it ultimately reducible to psychological — even psychobiological or neurological — considerations? My suggestion is that we tend to pose this question in an unfortunate way. Following Manners and Kaplan I would argue that reality does *not* come in "levels" such that there are psychological phenomena in the world that should be studied by psychology and then different phenomena — call them social or cultural — which are to be studied by anthropology, sociology, and history. Rather, all of us who are doing social science are concerned with one basic kind of reality, which you can call human phenomena, human affairs, human behavior, or what have you. The distinction between the psychological and culturological is not a distinction of chunks of reality but rather a distinction in perspective on the single reality which is human affairs. If you ask the questions, "Why do people behave the way they do? What are the reasons why people do particular things, say particular things?" then you are asking what I call individual questions. Questions about the purposes, the motives, the drives, the intentions that people have for acting as they do. Answers to such questions reveal a particular kind of meaning in human affairs. I call it "intentional" meaning, because it concerns motives, drives, intentions, and so on. On the other hand, there is another kind of question that you can ask of human affairs. That is, "What is the nature of the things that people do?" when such things are viewed in and of themselves. Instead of asking why so-and-so has a certain belief, you ask about the nature of the belief that so-and-so has. Then you are asking what I call "institutional questions." You are asking about systems of belief, patterns of behavior, in their own right. These questions are not concerned with what motivates people to conform to rules or to break them, but rather with the structure, the grammar of the rules themselves. The kinds of responses one gives to institutional questions relate also to meaning, but this time that meaning is a matter of implications. The meaning of customs, norms, rules, beliefs is found in their systematic relations to or implications for other beliefs, customs, etc. Hence answers to institutional questions demonstrate the structure, the logical interconnections of social and cultural institutions.

The upshot of all this is that psychology and cultural anthropology do not refer to distinct levels of reality, and one is not reducible to the other. They represent different perspectives on the single reality of human affairs. Either type of question can be asked independently but, of course, a full understanding of human behavior requires both perspectives.

PANEL DISCUSSION by *Lorand B. Szalay*

The differences between Dr. Haydu's and my position are mainly semantic, terminological. I am inclined to speak of psychological meaning. By that I mean something similar to Dr Haydu's meaning of "experiential forms." I consider the distinction between two types of meanings — between lexical meaning and psychological meaning — particularly important. The lexical meanings are based on convention; they are shown in the dictionaries and lexica. Dictionaries tell us equivalent words which are used by two language communities in reference to the same category of objects or events. The normative lexica tell us by definition what a word means or is supposed to refer to. These convention-based referents, these logical definitions are quite different from what I mean by psychological meaning, or what Dr. Haydu has in mind when he speaks of experiential forms. To illustrate the difference, I will use a simple example. The lexical meaning of drug is clear. It involves a chemical substance with medical effects. Yet it is similarly clear that various people have quite different meanings for drug. Christian Scientists do not have the same meaning for drug, for instance, as do drug addicts. These different groups have different psychological reactions, perceptions, evaluations attached to drug, although in a strictly lexical sense drug for both groups refers to the same category of objects. The psychological meaning or coding reactions can be, and frequently are, drastically different even though the referents of a word are the same based on convention. The psychological meaning reactions of various groups are frequently different because they are influenced by different beliefs, experiences, affects. These meanings are frequently loaded with emotion, permeated by different values, and bear on different concerns and interests.

The importance of this psychological meaning reaction follows from its role in communication, in human choices and decisions, and from its influences on human behavior. This psychological meaning reaction will decide whether a communication will be noticed, listened to and followed, ignored or rejected. Whether a drug addict will listen to a communication on drugs, and what action he will take will be determined by his psychological rather than lexical meaning.

This practical importance of psychological meaning in human behavior and in interpersonal relations suggests that it deserves particular interest. First of all, its identity requires full recognition of its distinct qualities which clearly differentiate it from lexical meaning and with which it is still widely confused.

Particularly important is the fact that psychological meaning is a subjective reaction; it is influenced by the person's mental state, past experiences, emotions, beliefs, value system. It has perceptual as well as affective components; it depends little on logic. Rather it depends on subjective salience of its main components of interpretation, some of which may be void of logical, rational foundation. The salient components determine how people react to communications, what alternative behavior they may choose. Word meanings are units of thought processes as well as units of our language. This duality, which combines the covertness of the thought process with observable manifestations of the language, provides useful opportunities for making inferences on the covert process and its composition by using information offered by the overt language behavior.

My research over the last ten years has been focused on the empirical assessment of the psychological meaning reaction. I have developed a research method that uses verbal associations as a category of language behavior exceptionally rich in information value. In view of the little time available, I would prefer not to discuss here Associative Group Analysis (AGA), as this method is called. Rather I would refer you to the article "Verbal associations in the analysis of subjective culture" (Szalay and Maday 1973).

As a psychologist with multidisciplinary orientation, I am particularly interested in comparative cross-cultural studies and in the potential utility of this approach for my anthropologist colleagues who are searching for sensitive analytic techniques.

Working with verbal associations obtained from various culture groups, I came to the conclusion that it is possible to map certain important psychological dimensions of culturally characteristic meaning systems through associations. In line with the conceptualization developed by Osgood et al. (1957), Triandis and others, subjective culture may be approached as the meaning system characteristic of the people representing the culture. To reconstruct the meaning system we may rely on the assessment of the psychological meanings that are salient and representative of main semantic domains. In this approach not only cultural meanings of single themes may be reconstructed and compared between cultures but wider semantic domains represented by dozens of themes as well. Our paper compares U.S. and Slovenian meaning systems in the domains of political and social values. With a slightly different terminology, we may speak of a comparative analysis of the sociopolitical frame of reference of U.S. and Slovenian students.

I would like to discuss now a couple of old ideas and perhaps some extensions of them. It seems to me that the natives of every culture would conceive experiences differently from those of other cultures. The Whorfian hypothesis would state that perception, perhaps also the nature of experience, is determined by the nature of the language utilized. Why not consider the hypothesis that the semiotic conception of that culture determines the experience forms? In other words, does the way in which people perceive nature depend on the terminology they are provided by their language? Do the words available to us determine our experience? Has there been any work done on any type of theory that would tend to synthesize these different conceptions of experience due to both individual and cultural variability in the semiotic ecology of the culture?

My personal opinion is that the theory of linguistic determinism is far fetched, but I fully subscribe to the idea of linguistic relativism as it is promoted by Whorf, Sapir, and others. It is my impression that the misunderstanding of the essential thesis of the Whorfian position has produced a great deal of debate. The Whorfian hypothesis should not be interpreted to mean that different languages, different vocabularies, imply automatically different world outlooks. The central idea is merely that our vocabulary has the potential to influence our perception. While the influences of words with concrete referents may be negligible, their effects can be expected to increase in proportion with the level of abstraction. Particularly intriguing is the question of how concepts in the social relations influence how we perceive social realities. Since social realities are not concrete — they involve, for instance, people's relationship to each other, their relationship to social institutions, community, society, government — they are open to interpretation. I suspect that the type of interpretations we bring to work depend to a large extent on the terminology we adopt and use in describing social relations. Unfortunately, in this area of social relations, the impact of terminology on thought processes cannot be easily tested. The experiments conducted with color perception by Brown and Lenneberg have suggested, however, that even in this relatively concrete field, terminology does influence perception. In the field of social relations, I would expect an even greater impact. As a social psychologist, I am particularly interested in the influences of terminology, in the influences of concepts and conceptual systems on social attitudes, on the shaping of social relations, on the perception of social reality. I consider this field particularly important because of its potential implications. I hope that with empirical methods of high analytic sensitivity, this problem area will become increasingly researchable in the future.

With respect to our approaches, Dr. Haydu and I show a distinct difference in emphasis. He emphasizes the individuality of experience; my emphasis is on the collective portion of the experience. I have to point out, however, that my focus is not a matter of principle but largely a consequence of practical, pragmatic methodological considerations. I do not doubt the importance of the personal experience. As a social psychologist, however, the type of indepth analysis that I am particularly intrigued with represents a very expensive approach if it is focused on a single person. Furthermore, I am particularly intrigued by the experience of cultural differences. Cultural differences are particularly important because they predispose the relationship of entire collectives, cultures, nations. I assume that these cultural group characteristics can be assessed as shared individual experiences expressed as verbal association behavior focused on the shared portion of the individual experiences characteristic of one culture group. By comparing this with the shared individual experiences characteristic of another culture group, we approach the problem of cultural differences at the level of collective experiences. These collective experiences do not ignore the individual experience, but merely consider the shared portion of the individual experience. This strategy, however, can be defended only on the basis of relative priority and economy. It promises knowledge on the group, on the collective. It embraces only the shared portion of the individual experience and helps to understand the individual only to the extent that he or she is representative of a culture or a collective. This approach helps us understand an American compared to a Chinese, or the American farmer in comparison to an American intellectual, but it does not help to understand one particular American in comparison to any other. In defending this research strategy, I am not denying the importance of the idiosyncratic individual experience; I merely assign to it a lower order of priority for reasons of economic constraints, social relevance, and social applicability of knowledge.

REFERENCES

LENNEBERG, E. H., J. M. ROBERTS.

1956 The language of experience, a study in methodology. *Indiana Publications in Anthropology and Linguistics, Memoir 13.*

OSGOOD, CHARLES E., G. SUCI, P. H. TANNENBAUM

1957 *The measurement of meaning.* Urbana, Ill.: University of Illinois Press.

SAPIR, E.

1921 *Language.* New York: Harcourt, Brace and World.

SZALAY, L. B., B. C. MADAY

1973 Verbal associations in the analysis of subjective culture. *Current Anthropology* 14:33–50.

WHORF, B. L.

1956 "Language: plan and conception of arrangement," in *Language, thought and reality.* Edited by J. B. Carroll. Cambridge, Mass.: MIT Press.

PANEL DISCUSSION by *C. R. Welte*

My paper is on the subject of categories of values, and in it I developed a taxonomy and a nomenclature for value studies. However, I find Dr. Haydu's term "experience forms" very helpful in clarifying the nature of values, so let me try to relate values as affective symbols to the idea of experience forms.

But before I do that I must point out that when I refer to symbols, I am not referring to labels or conventional symbols. You see, there are two distinct processes that are both called symbolization: one in which the meaning is assigned and the other in which the meaning is experienced. It is in this latter sense that I will refer to symbols and symbolization. So, I will give you an example of a type of experience form that I think is the basis of an extremely important but overlooked category of human values. The experiences I refer to are those undergone by all infants in the first year of life. They include experiences of trusting, activity, and conation. The infant experiences trust and the reward and return of trust by the mother. Such experiences take on deep symbolic significance for the infant because they are related to self preservation or survival. Simply put, the merged memories of these experiences form a symbol whose meaning is survival. The symbol is a value or an experience form and, as Dr. Haydu puts it, "it searches for fulfillment, looking for its congenial 'object', or 'action'." As new fulfillments are found, the symbol becomes more generalized, i.e., our values become enriched, and trust extends to other people, to the environment, and to the self. These values in their fully developed form are often given the labels "empathy," "harmony with nature," "authenticity," and so forth. The words are not the values, the values are the experience forms — internal symbols — or in my terminology, "affective, symbolic elements of underlying cognitive structure."

Now, the type of experience form or value that I have been talking about, the kind that originates in universal infant experiences, is not a part of culture or a cultural value because it is not due to enculturation. Rather, it is due to the infant's capacity for symbolization being applied to common infant experiences in the very first months of life. Thus, we have a category of values that develops in all humans, i.e., a category we can call pan-human values. As I show in my paper, pan-human values form a yardstick for cultural values, and if my hypothesis is correct, they are also a vital factor in our search for the nature of man.

Biographical Notes

HOMER GARNER BARNETT (1906–) is Professor Emeritus of Anthropology at the University of Oregon. He was educated at Stanford University and the University of California in Berkeley. He has served as Staff Anthropologist for the United States Trust Territory of Pacific Islands, adviser to the Netherlands New Guinea Government, and as a member of the Research Council of the South Pacific Commission. He has conducted research in the Palau Islands, New Guinea, and among Northwest Coast Indians in the United States and Canada. He is the author of *Palauan society* (1949), *Innovation: The basis of cultural change* (1953), *The coast Salish of British Columbia* (1955), *Anthropology in administration* (1956), *Indian shakers* (1957), and *Being a Palauan* (1960).

JEREMY BOISSEVAIN (1928–) is presently Professor of Social Anthropology at the University of Amsterdam. He obtained his B.A. in 1952 (Haverford College, Pa.) and his Ph.D. in 1962 (London School of Economics). In between he served as CARE Mission Chief in the Philippines, Japan, India, and Malta. He has carried out research in Malta, Sicily, and Montreal and taught at the Universities of Montreal, Sussex and Malta and at the Institute of Development Studies, Brighton. His publications include *Saints and fireworks* (1965), *Hal Farrug* (1969), *The Italians of Montreal* (1970), and *Friends of friends* (1974). He also edited *Network analysis* (1973) and *Beyond the community* (1975). At present he is studying the impact of tourism in the Mediterranean region.

DWIGHT BOLINGER (1907–) is Professor Emeritus of Romance Languages and Literatures, Harvard University. He was educated at Washburn College, the University of Kansas, and the University of Wisconsin, and has held teaching posts at Washburn University, the University of Southern California, and the University of Colorado. He is continuing research in linguistics and has among recent publications *Degree words* (1972), *Aspects of language*, 2nd ed. (1975), and *Meaning and form* (1977).

LAURA B. DELIND (1947–) is currently completing her doctoral dissertation in anthropology at Michigan State University. Her fieldwork was conducted in a rural Michigan community and considered the changing patterns of behavioral organization and decision-making as a function of population size and diversity. Since 1976 she has pursued problems of rural community change as an instructor and research specialist within the Department of Anthropology. Her most recent publications include "The small U.S. community and anthropology research" (1975) and "An adaptive approach to behavior and organizational development: the case of the Leisureville fire department" (forthcoming).

DENIS DUTTON (1944–) was born in Los Angeles and is Assistant Professor of Philosophy at the University of Michigan — Dearborn. His degrees are from the University of California, Santa Barbara (B.A., 1966; Ph.D., 1974). He served as a Peace Corps Volunteer in the state of Andhra Pradesh, India, 1966–1968. His major research interests are aesthetics and the philosophy of the social sciences. Articles by him in these fields have appeared in a number of philosophy journals, and his article "Art, behavior, and the anthropologists" was published in *Current Anthropology* in 1977. He is editor of the journal *Philosophy and Literature.*

F. ALLAN HANSON (1939–) Professor of Anthropology at the University of Kansas, was educated at Princeton and Chicago and has done postdoctoral studies at the Universities of Oxford, Pittsburgh, and Auckland. He has conducted anthropological research on Tahiti and Rapa (French Polynesia) and in New Zealand. Major publications stemming from his Polynesian researches are *Rapan lifeways: society and history on a Polynesian Island* (1970) and *Bibliographie de Rapa, Polynésie Française* (1973, with Patrick O'Reilly). His interest in the philosophy of social science has resulted in *Meaning in culture* (1975).

GEORGE G. HAYDU (1911–) is presently Director of the Center for the Study of Integrative Transformations, Bayside, N.Y. His special interests include: the genesis, competence and transformations of experience forms. His publications include: "Psychotherapy, enculturation and indoctrination: Similarities and distinctions" (1973), "Cerebral organization and the integration of experience" (1972), and *The architecture of sanity* (1958).

JULIUS E. HEUSCHER (1918–) received his medical degrees from the University of California in San Francisco and from the University of Basle where he also completed his psychiatric training. He is currently in private practice and teaches at the Stanford University as well as at the Pacific Graduate School of Psychology. His interest in phenomenology and existential philosophy has influenced both his psychotherapeutic orientation and his investigation of folklore. This is reflected in numerous articles in his recent book, *A psychiatric study of myths and fairy tales* and in a forthcoming book, *Existentialism and folklore.*

BEVERLEY KILMAN. No biographical data available.

GEORGE MILLS (1919–) was Professor of Anthropology at Lake Forest College. He was educated at Dartmouth College and Harvard University. Among his publications are: *Navaho art and culture; Go I know not where, bring back I know not what*; *People of the saints*; *Broken present* (with Barriss Mills); and *Trip*.

GORDON E. MOSS (1937–) is currently Associate Professor of Sociology at Eastern Michigan University. He was educated at Brigham Young University, University of Utah, Rutgers, and the State University of New York at Buffalo. He is the author of *Illness, immunity and social interaction: the dynamics of biosocial resonation*, and co-author of *Growing old: an exploration of our treatment of and resources for the aging*. His published articles include, "Biosocial resonation: a conceptual model of the links between social behavior and physical illness."

VID PECJAK (1929–) is a Professor of Psychology at the University of Ljubljana, Yugoslavia. He received his Ph.D. in psychology from the same university. He was a Visiting Professor at the University of Illinois, the University of Hawaii, and Monash University, Australia. His fields are the psychology of cognition, psycholinguistics, and cross-cultural psychology.

JOHN OWEN REGAN (1931–) is currently a professor at Claremont Graduate School in the departments of Education and International Relations. His teaching and research is in the area of applied anthropological linguistics. His field work centered on early childhood populations in English speaking communities.

His "Childspeak" ethnography of communication data corpus of child–adult interaction is organized within a system based on the Smith–Trager anthropological linguistic school. He is director of the Discourse Analysis Commission of the Association Internationale de Linguistique Appliquée among the projects of which is the development of a taxonomy of applied linguistic discourse analysis systems and an observational model for the study of children's communicative development.

CALVIN O. SCHRAG (1928–) received his formal education at Bethel College (B.A.), Yale University (B.D.), Heidelberg University (Fulbright Scholar), and Harvard University (Ph.D.). He is currently Professor of Philosophy at Purdue University. He has taught at the University of Illinois, Northwestern University, and Indiana University; and he was a Guggenheim Fellow at Freiburg University, Germany, in 1965–1966. He is currently Executive-Secretary of the Society for Phenomenology and Existential Philosophy and Secretary-General of the International Society for Phenomenology and the Human Sciences. He is co-editor of the International philosophical quarterly, *Man and World.* He is the author of *Existence and freedom: towards an ontology of human finitude* (1961) and *Experience and being: prolegomena to a future ontology* (1969). He is co-editor and contributing author of *Patterns of the life-world* (1970). More recently he has authored "The crisis in the human sciences" (1975); "Praxis and structure: conflicting models in the science of man" (1975); and "The topology of hope" (1977).

LORAND B. SZALAY (1921–) was born in Budapest. He is a social psychologist, the Director of the Institute of Comparative Social and Cultural Studies, Silver Spring, Maryland, and teaches at the University of Maryland. He received his M.A. level diploma in modern languages at the Academy of Foreign Trade and Languages in Budapest (1950) and his Ph.D. in psychology at the University of Vienna (1961), he then undertook post-doctoral studies at the University of Illinois (1961–1962). While Associate Professor in research with The American University in Washington, D.C., he has developed the Associative Group Analysis method which is being used in numerous domestic and international applications. His main research interests are in psychocultural analysis, intercultural communication, socialization, and alienation. He has authored and coauthored over fifty professional publications, articles and books. He is a member of the American Psychological Association, American Sociological Association, D.C., and the International Studies Association.

ROBERT W. THATCHER (1942–) is currently Associate Professor in the Department of Psychiatry and Brain Research Laboratories at New York University School of Medicine. He was educated at the Universities of Oregon and Waterloo (Canada) and has held posts at Indiana University, Albert Einstein College of Medicine, and New York Medical College. His publications include *Functional neuroscience, Vol. 1: Foundations of cognitive processes* and numerous articles in basic research and applied clinical neuroscience.

CECIL R. WELTE (1915 –) was born in Philadelphia, Pa. After an engineering education he spent twenty years as a line officer in the U. S. Navy. He then took an M.A. degree in Mesoamerican Anthropology at Mexico City College, with the thesis "Aztec value orientations" (1962) which was published in the *Actas* of the XXXVth International Congress of Americanists. His primary theoretical interest is in the varieties of values that are involved in human motivation, and in relating them to the culture history of the indigenous people of southern Mexico. He considers the semantic base for values to be crucial. His studies in the theory of semantics resulted in "Levels of integration" published in the *General Semantics Bulletin* (1964). Since 1964 he has resided in Oaxaca, where he established and maintains the Oficina de Estudios de Humanidad del Valle de Oaxaca. Most of the work of the office is privately printed. He is a member of the Consejo Ejecutivo of the Museo Frissell de Arte Zapoteca.

Index of Names

Index of Subjects